MIND SHIFT

JOHN PARRINGTON

MIND SHIFT

HOW CULTURE TRANSFORMED THE HUMAN BRAIN

OXFORD

UNIVERSITY PRESS

OXFORD
UNIVERSITY PRESS

Great Clarendon Street, Oxford, OX2 6DP,
United Kingdom

Oxford University Press is a department of the University of Oxford.
It furthers the University's objective of excellence in research, scholarship,
and education by publishing worldwide. Oxford is a registered trade mark of
Oxford University Press in the UK and in certain other countries

First Edition published in 2021

Impression: 1

Published in the United States of America by Oxford University Press
198 Madison Avenue, New York, NY 10016, United States of America

British Library Cataloguing in Publication Data
Data available

Library of Congress Control Number: 2020948532

ISBN 978-0-19-880163-4

Printed and bound in Great Britain by
Clays Ltd, Elcograf S.p.A.

ACKNOWLEDGEMENTS

I would like to thank a number of people who have helped bring this, my third book, to fruition. I owe a particular debt to Latha Menon, my editor at Oxford University Press, who had a great deal of input into the book. Latha's critical comments had a major impact in pushing me to go that 'extra mile' in the writing and her encouragement at all stages made a huge difference to me during the five years that I have been working on this project. I would also like to thank Jenny Nugee of the OUP editorial team, for her help on a multitude of practical matters, and Charles Lauder and Saraswathi Ethiraju of SPi Global, for meticulous copy-editing of the book. Quite a number of people helped immensely with the writing of the book through their critical comments; I would like to thank Jonathan Bate, David Crane, Jane Hardy, Marnie Holborow, George Paizis, Naomi Rokotnitz, John Rose, Margarida Ruas, and Jozsef Somogyi, as well as an anonymous reviewer, for their comments in this regard. I would also like to thank Margarida Ruas for producing the superb set of line drawings for the book. I also owe thanks to Anthony Morgan for producing the author photo for the book cover. For their expert assistance with marketing and publicity, and answers to many questions on this front, I would like to thank Anna Gell and Kate Roche of OUP. I owe thanks also to friends and colleagues who have indulged me in my speculations about some of the themes of this book. I also owe a great debt of thanks is to my family, for providing such a warm and lovely home environment that meant so much to me during the long hours spent on researching and writing, and for putting up with the time I spent on this project. Finally, I would like to dedicate this book to the memory of the family members who are mentioned in it, but are no longer here to provide me with their love, inspiration and vision: my grandmother, father, mother and sister. You taught me so much and if thought can be viewed as a kind of dialogue, then I will be always grateful for your role in my own thoughts.

Note on copyright material

I am grateful for permission to include the following copyright material in this book:

Epigraph from William Golding, Pincher Martin. Published by Faber and Faber Ltd.

The publisher and author have made every effort to trace and contact all copyright holders before publication. If notified, the publisher will be pleased to rectify any errors or omissions at the earliest opportunity.

CONTENTS

CONTENTS

PART IV MIND IN TROUBLE

PART V THE SOCIAL MIND

PART VI MIND AND CULTURE

PART VII THE FUTURE OF MIND

'[Man] is a freak, an ejected foetus robbed of his natural development, thrown out into the world with a naked covering of parchment, with too little room for his teeth and a soft bulging skull like a bubble. But nature stirs a pudding there and sets a thunderstorm flickering inside...'

William Golding

INTRODUCTION

The Universe, according to estimates, is about ninety-three billion light years across and contains over a hundred billion galaxies, each with hundreds of billions of stars.[1] Each year we learn more about cosmic phenomena such as black holes and colliding galaxies through increasingly powerful telescopes. Yet to find the most complex object in the known Universe, we need look no further than the space within our own skulls. Weighing about a kilogram and a half, with the appearance and consistency of cold porridge, the human brain nevertheless has about as many nerve cells or neurons—around a hundred billion—as there are stars in the Milky Way.[2] Since each neuron can connect with around ten thousand others through cellular structures called synapses, there are about a hundred trillion such connections in the brain. And as the brain has hundreds of different types of neurons, its complexity is far greater than can be described by simply focusing on the total number of cells or their interconnections.

The power of the human brain is displayed in the wonders of modern civilization. As a species, we are distinguished by our ability to systematically shape and transform the world around us via technology, which itself we continually improve. By this means in 40,000 years we have gone from living as hunter-gatherers scratching a living from the earth to sending astronauts to the Moon and unmanned probes to the far reaches of the Solar System and beyond.[3] Such technological advances are based upon our species' ability to continually discover more about the natural world through science, with each new generation.

Of course if human culture were only concerned with science and technology it would be dull indeed.[4] To the human mind we also owe the literature of William Shakespeare and Emily Brontë, the music of Wolfgang Amadeus Mozart and Billie Holliday, and the visual art of Pablo Picasso and Paula Rego. Yet despite the human brain's capacity for such intellectual and technological feats, we still know astonishingly little about how it achieves them. This deficit in understanding is a problem not only because it means we lack basic knowledge of the biological factors that underlie our human uniqueness, but also because, for all its amazing capabilities, the human mind seems particularly prone to dysfunction.

Currently a quarter of the population of Britain suffers from a mental disorder, and the proportion is similar in other developed countries.[5] Some would argue that this reflects a tendency to 'medicalize' what might once have been considered to be within the spectrum of normal human behaviour. But it would surely be a mistake to underestimate the pain and suffering caused by conditions like schizophrenia, bipolar disorder, and clinical depression. Although drugs can alleviate many of the symptoms of these conditions, even some psychiatrists question whether such drugs treat the underlying causes of these disorders or merely mask the symptoms. In addition, a significant proportion of people fail to respond to such treatments for reasons that are often unclear, but which can have tragic consequences for sufferers and their loved ones.[6] I speak here from bitter experience, having recently lost my sister to a severe form of depression that spiralled out of control, being apparently unresponsive to a variety of different types of drugs and psychotherapies, resulting finally in her suicide by hanging, aged just 53. Sadly, there is a history of such matters in my family, adding a personal resonance in my case to the question of whether mental disorders are largely a product of biology or environment. In fact, as we shall see, the idea that one can separate 'nature' or 'nurture' in this way betrays a misunderstanding of both biological mechanisms and the unique role that society plays in the human condition.

Still, some would argue there is good reason to be optimistic about the prospect of developing new and better treatments for mental disorders in the not-so-distant future. Such optimism is based on our increasing potential to study how the brain works in various important new ways thanks to recent technological innovations. While methods for measuring the electrical and chemical changes that occur in neurons during nerve impulses have been available for some years, it is now possible to genetically label living cells using fluorescent proteins so that such measurements can be carried out in the brain of a live animal. One approach, called 'brainbow', mixes fluorescent colours to create a palette of 90 different shades, making it possible to distinguish many different neurons and their interconnections.[7]

We also have the DNA sequence of not only the human genome, but those of mice and our closest animal relative, the chimpanzee, and even of extinct proto-humans like Neanderthals.[8] While this only allows comparison at a whole organism level, increasingly sophisticated methods for taking a 'snapshot' of the genes turned on or off in different brain cells make it possible to compare such patterns of gene 'expression' in our brains with those of other species. Meanwhile studies are underway to create a 'connectome' of the human brain.[9] Just as the Human Genome Project mapped all the genes in our genome, so the Human Connectome Project aims to identify all the connections between different brain cells.

In addition, exciting new technologies now allow us to study the functional roles of individual genes and cells in the brain. For instance, optogenetics uses genetic engineering to make neurons in an animal brain become responsive to light.[10] By beaming laser light into the living brain using fibre-optic cables, it is possible to stimulate—or supress—the activity of such neurons in the brains of experimental animals, and investigate their roles. And new forms of 'gene editing'—discussed in my book *Redesigning Life* and later in this book—now mean that the function of specific genes in the brain can be investigated in practically any mammalian species, including our primate cousins, whose brains are most similar to ours in terms of size and structure.[11] Gene editing makes it possible to totally eliminate a gene's expression, or subtly modify its properties, akin to what happens in many human diseases, and then study the effects on

3

brain function. This allows the gene's normal function to be ascertained, but can also identify potential genetic links with specific mental disorders.

Accompanying these technological advances are important recent conceptual shifts in our understanding of how the brain works. For instance, glial cells—the other main type of brain cell besides neurons—which were formerly thought to only play a supporting role, are increasingly recognized as far more active in transmitting information through the brain.[12] Another important shift in our understanding concerns the role of the 'non-coding' genome in brain function. Until recently, the genome was defined primarily as the sum of the genes—the molecular entities that encode the proteins that form the building blocks of each living cell but also provide the motors and other molecular machinery that drive its activity. Yet genes account for less than 2 per cent of the genome, and recently there has been a realization that anything between an additional 8 and 80 per cent of the genome may have important roles too, the exact proportion with functional roles being a matter of lively debate.[13]

Amazingly, this portion of the genome—previously often termed 'junk' DNA—contains around four million switches that regulate the 22,000 or so protein-coding genes. It also produces many different types of RNA—DNA's chemical cousin. Previously RNA was viewed merely as an intermediary between DNA and protein, but regulatory forms of the molecule have recently been shown to have key roles in their own right. One is regulating chemical modifications of the DNA genome and the proteins associated with it, in response to environmental signals. Such 'epigenetic' changes mean that the genome is far more affected by cellular activity, and that of the body as a whole than had been suspected.[14]

Perhaps most surprisingly, regulatory RNAs have been shown to be secreted by both neurons and glial cells and to travel to other cells in the brain, where they can then directly affect the functions of such cells.[15] Indeed, recent studies suggest that RNAs generated by brain cells in response to stress can even travel via the blood to sperm cells, and are thereby transmitted to the embryo, carrying with them the ability to transmit stress and anxiety to the next generation.[16] If that sounds somewhat worrying, there is increasing evidence that more positive life events can beneficially affect the brain through epigenetic mechanisms.[17]

Yet while focusing on the molecular and cellular biology of the human brain is a crucial aspect of understanding how our minds work, it is unlikely to be enough on its own, for our brains are shaped by our social environment to a degree unique to our species. In this book, I argue that society radically restructures the human brain within an individual person's lifetime, and that it has also played a central role in the past history of our species, by shaping brain evolution. So if we are truly to understand the human mind we must explain how the biological object that is the brain has become infused with this social influence. But in so doing, we must also steer a path between two overly polarized views of the human mind.

The first viewpoint was expressed by Jim Watson, co-discoverer of the famous double helix structure of DNA, who said of the Human Genome Project that 'we used to think that our fate was in our stars, but now we know that, in large measure, our fate is in our genes.'[18] This viewpoint can be applied to humans both as a species and as individuals. At the species level, it is linked with the idea that human behaviour and society are direct readouts of our genetic code, which is why the genome is sometimes referred to as the 'blueprint' of the species. Such a view may also be associated with the belief that supposedly universal human characteristics like competition, selfishness, sexism, homophobia, racism, and even a willingness to make war on other nations are intimately linked to our biology, and not just tendencies specific to the particular type of society in which we currently live.

At the individual level, this viewpoint seeks to explain differences between people—expressed via their personality, intellectual capacity, or sexuality—primarily at the level of genetics. For Jim Watson, 'If someone's liver doesn't work, we blame it on the genes; if someone's brain doesn't work properly, we blame the school. It's actually more humane to think of the condition as genetic. For instance, you don't want to say that someone is born unpleasant, but sometimes that might be true.'[19]

Showing how widespread such views are among some sections of the scientific establishment, Daniel Koshland, former editor of *Science* magazine, when asked in 1989 why so much money was being invested in the Human Genome Project rather than used to help the homeless, replied, 'What these people don't realize is that the homeless are [mentally] impaired…Indeed, no group will benefit more from the application of human genetics.'[20] Here, current social problems are reduced to a problem of the individual, rooted in a defective biology. But Koshland's statement also reflects the optimistic view that it will be possible to identify a clear link between genes and mental disorders, thereby leading to revolutionary new drug treatments for such disorders.

The second viewpoint rejects the idea that human behaviour and society are primarily a matter of biology, but makes the equally bold assumption that genetics has little or no influence on how an individual turns out in life. Such a view was expressed forcibly by John Watson, founder of the behaviourist movement in psychology, who in 1930 said, 'Give me a dozen healthy infants, well-formed, and my own specified world to bring them up in and I'll guarantee to take any one at random and train him to become any type of specialist I might select—doctor, lawyer, artist, merchant-chief and, yes, even beggar-man and thief, regardless of his talents, penchants, tendencies, abilities, vocations and the race of his ancestors.'[21]

More recently, in 2013, the British Psychology Society (BPS) issued a statement attacking what it calls the 'biomedical model' of mental disorders, which many BPS members see as the dominant standpoint of most psychiatrists and one that views such disorders primarily as illnesses to be treated using drugs. Instead Lucy Johnstone, a clinical psychologist who helped draw up the statement, rejects the idea that mental disorders have any biological basis, saying that 'On the contrary, there is now overwhelming evidence that people suffer breakdown as a result of a complex mix of social and psychological circumstances—bereavement and loss, poverty and discrimination, trauma and abuse.'[22]

This statement is interesting because it suggests that psychiatrists and clinical psychologists—the two main sets of health professionals that treat mental disorders—fundamentally disagree about the basis for mental disorders, despite having to work together to diagnose and treat such disorders. So what is the best way to resolve this dilemma? In this book I will make the case that focusing solely on biology or environment alone is unlikely to provide us with a proper picture of how the mind works, or explain why some individuals rather than others succumb to mental disorders. I will argue that purely biomedical, or psychological, views of the mind and its disorders both suffer from major flaws that restrict our understanding of how the human brain and mind work, and also adversely influence our ability to properly diagnose and treat serious mental disorders.

A central problem with the view that mental disorders are primarily due to differences in an individual's DNA has become apparent following attempts to define such clear genetic links in the wake of the Human Genome Project. At the project's conclusion, Daniel Koshland was far from alone in believing that clear genetic links for 'illnesses such as manic depression, Alzheimer's, schizophrenia, and heart disease' would soon be uncovered, providing new targets for drug design.[23] To identify such links, over recent decades, millions of dollars have been invested in so-called 'genome-wide association studies' (GWAS).[24]

Yet while GWAS have identified over a hundred links between different regions of the genome and schizophrenia, the vast majority are not even in the genes themselves but in non-protein-coding regions, and each apparently contributes only a tiny amount to the chance of succumbing to this disorder.[25] This seems to make a mockery of the idea that just a few 'schizophrenia genes' strongly influence an individual's chance of succumbing to this disorder. What is more, this scenario is being repeated for other common mental disorders.

As we shall see, another explanation for these findings is that rather than many small genetic differences contributing to a disorder, instead particular affected individuals may have genetic differences that contribute strongly to their chance of getting a disorder, but are rare in the population as a whole. But it remains far from clear whether this is true for a disorder as a whole, or only for a proportion of affected individuals. Either way, both scenarios pose potential problems for the development of new drug treatments, for if many genetic differences, each with a tiny effect, contribute to a disorder like schizophrenia, how could these be targeted simultaneously with drugs? Or if rare gene variants are the key to understanding mental illness, each different in a specific individual, then could targeting such variants really form a viable economic strategy? Perhaps most importantly, a focus on genetic differences alone neglects the fact that these may only create a predisposition in a highly specific set of environmental circumstances, so that they are far from deterministic in their action, and may even be irrelevant in most other circumstances.

Indeed, the biomedical model as a whole, and therapies designed around it, may require a major rethink as we learn more about the brain's intricate structure, and the way that genes—and mutations in these genes—affect its overall function. A still common view in psychiatry is that serious mental disorders are due to an imbalance in neurotransmitters—the chemicals that transmit signals between neurons—in the brain.[26] So schizophrenia is believed to be caused by an excess of dopamine, while depression is due to a lack of another neurotransmitter, serotonin, also known as 5-HT. Bipolar disorder presents a challenge to this view since sufferers fluctuate between depression and psychotic mania, as this disorder's previous name—manic depression—indicates. Here the idea is that the fluctuating moods are due to cyclical changes in serotonin, dopamine, and a third neurotransmitter, noradrenaline.

One problem with viewing the human brain as a vat of chemicals, with excesses or deficiencies of particular chemicals causing a particular disorder, is that this contradicts new insights about its intricate structure. Although neurotransmitters play key roles in normal brain function, so that alterations in their action may underlie particular mental disorders, such changes only make sense in the context of the detailed neural circuits that constitute the functional architecture of the brain and its sub-regions. The chemical imbalance viewpoint may also be based on a faulty interpretation of how some drugs used to treat mental disorders actually work. Selective serotonin re-uptake inhibitors (SSRIs), of which the most well known is the drug Prozac, are commonly used to treat depression.[27] SSRIs increase the concentration of serotonin in the brain by blocking its re-uptake by neurons, which has led to the claim that depression is caused by a serotonin deficiency.

Yet recent evidence suggests that SSRIs may also stimulate the growth of new neurons.[28] Such neurogenesis is becoming recognized as a very important process in the adult human brain, in contrast to previous beliefs that it only occurred in the embryo, foetus, and child. Defects in neurogenesis may therefore underlie some types of depression, which is why SSRIs can be effective as a treatment. But lest we get too carried away with substituting one explanation for depression with another, it is important to note that a significant number of seriously depressed individuals fail to respond to SSRIs, my sister being a tragic example in my own family. Coupled with the fact that genetic evidence is pointing to a variety of underlying biological causes for depression, it seems likely that we will need to revise ideas that view such a biological link in an overly simplistic and unitary fashion.

If these are some of the problems with the biomedical model of mental disorders, what about the idea that such disorders are solely socially and psychologically based? I have already mentioned the behaviourist movement in psychology, which was pioneered by John Watson and later by B. F. Skinner, becoming particularly influential in the mid-twentieth century. Behaviourism was initiated by the Russian psychologist Ivan Pavlov's observation that dogs salivate not only when presented with food, but also when in the presence of a human laboratory assistant whom the dogs associate with bringing that food.[29] This led to Pavlov's famous experiment in which he showed, by ringing the bell at the same time as the food was presented, that a dog could eventually be trained to salivate in response to a bell alone.

Behaviourism developed into the view that all behaviour, including that of humans, is due to conditioned reflexes.[30] Behaviourists believed that the brain could be treated as a 'black box', whose specific biology was irrelevant to understanding how the stimuli that entered it resulted in subsequent actions. Eventually there was a backlash against the idea that all human behaviour can be seen as equivalent to rats pressing levers to get a reward. Yet the influence of behaviourism on psychology today is more pervasive than is often realized.

The dominant current view in psychology is 'cognitive behaviourism'.[31] Like classical behaviourism, this sees the human mind as something that processes incoming information and generates a response. However, this viewpoint does not believe that the specific structure of the mind can be totally ignored; instead, it sees the mind as similar to a computer, and with humans having a bigger brain in proportion to body size than other animals, the extra information processing that occurs within our brains is assumed to explain the distinctive features of human consciousness. Based on this model, the standard method of treatment used by many psychologists today is 'cognitive behavioural therapy' (CBT).[32] This views people with a disorder like depression as being trapped in a vicious cycle of negative thoughts and feelings. It does not ignore the fact that depression may reflect real dilemmas in a person's life, but argues that the best way out of the depressive state is to break down problems that seem overwhelming into smaller parts, allowing exit from a pathological cycle.

In contrast to Freudian psychoanalysis, which looks for the source of mental illness in the past, CBT is more focused on the present. As an approach for treating mental disorders, CBT is very influential, being the main form of therapy used by the British National Health Service (NHS).[33] To some extent this is because, unlike psychoanalysis with its expensive therapy sessions extending often over several years, led by practitioners who are required to have undergone extensive training, CBT can be carried out by less skilled practitioners, over a much shorter space of time. Because of this apparent cost-effectiveness, CBT has sold itself as a pragmatic, scientific approach to the treatment of mental disorders. And indeed I have benefited from this approach myself in that it has helped me deal with the effects of various traumatic deaths in my family, as well as with more mundane life incidents.

Yet recently doubts have been raised about the value of this type of therapy, at least for serious mental disorders. So Tom Johnsen and Oddgeir Friborg of the University of Tromsø in Norway conducted a 'meta-analysis'—an approach that uses statistics to integrate the findings of multiple independent studies—to assess 70 studies, conducted between 1977 and 2014, that looked at the use of CBT in treating depression.[34] This analysis indicated that the therapy is only half as effective at treating this disorder now as when it first began to be used in the 1970s. So what could be responsible for this startling decline?

One explanation is that cuts in healthcare budgets mean the quality of therapy has decreased, as the length of sessions is reduced or less experienced therapists are employed. But another explanation is that early publicity around this approach made it seem like a miracle cure, so it may have functioned like one initially by virtue of the placebo effect, whereby a patient experiences an improvement in their condition due to personal expectations, rather than because of the treatment itself. Now expectations are more realistic, the effectiveness of CBT has fallen as a consequence.

For psychotic disorders like schizophrenia, the situation is even more problematic. Although initial meta-analyses suggested that around 50 per cent of schizophrenics showed a significant reduction of symptoms following CBT, more recent ones indicate that only 5 per cent of people benefited significantly compared to untreated individuals. Most troubling was a meta-analysis published in the *British Journal of Psychiatry* in 2014.[35] The largest of its kind, this examined the effects of CBT in 3,000 schizo-phrenics. The analysis revealed a 'small' therapeutic effect in reducing symptoms like delusions and hallucinations, yet even this minor effect disappeared in studies conducted 'blind', that is when effects were assessed without knowledge of whether an individual had undergone therapy or was left untreated.

The failure of both pharmacological and 'talking cure' therapies to effectively treat many cases of serious mental disorder comes with a high human cost. Figures released by the British NHS in 2016 showed that the number of annual deaths of mental health patients in England had risen by 21 per cent over the previous three years.[36] The number of mentally distressed patients who killed themselves, or tried to do so, rose by 26 per cent over the same period. This surge in deaths may be partially due to cuts in the health services and support networks available to help such people. According to Paul Farmer, chief executive of Mind (a British mental health charity) and chair of NHS England's mental health taskforce, 'Suicides among people in touch with crisis resolution and home treatment teams—which are there to support people in crisis to stay in their own homes rather than be admitted to hospital—have increased significantly. These teams have in recent years been starved of funding and in some cases have been disbanded altogether or else merged into community teams, losing their specialist function.'[37] However, Simon Wessely, president of the Royal College of Psychiatrists, believes that the apparent dramatic increases may be more a reflection of healthcare staff being more likely to report such events.[38]

What is clear is that the incidence of reported mental disorder is rising at the same time as healthcare services seem increasingly unable to cope with increased demand. In Britain, the number of people seeking help for mental health problems increased by more than 40 per cent between 2005 and 2015, and the number of prescriptions for antidepressant drugs rose by more than 100 per cent during this period.[39] Meanwhile, the number of specialist mental health nurses fell by more than 10 per cent over that time. While such figures may seem like mere numbers to many people, to those grieving the loss of a loved one who ended their life following a mental disorder, they serve as a grim reminder of how different things might have been with better resources. In the case of my sister's suicide, the coroner concluded that she was 'discharged too informally by those responsible for her with no paperwork and no advice to the family on how to help her.'[40] Of course it is easy with hindsight to imagine what might have been, but there seems little doubt that the erosion of NHS services has had a negative impact on mental health diagnosis and treatment.

In the USA, the lack of a publicly funded health service like the NHS means that the situation for people with mental disorders is even worse.[41] Until recently, even those with a private health insurance plan were not covered for mental illness. In November 2013 President Barack Obama's health reforms forced insurance companies to cover mental disorders in the same way as other medical conditions.[42] Yet the legacy of this lack of cover means that the infrastructure to deal with mental disorders remains woefully inadequate. Paul Appelbaum, a psychiatrist and expert on legal issues in medicine at Columbia University, New York, has noted that 'right to care does not mean access to treatment. Tens of millions of people who did not have insurance coverage may now be prompted to seek mental health treatment. And the capacity just isn't there to treat them. There really is no mental health system in the US.'[43] As a consequence, more than half a million US citizens with serious mental illness are falling through the cracks of society, leading Tim Murphy, a child psychologist and Republican congressman, to remark that 'we have replaced the hospital bed with the jail cell, the homeless shelter and the coffin. How is that compassionate?'[44]

While US states have been reducing hospital bed places for decades, the worst cuts occurred following the recession that began in 2008, with $5 billion disappearing from mental health services budgets from 2009 to 2012.[45] In the same period, 4,500 public psychiatric hospital beds—nearly 10 per cent of the total supply—were lost in the USA. According to the 2012 US National Survey on Drug Use and Health, nearly 40 per cent of adults with 'severe' mental illness received no treatment at all.[46] One possible consequence of this situation is mass shooting incidents, reports of which now regularly punctuate the US news. Yet a less reported but ultimately far greater loss is the number of Americans who commit suicide each year, with, for instance, almost 45,000 people taking their own lives in the USA in 2016—a fate that claims the lives of more citizens than car accidents, prostate cancer, or homicides, according to the US Centers for Disease Control and Prevention.[47] Worryingly, the US suicide rate increased 33 per cent from 1999 to 2017, with rates rising particularly sharply since 2006.[48] So while the US suicide rate increased by about 1 per cent a year from 2000 to 2006, that rate doubled from 2006 to 2016.[49] And although suicide is the starkest indicator of mental distress, there are others; so drug overdoses claimed 70,000 US lives in 2017. Because of this, US life expectancy, perhaps the broadest measure of a nation's health, fell from 2016 to 2019, in part because of the rise in drug overdoses and suicides.[50] This was the biggest three-year drop since the years 1915 to 1918.

Mental health problems cause misery for sufferers and their families, but also have an economic impact; for instance they cost the British economy £110 billion a year, according to a 2009 report by the Royal College of Psychiatrists, the London School of Economics, and the NHS Mental Health Network.[51] Obviously, a better resourced health service would be a key step to helping people with mental disorders, but improving the treatments—based on both drugs and psychotherapy—available for such people is also critical. Ultimately, the possibility of such an improvement rests on a better understanding of how the human mind works, and how this relates to human brain function, topics with which this book will be concerned. Here though we face a problem, for while there are many different viewpoints about these subjects, what is lacking is a unifying framework that binds them together.

This problem is not new. In 1934, comparing viewpoints about the mind at that time, the Russian psychologist Lev Vygotsky stated that 'there exist many psychologies, but [no] unified psychology.'[52] At the time, three particularly influential views were introspective psychology, behaviourism, and psychoanalysis. Introspective psychology, pioneered by individuals like William James (brother of novelist Henry James) sought to understand the human mind by questioning individuals about their innermost thoughts. Behaviourism, as already mentioned, was stimulated by Ivan Pavlov's discovery of conditional reflexes. The third viewpoint, psychoanalysis, developed from Sigmund Freud's discovery that the neurotic individuals he analysed seemed to be supressing important thoughts and desires, particularly those relating to sex, leading to the idea of an unconscious mind coexisting alongside the conscious one.

Vygotsky believed that James, Pavlov, and Freud had all provided important insights, but he pointed to two problematic tendencies within the schools of thought that grew from these insights. The first was for each to develop into an all-encompassing view of the mind that tended to exclude other viewpoints; the second was for this world view to collapse under the weight of its own internal contradictions.[53] The first tendency was illustrated by the way behaviourism developed into the idea that all human behaviour is merely a series of conditioned reflexes. In contrast, Freudian psychoanalysis viewed an individual's behaviour, and even human society as a whole, as a battle between the conscious and unconscious mind. Meanwhile, introspective psychology argued that only the individual can truly assess his or her innermost feelings. These mutually exclusive ways of viewing the mind meant that 'any behavioural or mental act being expressed in terms of these three systems would acquire three entirely different meanings.'[54] The second tendency involved a view of the mind expanding to the point that it became almost a parody of the original insight. So Freud's view that unconscious impulses are primarily due to repressed sexuality led to his eventually trying to explain all of human society in this way, so much so that he came to see 'communism and totem, religion and Dostoevsky's writings, occultism and commercials, myth and Leonardo da Vinci's inventions [as all] just a libido in disguise and nothing else.'[55]

Given that Vygotsky was writing in the early twentieth century, one might wonder how relevant his arguments are today. Yet almost a century later, it is far from clear whether contemporary mainstream viewpoints about the mind are any more unified. As I have already highlighted, the two views of the mind that dominate clinical practice—the biomedical model espoused by most psychiatrists, and the cognitive behavioural one adhered to by the majority of clinical psychologists—are based on perspectives that seem to share little common ground. So does this mean that a unified explanation of how the mind works, encompassing both biological and social aspects of human consciousness, will always be outside our reach? Certainly this is the view of some commentators; so Stuart Sutherland, a psychologist at the University of Sussex, stated in 1989 that 'consciousness is a fascinating but elusive phenomenon: it is impossible to specify what it is, what it does, or why it evolved.'[56] Yet it is my view that this is not so, and in the rest of this book, I will explain why.

Building on Vygotsky's original insights, but linking these for the first time to the findings of twenty-first-century neuroscience, I will argue that the human mind has undergone a qualitative transformation, enlarging the brain, giving it greater flexibility and enabling higher functions such as imagination, and that this shift has been driven by tool use, but especially by the development of one remarkable tool—language. I will show how the complex social interaction brought by language opened up the possibility of shared conceptual worlds, later enriched with music, art, and literature. This 'mind shift' has enabled modern humans to leap rapidly beyond all other species, and generated an exceptional human consciousness, a sense of self that arises as a product of our brain biology and the social interactions we experience. As a consequence, our minds, even those of identical twins, are unique because they are the result of this remarkably plastic brain, exquisitely shaped and tuned by the social and cultural environment in which we grew up and to which we continue to respond through life. However, suffering, delusions, and despair can occur when this symphonic brain activity goes wrong. This then is the book's theme, and it will require re-examining what we mean by consciousness, but also what it means to be human. This I will seek to do. But before doing so, let us first consider previous ideas put forward to explain how the human mind works, and why I believe we need a very different approach.

PART I

ORIGINS OF MIND

CHAPTER 1

MIND AND MATTER

W hat is the basis of human thought? How does human consciousness differ from that of an animal? Why do different people have
such distinct types of personalities? Is the human mind separate from the
body, or intimately linked to it? Fundamental questions such as these
have occupied the minds of philosophers at least since the ancient Greeks.
For instance Aristotle believed that consciousness exists as a continuum
of different types of 'souls' (Figure 1).[1] According to him, plants have a
vegetative or nutritive soul, which controls their growth, nutrition, and
reproduction, while animals have such properties too, but also possess
the powers of perception and locomotion. Aristotle thought that any
organism able to feel can experience pleasure; therefore animals also have
desires. However, he saw humans as unique in having a 'rational' soul,
which gives our species the power of reason and thought. Aristotle's ideas
shaped discussion about consciousness for almost two thousand years.
But they did not address the question of why humans are different from
other animals, or how we might begin to understand animal, but also
human, consciousness in a scientific manner.

In fact, in the hands of the Catholic Church, such ideas were mainly
used to consolidate the idea that the human soul was outside the bounds
of science. This was unfortunate because Aristotle was one of the first
individuals to investigate nature by carrying out experiments. However,
such an approach was seen as highly controversial in medieval times if it
began to directly challenge existing ideas sanctioned by the Church.

With such sanctions against those who advanced new theories about
the structure of the Universe, it is perhaps not surprising that speculation

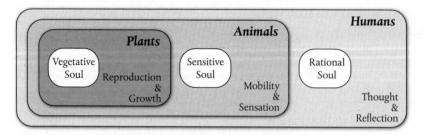

Figure 1 Aristotle's three types of soul.

about the material basis of human consciousness—the realm of the 'immortal soul'—could be even more dangerous. However, as the seventeenth century progressed, even here change was afoot, as new ideas were put forward about the human body, and how it might function. And the person who really began to change our views in this respect was an Englishman—William Harvey.

Born in Folkestone, Kent in 1578, Harvey was educated in Cambridge, but his most formative experience was his later studies at the University of Padua in Italy.[2] Uniquely for the time, students at Padua learned about the human body by dissection and observation, rather than simply regurgitating the views of ancient thinkers. Returning to England in 1602, Harvey became a medical doctor, but he also began studying the heart and blood vessels, performing experiments on animals as well as observing patients.[3] Such studies led Harvey to propose that the heart pumps blood in a one-way circulatory system.

The realization that the heart works like a pump transformed our understanding of this organ, but also helped stimulate the idea that the whole human body can be viewed as a machine.[4] In particular, the philosopher René Descartes became a major exponent of this viewpoint. Born in La Haye en Touraine, France, in 1596, Descartes refused to be bound by previous thinkers' ideas, arguing that he would investigate key topics 'as if no one had written on these matters before.'[5] He proposed that many 'motions occurring inside us do not depend in any way on the mind. These include heartbeat, digestion, nutrition, respiration when we are asleep, and also such waking actions as walking, singing, and the like'.[6] This was bold thinking, for it suggested that not only bodily functions but also aspects of human behaviour can be viewed as analogous to the work-ings of a machine. But Descartes was careful to note that higher mental functions such as conscious awareness, free will, and personality cannot be explained in this manner, saying that 'on the one hand I have a clear and distinct idea of myself, in so far as I am simply a thinking, non-extended thing [that is, a mind], and on the other hand I have a distinct idea of body, in so far as this is simply an extended, non-thinking thing.'[7] This led to his famous statement, 'I think, therefore I am.'[8]

Although Descartes believed that the soul can influence the body through a point of contact he arbitrarily located in the pineal gland, he argued that its supernatural form made it impossible to understand by scientific methods. This separation between body and mind left open the possibility that the mind has an immortal nature, as the soul, in keeping with the idea of an afterlife. Descartes' standpoint made some sense given the limited understanding of brain function at that time. It may also have been a pragmatic position given the Church's persecution of individuals for putting forward scientific views about the physical universe, let alone questioning the existence of an immortal soul. But the effects of this separation have clouded understanding of the human mind—and how it works—to the present day.

One of the first people to challenge the idea that the mind could not be studied scientifically like the rest of the human body was the philosopher John Locke. Born in Somerset in 1632, Locke grew up during the English Civil War of 1642–9;[9] indeed his father was a captain in the parliamentary army, and as a student at Westminster School in London, Locke was deeply affected by hearing the groan of the crowd at Charles I's execution, which took place at nearby Whitehall in January 1649.[10] The social revolu-tion that accompanied the civil war transformed British society by overturning the old feudal order, allowing new ideas about science and society to blossom.[11] As a scholar at Christ Church in Oxford, Locke turned his attention to the human mind; his solution to the problem of understanding how the mind worked was to view it as a 'blank slate', expressed by his statement, 'Let us then suppose the mind to be, as we say, white paper void of all characters, without any ideas. How comes it to be furnished?...To this I answer, in one word, from experience.'[12]

Such a view explained how ideas could change in society, for an individual's changing experience could also change their consciousness. This was an important principle in post-revolutionary British society in which, in theory at least, a person's rise in society was now based on merit, and not upon their position in the feudal order. Yet Locke did not completely break from the dualism that Descartes had introduced into discussions about human consciousness. Acknowledging that there is surely more to the human mind than simply experience, he proposed that there must also be some material factor that binds an individual's life story into a sense of self and personal identity.[13] However, he skirted the issue of how such a unifying influence might arise in the brain by stating that God had made humans this way, so in this respect his proposal was not much of an advance on Descartes' model, in that it left the higher aspects of human consciousness as an unknowable entity.

While Locke was prepared to prod at the boundaries of the dualism introduced by Descartes in his discussions about human consciousness, the real iconoclast in this respect was the Scottish philosopher David Hume. Born in Edinburgh in 1711, Hume took the blank slate view of the mind to its logical conclusion, by saying that there really is nothing more to human consciousness than experience. Addressing the issue of how this could ever result in a unified sense of personal identity, he proposed that this is to some extent an illusion. Instead, for Hume, the self is nothing but a collection of diverse experiences. However, recognizing, like Locke, that something must bind such experiences together for consciousness to form a coherent whole, he proposed that 'there is a secret tie or union among particular ideas, which causes the mind to conjoin them more frequently together, and makes the one, upon its appearance, introduce the other.'[14] As an attempt to explain human consciousness purely by materialist principles, and without any need for a supernatural soul, it was an important step forward. Yet by failing to properly explain how ideas cohere into a unified whole, Hume was unable to complete the revolution in thinking about the mind that he had initiated.

Despite the deficits in Locke and Hume's views of human consciousness, the blank slate view of the mind is still very influential in psychology, for, as already mentioned, it is central to behaviourism. Indeed, behaviourists like John Watson and B. F. Skinner followed Locke in ignoring the question of how experience might generate a sense of personal identity in a human brain treated as a blank slate. As one critic has put it, this view assumes that 'children are born with brains of soft clay, their mental makeup...infinitely mouldable'.[15]

Yet as should be obvious to any parent with two or more children, siblings can grow up in the same family environment but display completely different personalities, interests, and degrees of motivation. And treating the mind as a blank slate also fails to explain why humans differ from animals, and why some people thrive in a challenging environment, while others in the same environment succumb to mental disorders.

In the face of such criticisms, 'cognitive behaviourism' has sought to create a more sophisticated way of understanding the mind. Rather than basing itself on the simple stimulus–response circuit first identified by Pavlov, cognitive behaviourism sees the human mind as a very sophisticated information-processing device, or computer.[16] By analogy with computers, this view of the mind sees it as a device that receives information from the outside world, processes and stores this information, and then uses it to produce a response, either immediately or in the future. The particular power of human minds is then seen to arise from the fact that our relatively bigger brains endow us with a much greater computing power than the brains of other animals. But there are also a number of limitations to this view.[17]

For a start, by focusing primarily on the logical aspects of cognitive processing, it neglects the emotional, creative, and social side of human thought. It also does not explain why actual computers, despite having vastly greater number-crunching abilities than a human brain, show no signs of developing the conscious awareness of the human mind, although I will return later to the question of whether computers might achieve this ability in the future. Moreover, this model rarely engages with the question of how the different neurons in a living human brain come together to create such a super-powerful computing device, or why there seems to have been a qualitative leap in our consciousness compared to those of other species, including our closest cousins, chimpanzees. Linking all these issues is a problem with both traditional behaviourism and its more modern cognitive version, which is that the mind is largely treated as an abstraction. It remains unclear how such a mind might relate to actual brain structure and function. Because of this, neither approach seems to take us much further than the unknowable soul of Descartes' model, in terms of leading to an understanding of how consciousness takes form in a human brain.

Given that we know vastly more about the biology of the brain than Descartes did, can we use such knowledge to better understand the mind? A major advance came with the discovery that all life can be viewed as a flow of information from the DNA of the genes to the proteins which act as the building blocks of each cell, and the body as a whole. Although DNA was discovered by the Swiss biologist Friedrich Miescher in the 1860s, it took almost a hundred years for scientists to realize its unique role in the life process.[18] This realization came with the theoretical elucidation of its 3D structure by Jim Watson and Francis Crick in 1953, building on the experimental studies of Rosalind Franklin and Maurice Wilkins.[19]

The discovery of the famous double helix structure of DNA was a major breakthrough for two reasons.[20] First, it showed how the 'molecule of life' reproduced itself at each cell division. Essentially, the helix breaks in two, and a new double helix is formed from each single strand of DNA. Second, the DNA structure showed how the genetic material might act as a 'blueprint' for the organism. The realization that DNA could be viewed as a chain made of four letters—the DNA 'bases' A, C, G, and T—that acted as a linear 'code' led to the discovery that three bases of DNA code for a

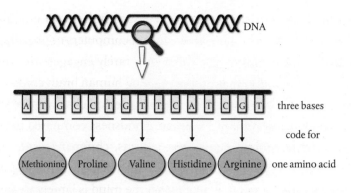

Figure 2 Genetic code from DNA to protein.

particular amino acid, the basic units that when linked together form proteins (Figure 2). For instance, GTT codes for the amino acid valine, CCT for proline, and so on for all twenty amino acids. Working out the genetic code was a key scientific achievement of the 1960s.[21] Finally in 1977 Fred Sanger developed a method for 'sequencing'—that is, reading—all the letters of a genome. Following a colossal effort involving scientists across the globe, the human genome was fully sequenced in April 2003.[22]

The central importance of DNA to life—and the fact that differences in genomes ultimately define to which species an individual organism belongs—led to claims that possession of the human genome sequence would have a major impact on our understanding of what it means to be human. The late Sir John Sulston, who led the Human Genome Project in Britain, said at the end of the project that 'we've now got to the point in human history where for the first time we are going to hold in our hands the set of instructions to make a human being.'[23] So given that a defining feature of being human is our conscious minds, how have advances in genetics affected our ability to understand the way our minds work? To properly address this question, it is worth restating a few basic principles of genetics.

Genetics was established by Gregor Mendel, a monk working in a monastery in Brno, once part of the Austro-Hungarian Empire but now in the Czech Republic. In 1865 Mendel showed that particular characteristics are passed down to future generations according to defined 'laws' of inheritance.[24] These laws stem from each characteristic being determined by 'factors'—now termed genes—that come in opposing pairs, with one member of the pair dominating the other. In the pea plants that Mendel studied, such pairs determine long versus short stems, or smooth versus wrinkled pods. In 'dominant' inheritance, a single copy of the dominant gene gives the plant a particular characteristic or trait, thereby making itself known in every generation. In 'recessive' inheritance, two copies of the recessive gene are required to produce the trait: parents with only one copy do not show the trait themselves, but a quarter of their offspring have both copies, and therefore display the trait.

Although Mendel studied pea plants, the laws of inheritance that he established were subsequently shown to be true for predispositions to particular human diseases (Figure 3).[25] Huntington's disease, associated first with jerky movements, then psychotic behaviour, and finally dementia and premature death, is a dominant genetic disorder. In contrast, cystic fibrosis, in which sticky mucus accumulates in the lungs and pancreas, stopping them functioning properly and eventually leading to an early death, is a recessive disorder. Because of this, it is passed on to future generations by 'carriers' who meet and have children without being aware that they then have a one-in-four chance of having a child with the disorder.

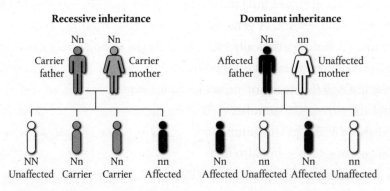

Figure 3 Inheritance of recessive and dominant characteristics.

Sometimes human behaviours like 'selfishness' or 'nationalism' are said to have a clear genetic basis. Such statements are based on the often unspoken idea that these conditions are inherited with the same mathematical precision that Mendel described for his pea plants, or is true of a disorder like Huntington's or cystic fibrosis. However, given that behaviours like selfishness are asserted to be a general characteristic of humans as a species, such associated genetic differences are presumably thought to have already become established in every living person, not just a few individuals. Take, for instance, the 'helping gene' that Mark Pagel of the University of Reading has proposed to code for an 'emotion that disposes people to be friendly.'[26] In fact this gene turns out to have a dual character, for while it makes us friendly to other individuals of our particular nation, towards people of a different country it becomes a 'xenophobic' gene that leads to distrust of foreigners and can even be an important factor in making people willing to go to war with those of another nationality. This is just one illustration of how claims about the influence of genes may be used to reinforce social stereotypes, such as the idea that people are naturally selfish, or jingoistic, why women are less likely to occupy the top jobs, or why men do not tend to like ironing.[27]

Critics of such claims have mainly focused on their reactionary conclusions about human nature, or women's position in society compared to men. But from a scientific point of view, a more fundamental problem is that such claims make little sense given how genes, and the genome as a whole, actually work. A gene only produces a particular protein when stimulated to do so, typically by a chemical signal in the cell that itself is activated by an agent like a hormone that arrives at the cell surface, attaches itself to a 'receptor' protein, and triggers a cellular response.[28] For instance, adrenaline stimulates receptors on heart, muscle, and liver cells, and thereby triggers an immediate response—the 'fright, flight, fight' reaction to a scare—but also induces longer-term changes in the activity of particular genes. Other cellular signals switch genes off, rather than turn them on.

Importantly, such signals activate specific proteins, themselves coded by genes. But this also means that genes can only influence the cell—and the wider body—as part of a network of proteins, also coded by genes. Not only do many proteins combine to form multi-protein 'complexes', but also they often work within 'signalling pathways', in which protein A influences the actions of protein B, which regulates protein C, and so on. This means that in an individual cell, we can only understand how genes affect the properties of that cell if we also consider the other proteins with which a particular protein interacts. And since proteins are encoded by genes, that means that genes themselves interact in complex ways; indeed biologists now commonly talk of 'gene regulatory networks'.[29]

All of this is important in considering how particular human behaviours, and human consciousness itself, might have evolved. Remember, genes themselves are never a direct target for the forces of natural selection; instead, those forces can only influence two aspects of an organism: its capacity for survival within a particular environment, and its subsequent ability to reproduce and pass on its genes to future generations. Of course, serious genetic abnormalities can influence this process. Tragically, boys with the muscle-wasting disorder Duchenne's muscular dystrophy (DMD) typically do not live longer than their teens, and are generally in a wheelchair long before this, such is the severity of the disease.[30] As a consequence, such boys never transmit their disorder to future generations. Instead, DMD is passed on by female 'carriers' who do not suffer from this condition, because it is due to a defect in a gene called dystrophin located on the X chromosome.

Since women have two X chromosomes, carriers have a normal copy of dystrophin which makes up for the loss of a functional other copy. However, half their male offspring have no such protection. DMD, like the blood-clotting disease haemophilia, is an 'X-linked recessive' disorder (Figure 4). In contrast, a dominant disease like Huntington's makes its effects known each generation. But because the symptoms only manifest themselves relatively late in life, people with this disorder often pass on

X-linked recessive inheritance

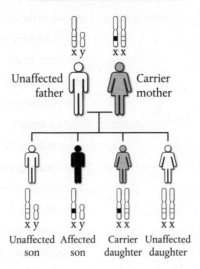

Figure 4 Inheritance of X-linked recessive characteristics.

this genetic defect to their children before they realize they have the disease and can transmit it.[31]

It has been possible to link disorders like cystic fibrosis, DMD, and Huntington's to single genes because the genetic changes associated with them have such dramatic effects. For instance, with DMD, no functional dystrophin protein is produced, while in Huntington's, the negative effects of the mutant huntingtin protein are so severe that a sufferer eventually dies from the condition. Yet there is growing evidence that more common conditions, whether bodily disorders like diabetes or heart disease, or those that affect the mind like depression or schizophrenia, have no such simple genetic link.

Why this is so is something that I will explore in greater detail in later chapters of this book, but it does suggest that claims about the genetic basis of human behaviour, and differences between human individuals, that are based on a simple extrapolation from rare single-gene disorders have overexaggerated the similarities between these disorders and the far more common ailments that can affect both the mind and body.

This also seems to be the case for other human characteristics often claimed to have a clear genetic basis, ranging from intelligence to sexual inclination. For instance, a recent study investigated nearly 80,000 children and adults in order to identify differences in DNA sequence across the human genome associated with 'general intelligence'—defined by an individual's IQ score and ability to give answers to brief touchscreen puzzles. The study identified 52 genomic regions that showed such an association.[32] However, these only account for about 5 per cent of the IQ scores among different people. Given that IQ tests themselves have been criticized for their class and cultural bias, that is hardly a ringing endorsement for a genetic basis for intelligence. Media reports in 1993 that a 'gay gene' predisposes men to becoming homosexual were misleading given that no gene had been discovered, only an association with a slight genetic difference in the X chromosome.[33] In fact, subsequent attempts to find an actual gene failed to identify such an entity.[34] None of this means genes are not involved in human behaviour—an aspect I will explore later in this book—but it does mean we need to move beyond simplistic views about their involvement.

Linked to the attempt to identify simple links between genes and human behaviour has been the claim that such genes map directly on to structures in the brain. Harvard University psychologist Steven Pinker has proposed that the human brain is an interacting community of distinct modules, each specialized for a particular function. Conveniently, each module is suggested to have a specific genetic basis.[35] Explaining how such genetically coded brain modules arose, Pinker has argued that each evolved as a response to problems faced by our hunter-gatherer ancestors as they sought to survive and prosper in a hostile environment. Yet neuroscientist Steven Rose of the British Open University has criticized Pinker on the grounds that the latter seems 'not very interested in actual brains, since his mental "modules" may or may not map onto actual neuronal ensembles.'[36] In fact recent studies of the brain confirm that the brain is modular in some respects, yet increasingly point to consciousness as a feature of the whole brain. This raises the question of how a brain that is partly modular in its structure might at the same time only function properly as an interconnected whole—something we will consider later in this book.

In summary, a major problem with much popular speculation about the biological roots of consciousness is that those who advocate a gene-based view of consciousness often appear to have little understanding of modern genetics, while speculation about how brain structures shape that consciousness often bears little resemblance to emerging knowledge about the complexity of an actual human brain. There is a common thread here, which is that idealized genes and brains have been substituted for real ones. Unfortunately, because of this tendency, it is not clear how much we have really advanced forwards from René Descartes and his belief that the human mind was an unknowable entity or, for that matter, the behaviourists with their view that the human mind could be treated as a black box.

In contrast, to understand human consciousness we need to understand real genes, real brains, and how these have evolved in humans compared to other species. And to do this means first of all re-examining the unique features that distinguish us from such species.

CHAPTER 2

TOOL AND SYMBOL

To understand the basis of human conscious awareness, a good starting point would be to establish what makes humanity unique compared to other species on our planet. To do so, however, means going against a trend in science over the past few centuries that has tended to downgrade the uniqueness of *Homo sapiens* and our place in the Universe, for entirely valid reasons. This trend began with Nicolaus Copernicus's demonstration in 1543 that instead of being at the centre of the Universe, the Earth is merely a satellite of the Sun, which we now know as just one star amongst trillions of others. A further blow to our egos came with Charles Darwin and Alfred Russel Wallace's theory of natural selection, most famously expounded in Darwin's *On the Origin of Species*, published in 1859, which showed that humans are only one of many branches on the tree of evolutionary change.[1] We now appreciate that while life itself has existed on Earth for 4 billion years, humans only diverged from our closest relatives, the chimpanzee, between 6 and 10 million years ago.[2] And comparisons of human and chimp genes have shown that these are 99 per cent similar in DNA sequence.[3]

Such discoveries have helped to undermine the idea, associated with thousands of years of religious dogma, that humanity is unique among the life forms on our planet, having been created in the image of some deity, or deities. Yet it surely does not require any belief in a supernatural being to recognize there is something quite distinctive about our species compared to others in the animal kingdom, including every other primate species.

Growing up in an advanced modern civilization, it is easy to forget how remarkable are the technologies that *Homo sapiens* has generated that help define a typical human life. And far from reaching a plateau, humanity's technological achievements are accumulating at an exponential rate. Not that everyone will agree on the value of all such technological developments. A growing problem for those seeking to publicize the most up-to-date images from Pluto or the moons of Jupiter is having to compete with the latest selfies from Kim Kardashian or Beyoncé. Yet fundamental or frivolous, nothing generated by any other species on Earth compares with our human powers of technological mastery and innovation.

Take for instance our closest animal relative, the chimpanzee. Recent studies have shown that chimps have a sophisticated social structure, are able to use natural objects as tools, and can construct habitats such as nests in the trees in which they live.[4] Yet nothing suggests that the way chimps live today is very different to how they have existed for the past few million years except, tragically, as a result of human activity, which is gradually eroding their habitats and pushing them to the edge of extinction.[5] In this respect, despite their genetic similarity with humans, chimps have far more in common with every other species on Earth, in being bound by the laws of natural selection. As Darwin and Wallace demonstrated, this allows organisms to radically change their character over time, but at such a slow rate that changes can only be measured over millions of years. In contrast, over the past 40,000 years, humans have gone from scratching a living from nature to building cities, cultivating crops and farm animals, and devising inventions which have taken humans beyond our planet, imaged individual atoms, and captured light from the farthest reaches of the Universe.

So how did this divergence between humans and other species arise? In addressing this question, we have two main options. One is to use archaeology to uncover fossilized human skeletons and cultural artefacts from past human, or human-like, societies. The other is to study human evolution through comparisons of different genomes—modern humans versus chimpanzees, other apes, and extinct proto-humans like Neanderthals. While DNA evidence has only been available over the past few decades, scientists have been drawing conclusions about human prehistory ever since the discovery of a Neanderthal skull in Germany in 1856—three years before Darwin published his *Origin of Species*.[6] Since then, a multitude of fossils have been discovered spanning several million years of human evolution, along with artefacts such as stone tools. Yet much remains uncertain about the exact pattern of human evolution, partly because many fossils exist merely as fragments, and partly because of misconceptions that have affected the interpretation of these fossil finds.

One such misconception led to that first Neanderthal skull being disregarded by Herman Schaaffhausen, the naturalist who initially studied it. Ignoring the skull's location next to extinct cave bear and mammoth bones, Schaaffhausen claimed it belonged to a member of one of the barbarian tribes mentioned by Roman historians.[7] Only some years later was the skull recognized as belonging to an extinct proto-human species. And showing that even the greatest scientists can make mistakes, another misconception in our view of human evolution came from none other than Charles Darwin. In his book *The Descent of Man*, published in 1871, Darwin considered which key distinguishing features of humanity— walking upright, speech, tool use, and a bigger brain—might have evolved first. Darwin plumped for a big brain.[8] He did recognize the importance of the other features, but saw these as occurring after the growth of the brain, and therefore as a consequence of this growth. Yet Darwin's order of events was mistaken. We now know that a distinctively large brain only became a feature of humanity several million years after our ancestors first embarked on their unique evolutionary path. Instead, what first set us on this path was when proto-humans first began walking on two feet. Next came a particular aptitude for using tools. Dramatic brain growth only occurred much later in our evolution.

Of course when Darwin was writing there was little fossil evidence to help illuminate human evolutionary history. For as well as the confusion about the exact relationship of Neanderthals to humans, there was no evidence at this time of more ancient proto-human fossils. However, archaeologist Bruce Trigger of McGill University in Montreal has suggested that Darwin was also constrained by a 'reluctance to challenge the primacy which the idealistic religious and philosophical thinking of his time accorded to rational thought as a motor in bringing about cultural change.'[9] In other words, Darwin may have been overly swayed by his intellectual's view of the importance of the mind compared to other human attributes such as our hands, with their particular skill at making and using tools.

Intriguingly, one individual at this time who did guess the correct sequence of human evolution was Friedrich Engels, which might seem surprising given that he is more usually known as a political thinker and activist, not a source of insights about the natural world. Yet in an essay he wrote in 1876 entitled 'The Part Played by Labour in the Transition from Ape to Man', Engels argued that human evolution was initiated when our ancestors began walking on two feet, which freed the hands to design and make tools.[10] Such tool use, carried out as a highly social activity, eventually gave birth to language, and this, together with shared labour activities, began to fuel the growth of the brain in a kind of positive feedback mechanism.

Engels' essay was remarkably prophetic given that such a pattern of events is now accepted as the correct sequence of human evolution. Unfortunately, the essay lay unpublished until after Engels' death in 1896, when it appeared in *Die Neue Zeit*, a German socialist newspaper.[11] There its insights were ignored by the professional scientific world, or, more likely, it was not even known about in those circles. This was unfortunate: guided by Darwin's proposal that our large brains were the first distinctive human feature to evolve, for many years archaeologists accepted erroneous findings and disregarded important clues in their bid to make the fossil evidence fit such a mistaken pattern of evolutionary change.

One famous piece of wrong-footing was a discovery made near Piltdown, in Surrey, England, in 1912. Here, Charles Dawson, an amateur archaeologist, identified a fossil skull with an ape-like jaw but a cranium of modern human-like proportions.[12] Fitting as it did the proposal that the first proto-humans had an ape-like posture and a large brain, the find led to newspaper headlines along the lines of 'Missing Link Found—Darwin's Theory Proved'. Yet the 'fossil' was a fake, something only realized in 1953 when analysis showed that the main part of the skull was only 500 years old and the jaw came from a modern-day orang-utan. Who propagated the hoax remains debatable, although recent evidence has implicated Dawson himself.[13] In the meantime, not long after the Piltdown 'discovery', a genuine archaeological find by Raymond Dart in South Africa in 1924 of the partial remains of an ape-like creature with a small brain, but which walked on two legs, was largely disbelieved.[14]

Only in 1974 when Donald Johanson discovered a complete 3.2-million-year-old female skeleton with an ape-sized brain and erect posture in Afar, Ethiopia, was Darwin's sequence finally laid to rest.[15] Johanson named the skeleton 'Lucy', after the Beatles' song *Lucy in the Sky with Diamonds*, which was playing in the archaeologists' camp at the time of the discovery. More seriously, the skeleton was proposed to be that of a new proto-human species, *Australopithecus afarensis*, meaning 'southern ape from Afar'.

Although Engels was ahead of his time in recognizing the general pattern of human evolution, in other ways his account is misleading in giving the impression of a smooth progression from ape to human. We now know the evolutionary process was far from such a linear ladder of progress, being more like a branching tree, involving multiple proto-human species—also known as hominids—many coexisting for long periods of time (Figure 5).[16]

Such complexity is exacerbated by proto-human remains often only consisting of fragments of bone, teeth, or bits of rock that may have, or not, been stone tools. This makes assigning a new fossil to a known proto-human species, or deciding whether it belongs to a totally new one, a process open to much discussion. As palaeontologist Chris Stringer of the Natural History Museum in London has noted, 'the field of human evolution is littered with abandoned ancestors and the theories that went with

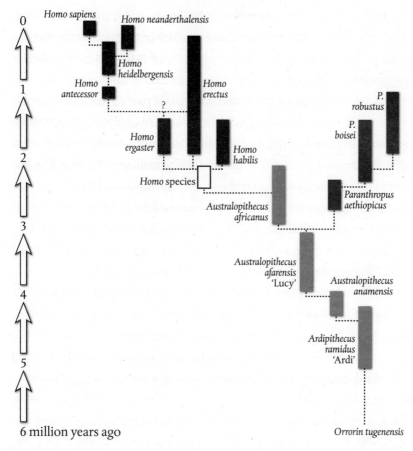

Figure 5 Evolution of humans from apes.

them...Failure to realize the complexities involved in trying to interpret a few fossils scattered sparsely through space and time has characterized the approach of even the most competent workers, resulting in naïve interpretations.'[17] Despite this diversity, the trend is that first identified by Engels, namely of upright walking freeing the hands to use and design tools. We should consider, then, at what point proto-humans first began using tools in a distinctive way compared to other apes, and how this might have led to other features like language capacity and a bigger brain.

Definitively answering the first question is complicated by the regular appearance of new evidence to challenge past views on this matter.

This capacity of nature to challenge accepted viewpoints is, of course, what gives science its power, but it can create much debate and disagreement in the process. For human evolution, an early consensus was established by Louis and Mary Leakey, who in 1964 discovered a 1.8-million-year-old skull of a species they named *Homo habilis*, or 'handy man', because the skull was found next to hand-crafted stone tools.[18] Giving the new species the prefix *Homo* signified it as the start of a more human-like line of evolution, in contrast to *Australopithecus afarensis*, which was bipedal, but has otherwise been thought to be far more ape than proto-human.

Yet recently Sonia Harmand and Jason Lewis, of Stoney Brook University in New York, have challenged this distinction, through their discovery of stone tools dated to 3.3 million years ago, raising the question of which proto-human species made such tools.[19] Other studies have uncovered butchered animal bones from this time that show the marks of precision stone tools.[20] An obvious candidate is *Australopithecus afarensis*, which lived from 3.9 to 2.9 million years ago. This pushes back the point in prehistory at which proto-humans may have designed and used tools. For palaeoanthropologist Henry Bunn of the University of Wisconsin, this implies that while proto-humans at this time 'had small australopithecine-sized brains...nevertheless they figured out how to cut through often tough hide to efficiently get the meat off the bones and break the bones open for the marrow.'[21]

What might have triggered this move into tool-making? The stone tools identified by Harmand and Lewis were made during a 'drying out' period in this African region between 2 and 3 million years ago that would have presented our ancestors with a much greater variety of habitats, such as woodlands and grasslands, than previous primates had known. An enhanced ability to use tools at this time could have been particularly fortuitous, in allowing proto-humans to be more adaptable and extract food from a greater range of areas.[22] Despite these steps towards the human condition, our ancestors in this prehistoric era appear to have still partially shared the tree-dwelling lifestyle of their primate cousins.

In addition, the tools they created were probably still quite ad hoc and largely disposable. Thomas Wynn, a palaeoanthropologist of the University of Colorado, believes that the new findings suggest that 'when you needed a stone tool and you didn't have one, you just made one, then dropped it.'[23]

A breakthrough in tool-making seems to have taken place in a later proto-human species—*Homo erectus*. According to Wynn, 'the technology [of this species] is really different, more sophisticated in a cognitive way than anything earlier hominids or chimpanzees could do—some see cognitive abilities to coordinate spatial and shape information that chimpanzees don't have.'[24] *Homo erectus* was far less based in the treetops, and the tools it created were much less disposable, pointing to a greater investment of labour in these objects, and also perhaps more of a sense of their use in different, future situations.

If sustained tool use was the initial defining feature of humanity, how did this translate into a capacity for language and accelerated brain development? Engels believed that increasing use and design of tools made possible by the bipedalism of our proto-human ancestors was a key step in the development of language. As he put it, communal tool use 'helped to bring the members of society together by increasing the cases of mutual support and joint activity…Men-in-the-making arrived at the point where they had something to say to each other.'[25] During this process, 'the reaction of labour and speech on the development of the brain and its attendant senses, of the increasing clarity of consciousness, power of abstraction and of conclusion, gave both labour and speech an ever renewed impulse to further development.'[26] So what is the evidence for such a process of positive feedback?

One problem in exploring this question is that, unlike fossils and stone tools, language leaves no direct imprint to be discovered by archaeologists, although it does make its presence known in the human body through changes in the larynx and in certain brain regions. Nevertheless, these are subtle changes that are difficult to detect in a fossilized skeleton. Instead, we must guess the likely development of language evolution by indirect means.

Thomas Morgan and colleagues at the University of California, Berkeley, recently explored how language as a means of interaction in modern humans affects their ability to learn how to create prehistoric stone tools.[27] Morgan's team divided volunteers into groups. The first group were given a stone and a finished product and just told to get on with the task. The second group learned primitive tool-making technology by watching an expert and then trying to replicate what they did, but with no interaction. In the third group, the toolmaker actively showed the volunteers what they were doing but without gestures, while in the fourth group, gesturing and pointing were allowed but talking was forbidden. In the fifth group, the toolmaker explained verbally whatever was necessary for volunteers to appreciate the production process. Finally, the quality of the tools produced by each group was assessed.

The study showed that volunteers told just to get on with the task produced poor results, but so did those who passively watched an instructor. But groups in which gesturing and talking were allowed produced far higher quality products, with gesturing doubling the quality and verbal teaching quadrupling it. This suggests that the successful spread of even the earliest known tool-making technology, more than two million years ago, required a capacity for teaching by gesture, and maybe the beginnings of spoken language.[28]

Another way of investigating the question is to image brain activity when people are making stone tools and compare this to activity when a person is engaging in language, as did Natalie Uomini and Georg Meyer of the University of Liverpool. They used functional transcranial Doppler ultrasonography (fTCD)—which measures blood flow to different brain regions and therefore their likely relative activity—to study ten expert toolmakers performing two different tasks.[29] The first task involved crafting an Acheulean hand axe, a prehistoric symmetrical tool that requires considerable planning and skill; the second silently imagining words beginning with a given letter. The volunteers interspersed such activity with non-skilled behaviour—banging two stones together for the first task and resting without thinking of words for the second. The researchers found that blood flow changes during the first ten seconds of each experimental period—when the volunteers were working out how to shape the stone or thinking up their first words—were very similar.

Uomini and Meyer think that their findings suggest that language and tool-making coevolved as early as 1.75 million years ago. Yet since they used a tool-making method known as the Late Acheulean, dating from half a million years ago, that put much greater emphasis on symmetry and aesthetic considerations than the Early Acheulean method of 1.75 million years ago, Michael Corballis, of the University of Auckland, New Zealand, believes that while the study does suggest coevolution between language and toolmaking, 'language itself emerged much later, but was built on circuits established during the Acheulean period.'[30]

Exactly how language first arose in humans may always remain a matter of speculation in the absence of direct evidence. What we can investigate directly is what makes language capacity unique to human beings. In fact not everyone accepts this uniqueness. One area of dispute is the extent to which humanity's extraordinary language capability is innate, rather than something learned. I have already mentioned the view of behaviourists like B. F. Skinner that all human behaviour is a series of conditioned reflexes. Skinner viewed language as no different in this respect. He believed that if a child says 'milk' and her mother subsequently gives her some, the child finds this rewarding, enhancing her language development. Evidence for this view came from studies like one by Doug Guess of Kansas University, who in 1968 claimed he had taught a girl with severe learning disabilities to make correct grammatical utterances using positive reinforcement of praise and food.[31] And, indeed, studies of chimpanzees in the 1960s and 1970s suggested that the ability to be taught language by such reinforcement might not even be unique to humans.

Although chimps have no vocal apparatus for making the sounds of human speech, Allen and Beatrix Gardner, psychologists at the University of Nevada, claimed to have taught a chimp, Washoe, to use American sign language for the deaf.[32] The Gardners raised Washoe like their own child; she wore clothes and sat with them at the dinner table. During her life

Washoe learned over 250 symbols and was even said to have coined completely novel words. For instance, after seeing a swan, Washoe made the two signs for 'water' and 'bird', suggesting she had grasped the link between the two. Hearing about this incident, Harvard University psychologist Roger Brown viewed it 'like getting an S.O.S. from outer space'.[33]

Washoe was not the only ape to display an apparent flair for learning sign language. A bonobo chimp, Kanzi, learned over 400 symbols, and was said to have invented novel words, referred to past versus present events, and understood others' points of view, all implying a more complex understanding of language than simply associating words with their objects.[34] And a gorilla, Koko, was reported to be able to understand grammar, something only humans were thought capable of doing. On being informed of the death of Robin Williams in 2014, it was said that Koko made the sign for 'cry', and 'looked very thoughtful', despite having only met the actor once in 2001, when Williams spent an afternoon with the celebrity ape.[35]

Despite the potency of such stories, major criticisms have been raised regarding the idea that language acquisition in humans is simply a matter of reinforcement. Notably in the 1960s Noam Chomsky of the Massachusetts Institute of Technology identified key problems with this view.[36] Chomsky highlighted a central feature of human language, namely its capacity for expressing an infinite number of ideas using a limited set of symbols. Such creativity is made possible not only by the size of a person's vocabulary, but also by how words are strung together in a precise order. Chomsky argued that the speed with which children learn new words, but also how to combine them in a grammatically precise fashion, means our species must have an innate biological capacity for complex language.[37]

But what about the claims that other primates can learn human language? In fact there is now evidence that experimenters who claimed amazing linguistic abilities for Washoe the chimp or Koko the gorilla may have overinterpreted their responses, and even unconsciously given the apes visual cues about what to say. Herbert Terrace, who studied the chimp Nim Chimpsky—this name being a dig at Chomsky—recognized this problem midway through the studies. According to him, 'previously, I had been so fixated on watching Nim's hands that I missed what his teachers had been doing on the side while in shadow. About one quarter of a second prior to Nim's signing, they were making the same or similar signs! This discovery downgraded what we thought were spontaneous utterances to imitative responses.'[38]

Researchers may also have selectively picked examples of ape signings because they were so keen to identify human-like language in these non-human species. The individuals who taught apes sign language invested huge amounts of professional but also personal time with their wards.[39] For instance, Penny Patterson, who worked with Koko the gorilla, devoted much of her life to this animal since she began working with her in 1971. It is pertinent to ask, therefore, whether such researchers became so devoted to their apes that like a proud cat- or dog-owner who swears their pet is more intelligent than the average, their judgement was swayed by close proximity to their experimental subjects.[40] For instance, did Koko really display sorrow on the death of Robin Williams, whom she met for one afternoon 13 years before, or was she merely mimicking the sorrowful attitude of her human carers?

Building on such concerns, critics have pointed to important differences between humans and these taught apes in their use of language. A central feature of human language is its spontaneous nature, whether small talk about the weather with a stranger, a toddler pointing out a cute dog to her parents, or the many fascinating questions that a book about human consciousness might prompt from its readers. In contrast, apes seem not to care for chit-chat. For instance studies of Kanzi show that only 4 per cent of his signs were comments, the other 96 per cent being purely functional, such as requests for food or toys.[41] Such criticisms have not gone down well with some researchers in this area. In an exchange between Penny Patterson and Herbert Terrace in the New York Review of Books in 1980, Terrace accused Patterson of only eliciting signed responses from Koko by asking her questions. Patterson at first defended her research methods, but eventually gave up on the debate, saying that her time would be 'much better spent conversing with the gorillas'.[42]

Since taught apes may often be merely mimicking the gestures of human researchers, and researchers may be showing selective bias in drawing conclusions about ape language capacity, another approach would be to study whether apes show a capacity for sophisticated forms of communication in the wild. Recently Catherine Hobaiter and colleagues at the University of St Andrews in Scotland investigated wild chimpanzees and claimed to have found that they can communicate 19 specific messages to one another with a 'lexicon' of 66 gestures. Hobaiter believes the chimps are similar to humans in having a communication system in which they deliberately sent a message to another individual. For her, 'the big message...is that there is another species out there that is meaningful in its communication, so that's not unique to humans.'[43] Yet while Hobaiter claims that some gestures are only used to convey one meaning, like leaf clipping, where a chimp takes small bites from leaves to elicit sexual attention, undermining this claim, other gestures were more ambiguous; for instance, a grab can apparently signify either 'Stop that', 'Climb on me', or 'Move away.'[44]

Susanne Shultz, a biologist at the University of Manchester, believes the researchers' attempts to fill in the gaps in our knowledge of the evolution of human language are commendable, but 'the vagueness of the gesture meanings suggest either that the chimps have little to communicate, or we are still missing a lot of the information contained in their gestures and actions. Moreover, the meanings seem to not go beyond what other less sophisticated animals convey with non-verbal communication.'[45] It therefore seems that fundamental differences remain in terms of language capacity between humans and apes.

Yet while the behaviourist view that human language acquisition is simply an accumulation of conditioned reflexes now looks incorrect, recent studies have also challenged Chomsky's view of a biological basis for a 'universal grammar' shared by all humans.[46] Instead, increasing evidence points to both human biology and the process of growing up in a specific human society as being factors of equal importance in the formation of language. I will return to this question later by looking at new studies that are beginning to identify genes that underpin human language capacity, as well as the regions of the brain that play important roles in language development, but also at evidence that the environment helps to shape human language. For now though, let us consider another aspect at the heart of what makes humans unique: the relationship between individual and society.

CHAPTER 3

INDIVIDUAL AND SOCIETY

—◦◦◦—

The relationship between the individual and society has been hotly disputed among philosophers and politicians through the ages. Aristotle believed that 'man is by nature a social animal … society is something that precedes the individual. Anyone who either cannot lead the common life or is so self-sufficient as not to need to, and therefore does not partake of society, is either a beast or a god.'[1] This was in line with the ancient Greek viewpoint that human civilization was inconceivable without the organization of individuals into a 'polis', or body of citizens. The Athenians established the idea of democratic government in such a society, although their concept of democracy was highly limited.

In feudal societies, a person was defined by their position in a rigid social hierarchy rather than as an individual judged on their merits. This system was justified by the idea that everyone had their place in society, ordained by God. The notion is expressed succinctly in the lines of the English hymn 'All Things Bright and Beautiful': 'The rich man in his castle, the poor man at his gate, God made them high and lowly, and ordered their estate.'[2] Ironically, when this hymn was written by Cecil Frances Alexander, wife of the Archbishop of Armagh, in 1848, the British feudal aristocracy had long relinquished their role as rulers of society—although not all their wealth and influence—to a new governing class, who now run our capitalist economic system. Asserting themselves first in the Italian Renaissance, and then coming to power in the Netherlands, England, and France, after bitterly fought revolutions, a rallying point for this new class was the primacy of the individual. René Descartes, with his statement 'I think, therefore I am',[3] and Jean Jacques Rousseau, who saw every 'man [as] born free and master of himself',[4] were important proponents of this individualist viewpoint.

The importance of the individual is taken for granted in modern capitalist society. Indeed, former British Prime Minister Margaret Thatcher even notoriously claimed in 1987 that 'there's no such thing as society. There are individual men and women and there are families.'[5] While many would disagree with this extreme position, the view that society is merely a collection of individuals driven by self-interest is a very common one.

Yet recent studies have questioned the idea that we are naturally solitary individuals. Instead they suggest that socializing with others is so central to our species that rejection—whether by a lover, friend, or even while playing a game with total strangers—is registered in the same brain regions as respond to physical pain. One such study, led by Matthew Lieberman at the University of California, Los Angeles, imaged the brains of volunteers playing an invented online game called Cyberball.[6] When invisible online 'players' deliberately stopped passing the e-ball to the volunteer, the brain region that lit up in the scanner was the one also activated by physical pain. Looking at scans from two studies side by side, Lieberman noted that 'without knowing which was an analysis of physical pain and which was an analysis of social pain, you wouldn't have been able to tell the difference.'[7]

Other studies have undermined the idea that human beings are inherently selfish, indicating instead that altruistic acts trigger activity in the 'reward' region of the brain that is stimulated when a person experiences pleasure. For instance, Jorge Moll and colleagues at the National Institutes of Health, near Washington, DC, showed that this region became more active when people gave $10 to charity than when they received the same amount.[8] And Tristen Inagaki and Naomi Eisenberger of the University of California, Los Angeles, have shown that comforting someone in distress is a powerful stimulus for this brain region.[9] They performed brain scans on women holding their boyfriend's hand, both when the couples were sitting together normally and when the man was given an electric shock. The brain's reward region was activated on both occasions, but even more so when the women were comforting their distressed partners. Eisenberger believes this shows that while people 'typically assume that the benefits of social support come from the support we *receive* from others...it now seems likely that some of the health benefits of social support actually come from the support we *provide* to others.'[10]

A particularly surprising observation was made when the brains of people not engaged in any activity were monitored. Instead of the brain being relatively inactive during these periods as one might expect, a 'default activity' was triggered, and this occurred in those brain regions involved in social interaction with other human beings. Lieberman believes this is because 'the default network directs us to think about other people's minds—their thoughts, feelings, and goals...Evolution has made a bet that the best thing for our brain to do in any spare moment is to get ready for what comes next in social terms.'[11]

Studies like these raise the question of how the human brain became so attuned to social cues in this way. Here there are two issues to consider. One is evidence that primates in general have evolved to be highly sensitive to social interactions with other members of their species, and this has been accompanied by enhanced brain growth in order to handle these more sophisticated interactions. Robin Dunbar of the University of Oxford has demonstrated a good correlation between relative brain size in a particular primate species and the extent of the social interactions of individuals of that species with each other.[12] This could be one reason why primates as a group have relatively bigger brains than other mammals. Because of this, the very fact that humans are primates may explain why we are so primed to seek social interaction, and for our brains to reward us for such interactions.

Yet while social interaction may be hardwired into our brains because of evolutionary changes in our primate ancestors, some features of our strong tendency towards social interaction may be specifically human. In particular, considering what I said previously about the uniquely human characteristics of our capacity for tool design and use, and language, how might these be linked to our skill, and desire for, social interaction?

One scientist particularly interested in this question was the Russian psychologist Lev Vygotsky who developed novel ideas about human consciousness starting in the early 1920s and then with increasingly important insights until his tragic death in 1934 at the age of only 37, from tuberculosis.

Such were the depth and breadth of Vygotsky's insights that he was named the 'Mozart of psychology' by psychologist Michael Cole of Rockefeller University in New York, when his writings were 'rediscovered' in the 1960s, having been suppressed as too subversive by Josef Stalin and later rulers of the Soviet Union.[13] Vygotsky's ideas have had an influence on modern psychology—particularly regarding the link between individual and society, and the development of consciousness during childhood—that continues to this day.

In seeking to understand the unique features of human conscious awareness, Vygotsky was influenced by Friedrich Engels' view of our species' evolution. Although Engels' essay on this topic was ignored by most scientists, in the newly formed Soviet Union there was much more receptiveness towards it, following its translation into Russian in 1927.[14] Vygotsky was struck by the idea that tool-making and use constituted the crucial initial step in human evolution, and he creatively elaborated on Engels' account of how this allowed human beings to first begin to shape the world around them. Vygotsky proposed that language, through becoming internalized as thought, could itself be viewed as a tool, in this case shaping consciousness. What language and tools have in common is that they both act as mediators, yet while technical tools are aimed at 'mastering, and triumphing over nature', in contrast language is an 'internal activity aimed at mastering oneself.'[15]

How might language play such a role in shaping individual consciousness? Vygotsky explored this question by studying the development of conscious awareness in children. When he began developing his ideas in the 1920s, a key figure in this area of study was the French psychologist Jean Piaget.[16] At this time it was commonly assumed that children are merely less competent thinkers than adults, but Piaget's studies indicated

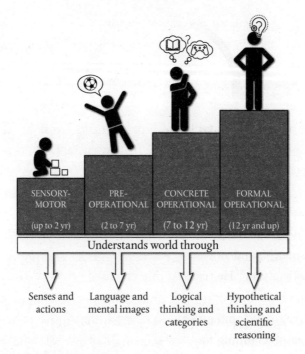

Figure 6 Piaget's stages of child development.

that young children think in strikingly different ways to adults, and go through discrete stages of development, marked by qualitative differences, rather than via a gradual increase in the number and complexity of behaviours, concepts, and ideas (Figure 6).[17] Piaget saw child development as a combination of an innate biological process and social influence through the learning environment, but his emphasis was definitely on the innate character of development.

While Vygotsky agreed with some aspects of Piaget's view of a child's mental development—such as the idea that this occurs in discrete stages—his conclusions about the main driving force behind the development of the child's mind were radically different.[18] In contrast to Piaget's notion that children's development necessarily precedes their learning, Vygotsky argued that learning is a necessary and universal aspect of the process of developing 'culturally organized', specifically human psychological functions.[19]

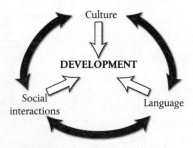

Figure 7 Vygotsky's theory of cognitive development.

One phenomenon Vygotsky focused upon was the tendency of young children to talk to themselves as they play. Piaget called this phenomenon 'egocentric speech', and noted its abbreviated form and grammatically fragmented structure.[20] He realized that it might guide the child's activities, but believed its insular nature and fragmented form indicated it was a relic of a past developmental stage, soon to be superseded by more mature forms of planning and self-control. In contrast, Vygotsky saw such speech as a vital stage in child development (Figure 7).[21] By studying how young children talk to themselves as they play, Vygotsky showed that egocentric speech guides a child's activities but is also a transitional phase later internalized as 'inner speech' which helps to create the thought processes of each individual.

Other important insights came from Vygotsky's investigations of how children begin to think conceptually.[22] His studies indicated that this is an active process, whereby the child seeks out the words and concepts that make sense of their everyday practical and social experience. Continual testing of the meaning of language against reality allows them to reach towards future knowledge and abilities, and this is apparent in the incredible creativity and questioning nature we typically associate with children.[23] One way Vygotsky studied concept formation was to ask children of different ages to sort 22 blocks of varying colours, shapes, heights, and sizes into groups.[24] The blocks were coloured red, blue, green, yellow, and white. But there were also four 'secret' categories, identified by nonsense words—'lag'—tall and fat blocks; 'bik'—small and fat blocks; 'mur'—tall and thin blocks; and 'cev'—small and thin—on the reverse of each block. The children were not told that these four words represented such complex concepts, nor did they know that the words referred to a combination of the height and the size of the object. The child was asked to group together all blocks that might belong to the same kind. After this sorting exercise, the experimenter picked a wrongly selected block, read its name, and encouraged the child to keep trying. Following each attempt, another wrongly selected block was picked up, and its name revealed. In this way, the child could discover to which characteristics the words referred.

Vygotsky called his test the 'double stimulation' method, since the child was stimulated both by the physical properties of the blocks and by the nonsense names.[25] The test was supposed to indicate the level of conceptual problem solving that the child had reached in their mental development. The overall finding of the study was that true conceptual problem solving was only achieved by older adolescents. But also interesting was what was revealed about the different stages of preconceptual thinking in children.

Vygotsky named the preconceptual stages syncretic, complex, and pseudo-conceptual. The youngest children grouped blocks syncretically, meaning they collected blocks at random, based on a vague idea that they belonged together.[26] Complex thinking was more advanced in that blocks were grouped by real similarity, but inconsistently. So in a 'collection complex', blocks were paired by complementarity: red with blue, tall with flat, and so on. In a 'chain complex', if the first block was red and round, the second one could be red and triangular, the third triangular and green. This type of thinking was distinguished from true conceptual thought by showing no hierarchical organization in the selections.

An intriguing phenomenon uncovered by the study was 'pseudo-conceptual' thinking.[27] Here, at first glance children appeared to be sorting blocks on a purely conceptual basis, selecting them for instance by a shared feature, like colour. The preconceptual nature of such a selection was only revealed when the experimenter turned over a block and exposed its inaccuracy according to the nonsense word describing it.

While a child using true conceptual reasoning responded by removing all selected blocks and starting again, one guided by pseudo-conceptual reasoning only removed the 'wrong' block and continued to insist on the appropriateness of the others they had selected. Vygotsky believed this revealed that 'the child's complexes...do not develop freely or spontaneously along lines demarcated by the child himself' but rather along lines preordained by the word meanings established in adult speech.[28] Only in the experiment was the child freed from the directing influence of the words of the adult language with their developed and stable meanings. In other words, a child grouping objects according to a common feature, such as a square shape, does not necessarily have an internal concept corresponding to that common feature. This raises the question of how a child progresses from pseudo-conceptual thinking to that guided by pure concepts.

A key distinction that Vygotsky made in this respect was between 'everyday' concepts and 'scientific' ones. With everyday concepts the child's attention 'is always centred on the object being represented and not on the thought that grasps it.'[29] In contrast, scientific concepts, 'with their quite different relationship to an object, are mediated through other concepts within their internal, hierarchical system of relationships.'[30] This means that a child can apply conceptual categorization without necessarily understanding it fully. The development of scientific concepts is particularly associated with formal schooling; however, Vygotsky believed that classroom teaching is most effective when it connects with the child's own spontaneous view of the world formed through their everyday experiences. It is through such an organic connection that a child is most likely to develop an understanding of the hierarchical basis of true conceptual categorization and not just apply it blindly.

This idea was explored in a 1994 study by Luis Moll and James Greenberg of the University of Arizona, while trying to encourage literacy in children of a Mexican community in Tucson, Arizona.[31] In the study, writing classes were geared to the children's everyday experiences outside the classroom and the learning experience mainly took the form of project work whose content was drawn from the community from which the children came. Moll and Greenberg's approach involved creating 'meaningful connections between academic and social life through the concrete learning activities of the students'.[32] Spelling and grammar were not ignored in the study, but the emphasis was on treating reading and writing as communicative and meaningful. With such an approach even the most uninterested or apparently incapable children made great improvements in their literary skills.[33]

Vygotsky was not the only thinker with such a view of the key role of language in mental development working in Russia in the 1920s and 1930s. Valentin Voloshinov, who also died young, in 1935, aged only 40, in this case after incarceration in a Stalinist 'Gulag' prison-camp,[34] came to similar conclusions during this period. Although Voloshinov and Vygotsky are not recorded as having met or read each other's work, certain themes 'in the air' in Russia at this time may have contributed to the large degree of overlap and complementarity in their approaches to the study of human consciousness.[35]

Voloshinov was more concerned with adult thought patterns than those occurring in children. He argued that people 'think and feel and desire with the help of words; without inner speech we would not become conscious of anything in ourselves. This process of inner speech is just as material as is outward speech.'[36] But he also believed that thought and language share the same transmission medium and have the same source, that is, society. He proposed that each person engages in 'horizontal' social relationships with other individuals in specific speech acts, and simultaneously in 'vertical' internal relationships between the outer world and their own psyche.[37] The mind is thus not an internal but a boundary phenomenon. Or as Voloshinov put it, 'individual consciousness is not the architect of the ideological superstructure, but only a tenant lodging in the social edifice of ideological signs.'[38]

Expressed in this way, it might appear that Voloshinov saw language, and therefore thought, as something imposed upon the individual, like the 'blank slate' view of the mind of the behaviourists. But this was far from the case, for he rather viewed language as a dynamic process, in which the individual plays a highly active role. And it is the social relations between the two individuals taking part in a verbal exchange that influence the way this exchange develops. As Voloshinov put it, 'the word is implicated in literally each and every act or contact between people—in collaboration on the job, in ideological exchanges, in the chance contacts of everyday life, in political relationships, and so on...The word has the capacity to register all the transitory, delicate, momentary phases of social change.'[39]

But how can this be so, given that all individuals in a particular society use the same words and the same language system? Voloshinov's explanation was that words used by groups with radically different circumstances and life activities become inflected with different and competing meanings as these groups struggle to express their life situations, their outlooks, and their aspirations. Because of this, for Voloshinov, language exists in 'a continuous process of becoming. Individuals do not receive a ready-made language at all, rather, they enter upon the stream of verbal communication.'[40]

Voloshinov's view of language is based on dialogue, but in a much deeper sense than we usually use this term in everyday speech. A key feature of any utterance produced by an individual is that it can only be understood in relationship to other utterances. In other words, understanding is a process in which the utterances of a listener come into contact with and confront those of a speaker. Or as Voloshinov expressed it, 'for each word of the utterance that we are in process of understanding, we...lay down a set of our own answering words. The greater their number and weight, the deeper and more substantial our understanding will be.'[41] A second way in which utterances can come into contact is through the practice of an individual taking over another's words or expressions. One familiar way in which this occurs is in parody, for instance when a speaker repeats the utterances of a well-known celebrity or politician by repeating such utterances with a different intonation or in a context different from the original one, in a way that can be humorous, sarcastic, or so on. It is through this multi-voicedness that parody achieves its effects. However, while parody may be a particularly obvious form of this phenomenon, Voloshinov believed that some element of multi-voicedness is an essential feature of practically every type of utterance.[42]

While Voloshinov viewed words as having multiple accents, he did not see this as a random process. Instead, verbal exchanges between individuals are governed by what he called 'speech genres'. These form a crucial link between the abstract level of a language system and the concrete richness of speech. Mikhail Bakhtin was another Russian philosopher of language who worked closely with Voloshinov in the 1920s and 1930s as part of an intellectual circle involving a variety of other individuals; indeed it has even been claimed, controversially, that Bakhtin was the actual author of the books published under Voloshinov's name.[43] Whatever the truth of this claim, Bakhtin continued to develop the concept of speech genres until his death in 1975. He noted that speech genres come in various different forms, and this range gives social interaction its range and dynamism. They might include 'social dialects, characteristic group behaviour, professional jargons, generic languages, languages of the authorities of various circles and of passing fashions, languages that serve the specific socio-political purposes of the day.'[44] Individuals can draw on different speech genres in their everyday life. So an administrative assistant might use a deferential genre while speaking with her line manager, a relaxed one while talking with workmates at lunch, and a more politicized genre at a meeting of her trade union branch.

Bakhtin noted that speech genres are so much a part of everyday speech that we generally use them unthinkingly. Or as he put it, 'our repertoire of oral (and written) speech genres is rich. We use them confidently and skilfully in practice, and it is quite possible for us not even to suspect their existence in theory.'[45] Bakhtin also saw speech genres as being grounded in reality, stating that 'genres correspond to typical situations of speech communication, typical themes, and, consequently, also to particular contacts between the meanings of words and actual concrete reality under certain typical circumstances.'[46]

Yet speech genres are also sensitive to changes in circumstances. For Richard Bauman of Indiana University, they represent a 'ready-made way of packaging speech' but also allow for 'creative, emergent, and even unique' performances.[47] Because of this, both the voice type of a speech genre and a concrete individual voice are simultaneously involved. This interaction occurs because, as Bakhtin put it, 'the word in language is half someone else's. It becomes "one's own" only when the speaker populates it with his own intentions, his own accent, when he appropriates the word, adapting it to his own semantic and expressive intentions.'[48] Crucially, being linked to social groups, speech genres are very sensitive to social change, in ways that I will explore later in this book.

While social interactions shape language, it is still individuals that make the utterances involved in concrete speech, and understand and respond to those made by other individuals. So how does language manifest itself in the individual mind? Voloshinov thought that the ideological character of language is also present in the individual. Antonia Larrain and Andrés Haye, psychologists at the University of Santiago in Chile, believe that this is because inner speech is a 'dialogue that consists of a constant negotiation and redefinition of ideological territories.'[49] The linguistic aspects of inner speech help the speaker to think and feel in a stable and recognizable way along with others, but in an ongoing human consciousness it is not the linguistic structure of speech but the 'dialogical dynamics internalized from social activity' that most characterize our inner thoughts.[50]

A challenging aspect of investigating the validity of this view and the general idea that inner speech is a central motor of consciousness is that processes that occur in the individual mind are very difficult to study. Because of this, those interested in investigating this issue further must resort to indirect methods of analysis. Vygotsky believed that since egocentric speech in children is an important transitory stage in the development of inner speech, clues about the latter's character might be found in the structure of egocentric speech.[51] Charles Fernyhough of the University of Durham has recently tried to monitor inner speech directly by getting adult volunteers to wear a device that bleeps at random intervals, at which point the wearer is prompted to note whatever was passing through their mind at the time.[52] This was repeated on six occasions. The subjects were then interviewed about the nature of their inner mental responses and their relation to spoken language.

By such routes, the emerging picture of inner speech is that it differs from external speech, in both structure and how it conveys meaning. It has a highly abbreviated quality because its context is already largely known to the individual in which it occurs. To some extent this is a feature of egocentric speech in children. As Vygotsky observed, a child talking to themselves 'already knows' what they are talking about and therefore has no need to name the subject; so only 'new information' is retained while 'old information' is simply presumed.[53] Yet inner speech is likely to be even more telegraphic and abbreviated than egocentric speech. Because of this, and as external speech requires tongues, lips, and voice boxes, Fernyhough estimates the likely speed of inner speech to be about 4,000 words per minute—ten times faster than verbal speech.[54] Other key differences include a predominance in inner speech of personal, private meaning over conventional, public meaning, and a tendency to agglutination—that is, for words to join to form more complex wholes.[55]

Based on his studies, Fernyhough believes there are different kinds of inner speech, which vary 'according to how compressed it is, how condensed. We think inner speech varies according to how much it's like a conversation between different points of view…And that fits with the idea that inner speech has a lot of different functions. It has a role in motivation…in emotional expression, it probably has a role in understanding ourselves as selves.'[56]

One situation in which inner speech clearly plays a role as a motivating influence is in sport. In a recent study by Amir Dana and colleagues at the Islamic Azad University in Borujerd, Iran, darts players were asked to aim at the board while inwardly voicing either positive or negative inner suggestions such as 'you can do it' or 'you're bound to miss'.[57] Those who spoke positively to themselves scored consistently higher. More generally, high-achieving athletes have been shown to talk more to themselves than less successful rivals.[58]

The idea that inner speech is essentially a dialogue raises many important questions. Fernyhough believes that 'we are all fragmented. There is no unitary self. We are all in pieces, struggling to create the illusion of a coherent "me" from moment to moment.'[59] But if this is the case, then how do most of us manage to navigate our way through life with the sense that we are each unique individuals, with our own goals and aspirations, through both successes and failures? And if the 'normal' human psyche is by nature fragmented, then how does this differ from individuals who by their behaviour and statements appear to have progressed significantly further than the norm in their state of mental fragmentation, such as people suffering from conditions like schizophrenia or multiple personality disorder?

If inner dialogue is central to human thought, where does this leave deaf individuals? Lacking a means of hearing external speech, are they also deficient in this aspect of consciousness? In fact deaf people do seem to have an inner dialogue, but one that can take a different form to inner speech. In a recent survey, a deaf person asked about this question stated that 'I have a "voice" in my head, but it is not sound-based...in my head, I either see ASL (American Sign Language) signs, or pictures, or sometimes printed words.'[60] The age at which hearing loss occurs seems to influence the exact modality of the inner dialogue. So one individual who lost their hearing aged two stated that they think in words, but without sound.[61] It seems that the inner dialogue that mediates human consciousness in this case still involves symbols, but not necessarily auditory ones. I will look at how consciousness is affected by loss of one or more of the senses later in this book, and indeed at some of the other questions raised in this chapter about the nature and role of our inner dialogue.

For now though, it is time to take a step away from psychology and turn to neuroscience. For one issue that neither Vygotsky nor Voloshinov got to address is how the features of the human mind that they believed distinguish human consciousness from that of other animals are influenced by each individual consciousness being also the product of a biological entity—the brain. One consequence of seeing mental functioning in human individuals as having its origins in social, communicative processes—and human thought as a kind of verbal dialogue—is that this should have the effect of transforming the different functions of our brains in a way that is unique to human beings compared to other species, even to such closely related primate species as chimps. But posing the matter in this way raises a conundrum, for how can humans and chimps be apparently so similar at the genetic level—99 per cent if protein-coding genes are compared—and yet our brains be so different at a functional level? It is time to look at what science is revealing about the human brain.

PART II

MIND AND BRAIN

CHAPTER 4

NERVE AND BRAIN

————◆————

The phenomenally complex structure that is the human brain is the product of a remarkable evolutionary process that has given rise to the conscious awareness that makes our species unique on Earth. So any attempt to understand the material basis of this uniqueness needs to engage with the structure and function of the brain.

The basic unit of the brain is the nerve cell, or neuron.[1] These cells are also the main units of the peripheral nervous system, which sends messages from the brain to the other tissues and organs that make up our bodies. Neurons are distinctive compared to other cells in being able to conduct electrical impulses, in the form of a flow of charged atoms (ions). These flows of ions occur thanks to protein pores, known as 'ion channels', found on the surface of neurons that allow particular ions to enter or exit the cell in a regulated fashion.[2]

The basic structure of neurons was first identified by Spanish biologist Santiago Ramón y Cajal in the late nineteenth century.[3] He used new methods for chemically staining cells to demonstrate that neurons have three main parts—a cell body, dendrites, and an axon. Cajal originally wanted to be an artist, and this artistic vision is apparent in the beautiful drawings he made—at a time when photography still had not made its entry into the biological sciences—of the structures of different neurons. As Cajal himself put it: 'Like the entomologist in pursuit of brightly coloured butterflies, my attention hunted, in the flower garden of the grey matter, cells with delicate and elegant forms, the mysterious butterflies of the soul, the beating of whose wings may someday—who knows?—clarify the secret of mental life.'[4]

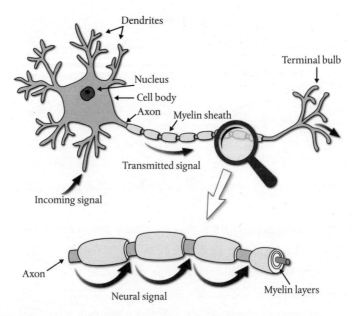

Dendrites

Terminal bulb

Nucleus

Cell body

Axon   Myelin sheath

Transmitted signal

Incoming signal

Axon

Neural signal

Myelin layers

Figure 8 The nerve cell and transmission of signal.

Building on Cajal's pioneering studies, subsequent investigations have shown that the neuron's cell body contains a nucleus—within which is the genome of the nerve cell—plus other important subcellular structures (Figure 8). But it is the dendrites and axon that are critical to the business of conveying electrical impulses from one neuron to the next.

It is through the dendrites that electrical activity is initiated in the nerve cell, and the axon is the structure that transmits the electrical signal to a neighbouring cell. In many human neurons, unlike the situation in simpler organisms, the flow of electricity along the nerve cell is aided by an insulating structure around the axon called the myelin sheath.[5] This only allows the electrical changes in the axon to occur at regular intervals where there is no sheath, forcing the impulse to 'jump' from gap to gap, which greatly speeds up its progress along the axon (Figure 8). It has been estimated that myelinated neurons can carry impulses up to 100 times faster than non-myelinated ones. This is particularly important given that some axons in the human body—for instance those in the spinal cord—can be a metre long.

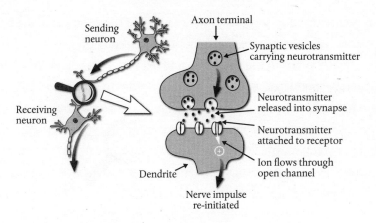

Figure 9 Synaptic transmission of nerve impulses.

A crucial property of the brain and nervous system is that neurons are connected together in chains. There is no direct electrical contact between adjacent neurons; instead, chemical messengers known as neurotransmitters are released into the narrow space between the axon of one neuron and the dendrite of the next—the synapse (Figure 9). By a process similar to a key fitting into a lock, neurotransmitters interact with 'receptor' regions of particular ion channels on the surface of the dendrite to open or close particular channels, altering the ionic composition of the neuron as a consequence.

How neurons conduct electrical impulses was shown by Alan Hodgkin and Andrew Huxley at the University of Cambridge in 1939.[6] They studied axons from giant squids, which were large enough that the researchers could impale them with electrodes, recording both electrical current and voltage. These studies showed that when unstimulated, the neuron has a negative electric charge. Activating neurotransmitters cause sodium ions to flow into the cell. Since sodium is positively charged, the neuron becomes more positive. At a certain point, it can become so positive that this triggers an explosive rise and fall of electrical charge called an action potential, the fall being due to positive potassium ions rushing out of the cell.[7] The action potential sweeps like a wave along the axon, and, due to the conducting effect of the myelin sheath, it can reach speeds of

250 miles per hour in some human neurons. At the axon's end, the action potential causes the release of neurotransmitter into a synapse, and this in turn activates or inhibits a neighbouring neuron via its dendrites. Once a neuron has conducted an action potential, protein pumps on its surface return its ionic composition to a resting level, at which point it can begin the process of electrical conduction once more.

Since an axon may have hundreds of branches, it can activate or inhibit the dendrites of many other neurons. As each neuron has many dendrites, whether the activation of a dendrite in a neighbouring neuron leads to an action potential depends on the information flowing into that neuron from other axons, and whether the neurotransmitters being released into each synapse are excitatory or inhibitory. In this way the behaviour of each neuron is influenced by the sum of the information flowing into it via all its dendrites.

This is the situation for all neurons, but there is much diversity in size and shape between these cells.[8] Multipolar neurons with the form described earlier—a cell body from which emerges a single long axon and a crown of shorter branching dendrites—are the most common type of neuron in our nervous system. But one type of multipolar neuron only has a single primary projection that functions as both axon and dendrites. And neurons involved in sensing pressure, touch, and pain have axons that split in two opposite directions, with one end heading for the skin, joints, and muscle, while the other connects with the spinal cord. The latter is a bundle of nerves running up and down the spine, and can be viewed as akin to an information superhighway, transmitting messages to and from the brain at rapid speed.

Neurons can be classified according to their functional role (Figure 10).[9] Sensory neurons supply information to the brain from sensory organs like the eyes, ears, nose, tongue, and skin. Motor neurons spread out from the brain and spinal cord to regulate the action of muscles, which they activate via a 'neuromuscular junction', a structure that works in a similar way to a synapse. Interneurons connect neurons in the brain; while some

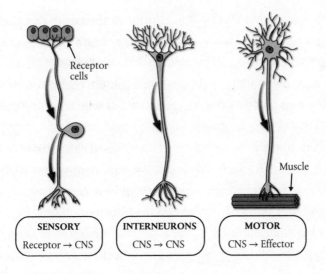

Figure 10 Three main classes of neurons.

link distant brain regions, others group neurons into smaller circuits of neighbouring cells. But descriptions of such major classes only give a flavour of the true complexity of the brain, which contains hundreds of neuronal types. Even a glance at some of the names assigned to nerve cells—cone, climbing fibre, crab-like, medium spiny, pyramidal, chandelier, and tripolar—gives a sense of the multitude of the variety of cell shapes found in the brain.

One class of neurons that have excited much interest recently gets their name not from their shape, but from their function. They were discovered by Giacomo Rizzolatti and colleagues at the University of Parma in Italy in 1992 when they implanted microelectrodes in the brains of monkeys to study their brain activity during different actions, such as the clutching of food.[10] One day, as a researcher reached for his own food, he noticed neurons activating in the monkeys' premotor cortex—the same region that showed activity when the animals made their own similar hand movement. Following up this intriguing observation, the group found that this activity was due to a class of neurons which they dubbed 'mirror neurons' that fire when a monkey performs an action, but also when it observes another individual making the same movement. Mirror neurons have since been shown to exist in the human brain.[11]

More controversially, some neuroscientists believe that our mirror neurons may allow us to simulate not only other people's actions, but also the intentions and emotions behind those actions. So Marco Iacoboni, of the University of California in Los Angeles, thinks that 'mirror neurons are the only brain cells we know of that seem specialized to code the actions of other people and also our own actions...without them, we would likely be blind to the actions, intentions and emotions of other people' and that mirror neurons let us understand others by providing some kind of inner imitation of the actions of other people, which in turn leads us to 'simulate' the intentions and emotions associated with those actions.[12]

Neurons are the most well-known cells in the brain but they are not the only type of cell in this organ. The other main type are the glial cells, also known as neuroglia (Figure 11).[13] Glia comes from the Greek word for glue, reflecting the view of these cells when they were first characterized as being merely a structural support for neurons in the brain. But further research showed that a class of neuroglia called oligodendrocytes supply nutrients to the neurons, and generate the myelin sheath that wraps such cells. Neuroglia have generally been considered very much secondary to neurons in terms of importance. Recently though, that view has been

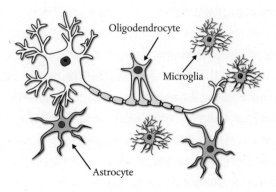

Figure 11 Neuron and neuroglial cells.

changing. For it turns out that glial cells can play a highly active role in brain cell communication, and perhaps in the development of human intelligence.

If asked to name the most intelligent person of the twentieth century, many people would probably identify Albert Einstein as their chosen figure of genius. When Einstein died in 1955 and left his body to be examined for the purposes of medical research, some people expected that his brain would turn out to be a lot bigger than those of less distinguished individuals. Yet a post-mortem showed that not only was Einstein's brain of average size, but also there was nothing unusual about the number or size of its neurons. But in the late 1980s, some scientists investigating Einstein's brain claimed that it had an unusually large number of neuroglia, especially in the association cortex, an area of the brain linked to imagination and complex thinking.[14] Quite what the significance of this finding about one, albeit very gifted, individual might be for our understanding of brain function is far from clear, but it is one of the discoveries that have challenged the idea of glial cells as merely supporting actors.

One reason why neuroglia were underestimated for so long was that they seemed to have no way of transmitting information in the brain. Neurons, as we have seen, carry information along their length via action potentials, and communicate with other neurons by releasing chemical messengers across connecting synapses. In contrast, neuroglia cannot generate action potentials. However, recent studies have shown that chemical communication is just as important for them. They turn out to have many of the same cell surface receptors as neurons, allowing them to communicate with both neurons and other neuroglia.[15]

One class of neuroglia, called astrocytes, play an important role in the formation of synapses.[16] If rodent neurons are cultured outside the body in the absence of astrocytes, they form few synaptic connections. But if astrocytes are added to the culture, the number of synapses rapidly rises, and synaptic activity increases tenfold. Astrocytes secrete chemicals that enhance synapse formation and directly regulate synaptic activity. The close physical association between synapses and astrocytes ensures that these are the first cells to respond to changes in synaptic activity during embryo development, but also in the adult brain.[17] Astrocytes are highly dynamic, constantly modulating their association with synapses—and thereby their influence—with the dynamic pattern being dependent on the state of the brain.

Another class of glial cells called microglia make up about 10 per cent of the cells in the brain.[18] For some years now these cells have been recognized as the brain's primary defenders against disease; they identify injured neurons and destroy them, and strip away defective synapses, thereby removing diseased cells that have a negative impact on the rest of the brain.[19] For decades, scientists thought of microglia only as immune cells and believed that they were quiet and passive in the absence of an infectious invader. Recently though, that idea has begun to change. One clue that microglia might have a wider role was the discovery that these are the fastest-moving cells in a healthy brain. For instance, recent studies have revealed that microglia can reach out to surrounding neurons and contact synapses in the absence of disease. These cells regulate the number of synaptic connections just as a gardener prunes a plant. This is important because, for reasons that are not fully understood, the brain begins with more synapses than it needs. Cornelius Gross, of the European Molecular Biology Laboratory in Heidelberg, who studies this process, has shown that microglia help sculpt the brain by eliminating unwanted synapses.[20] But how do the microglia know which synapses to remove and which to leave?

In fact, microglia can receive two types of message from neurons—one that identifies synapses that play an important role and should be preserved, another that highlights weaker ones that need pruning. Pruning is important, as shown by the demonstration by Gross's team that removing the receptor for a chemical named fractalkine from microglia leads to weak synaptic contacts caused by defective synaptic pruning in the hippocampus, a brain region involved in learning and memory. Gross believes this shows that during embryo development neurons 'call out' to microglia for assistance with pruning.[21] Yet it appears that there are also mechanisms in the brain to avoid overpruning, since another study, by Emily Lehrman and Beth Stevens of Boston's Children's Hospital, showed that useful synapses that are destined not to be pruned are identified by a protective protein 'tag' that deters microglia. If the receptor is eliminated genetically in a mouse, synapses are no longer protected, leading to excess engulfment by microglia and overpruning of neuronal connections.[22]

Recent studies of the role of glial cells in the brain are also revealing potentially important differences between humans and other species in the functions of these cells. They have shown that astrocytes reach dramatically longer distances in the human brain and propagate chemical signals faster than astrocytes in other species.[23] These features suggest an amplified role for human astrocytes in brain function. Other studies have shown that the genes that are switched on or off in human astrocytes are quite different from those in mice, so astrocytes may have evolved to play quite different roles in the two species.

Intriguingly, Steven Goldman and colleagues at the University of Rochester in New York have recently shown that baby mice whose brains were injected with human astrocyte precursor cells grew up to have better memories and learning abilities than their normal counterparts. This suggests that human astrocytes may have an enhanced ability to control synapses.[24] But future studies injecting chimpanzee or macaque glia into the mouse brain will be needed to determine whether the effects are due to properties unique to the human cells or ones that are common to those of all primates.

Another hint concerning the functions of astrocytes comes from studies of autism. Many genes implicated in some types of autism are expressed in astrocytes. Moreover, when researchers examined astrocytes in frontal brain regions in post-mortem samples from some individuals who had been diagnosed with autism, they found the astrocytes to be denser, with smaller cell bodies and fewer and shorter projections, than in unaffected individuals.[25] Changes in glial cells have also been claimed to play a role in dyslexia, stuttering, tone deafness, chronic pain, epilepsy, sleep disorders, and even pathological lying.[26]

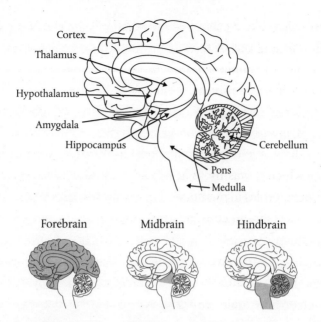

Figure 12 The different regions of the human brain.

Let us now move from the cellular structure of the brain, with its many types of neurons and glia, to its large-scale structure. The brain has three main parts: the forebrain, midbrain, and hindbrain (Figure 12).[27] The forebrain includes the cerebrum, which has two hemispheres and a highly folded surface 'cortex'—derived from the Latin word for bark, for indeed the cortex wraps around the brain like bark on a tree. The cortex is particularly enlarged in humans, in places having double the area it should for a primate of our size.[28] This is the part of the human brain thought to be involved in reasoning, planning, and problem solving. But, demonstrating how specific brain regions are involved in multiple functions, it also plays key roles in the regulation of movement, perception, visual processing, recognition of sounds, and speech.

Also buried within the forebrain are the various parts of the limbic system, which are more ancient from an evolutionary viewpoint and often called the 'emotional brain', reflecting their role in mediating different emotions, or even the 'lizard brain', to convey the fact that we share these brain regions with 'lower' organisms.[29] The limbic system contains the

thalamus which, like a central postal-sorting depot, channels information both into and out of the cortex. Then there is the hypothalamus,
which, as its name implies, lies underneath the thalamus, and regulates
thirst, hunger, sexual desire, reproduction, and the body clock. The amygdala is a region of the limbic system that plays a central role in processing
emotions; it is active in responses associated with fear, but also with
pleasure. Adjacent to this is the hippocampus, named after the Latin for
'seahorse', which it was thought by early researchers to resemble. This
plays important roles in memory—both in the formation of memories of
events and facts and in consolidating the transition from short-term to
long-term memory.

The midbrain is involved in functions such as vision, hearing, and
movements of the eyes and body.[30] Meanwhile, the hindbrain includes the
cerebellum, or 'little brain', so-called because like the cerebrum it has two
hemispheres and a highly folded surface.[31] This structure is involved in
the regulation and coordination of movement, posture, and balance;
however, as we shall see, there is increasing evidence that the cerebellum
may also play important roles in 'higher' mental processes, including
imagination and creativity. The hindbrain also contains the pons, which
controls consciousness and sleep, and the medulla, responsible for maintaining vital body functions, such as breathing and heart rate.[32]

While different brain regions have distinct functional roles, increasingly these are being shown to be linked in a highly coordinated way. One
way such coordination takes place is via electrical brain 'waves'—synchronized electrical pulses from masses of neurons communicating with
each other—whose character varies depending on the state of alertness of
the organism in which they occur.[33] These insights about brain waves
have coincided with a shift among some scientists away from a view that
reduces the brain to the behaviour of its individual components. For
instance, Earl Miller of the Massachusetts Institute of Technology points
out that previous theories about how the brain works tended to view it as
'a giant clock, and if you figure out each gear, you'll figure out the brain.'[34]
In contrast, Miller argues that the brain is better understood as 'networks

interacting in a very dynamic, fluid way', with oscillating brain waves, which he sees as 'the most powerful signal in the brain' central to the coordination of such networks.[35] Robert Knight of the University of California at Berkeley has similar views about this issue. He believes that 'you've got to have a way to get brain areas communicating. What oscillations do is provide a routing mechanism.'[36] Importantly, brain waves can carry out such routing extremely rapidly. Such waves may tune out extraneous information by temporarily shutting down unnecessary communication lines.

This basic structure of the human brain is shared by other mammals, and has many similarities with the brains of other animal groups like birds, reptiles, amphibians, and fish. But to understand human consciousness we need to identify the ways in which the human brain differs from those of other species. A striking feature of human evolution has been the rapid growth of our brains relative to other primates.[37] Over the past seven million years, the brains of our ancestors tripled in size, with most of this growth occurring in the past two million years.[38] One way to investigate the biological basis of this extraordinary growth is to study our proto-human ancestors. However, this investigation is complicated by the fact that brains rarely fossilize, and no fossilized proto-human brain tissue has been found. Instead scientists estimate ancient brain volumes from skull size, and studying the few rare fossils that have preserved natural casts of the interior of skulls with the imprints of the brain in the bone. In addition, methods like computerized tomography make it possible to scan the skull and create a digital brain representation.[39] Comparing such brain imprints and 3D representations from different fossil skulls can give us an idea about changes in proto-human brain structure.

Such analysis shows that for the first two-thirds of the period since proto-humans diverged from apes, the size of their brains was within the range of those of apes living today.[40] *Australopithecus afarensis*, the species of which the famous Lucy was a member, had brain volumes of 400 to 550 ml—about the size of a large orange—compared to chimp brains of around 400 ml, and gorilla brains between 500 and 700 ml.[41] Yet showing that size is not the only factor to consider, australopithecine brains show subtle changes in structure and shape compared with those of apes.[42] For instance in this species, the neocortex—the part of the brain involved in higher-order brain functions in modern humans—had begun to expand.

The most dramatic changes in brain size and structure occurred in the final phase of human evolutionary change.[43] Homo habilis, which appeared on Earth 1.9 million years ago and is particularly associated with stone-tool production and use, saw a modest increase in brain size, but also an apparent expansion of a region of the frontal cortex linked to language called Broca's area. The first fossil skulls of Homo erectus, which appeared 1.8 million years ago, had brains averaging just over 600 ml, in contrast to Homo sapiens, which has an average brain volume of around 1,400 ml.[44] Chet Sherwood, an anthropologist at George Washington University in Washington, DC, believes that there must have been a clear functional benefit for such a change. As he notes, 'brains don't just do that for no reason in evolution. Evolution is frugal and cost-effective, and brain tissue has extraordinary metabolic expense. There must have been some adaptive value to brain size increase.'[45]

Neanderthals, who lived between 400,000 and 40,000 years ago, had brains similar in size to those of modern humans. But Robin Dunbar and colleagues at the University of Oxford have suggested that Neanderthals devoted more of their brains to controlling their bodies and regulating their vision.[46] This was because these proto-humans had significantly wider shoulders, thicker bones, and larger eyes than early humans. The researchers believe that this latter difference could be because Neanderthals underwent their key stages of evolution in Europe, at a higher latitude, and therefore poorer light conditions, than Africa, where Homo sapiens originated. After 'correcting' for these differences, the amount of brain volume left for other tasks is significantly smaller for Neanderthals than for modern humans. So while the average brain volumes of the two groups is the same, at around 1,400 ml, the corrected average Neanderthal brain volume is just 1,134 ml, compared to 1,332 ml for humans.[47]

Further information about potential differences between human and Neanderthal brains has come from comparisons of brain imprints on skulls of the two species at different developmental stages. Philipp Gunz and Simon Neubauer at the Max Planck Institute in Leipzig used 3D imaging to study skulls of Neanderthals and humans that were either newborn or at different stages of childhood.[48] They found that Neanderthals had elongated braincases at the time of birth like modern human babies, but only the human ones change to a more globular shape in the first year of life. Neanderthals therefore reached large adult brain sizes through a developmental pathway different to that of modern humans. Moreover, the pattern seen in the Neanderthals is more similar to that of chimpanzees.

Gunz believes that the differences in the patterns of brain development that he and Neubauer have uncovered might contribute to cognitive differences between modern humans and Neanderthals.[49] And Robin Dunbar thinks his team's findings also suggest there was less capacity in the Neanderthal brain for higher cognition and social networking. As he puts it, 'having less brain available to manage the social world has profound implications for the Neanderthals' ability to maintain extended trading networks, and are likely also to have resulted in less well developed material culture—which, between them, may have left them more exposed than modern humans when facing the ecological challenges of the Ice Ages.'[50]

Although these arguments can seem persuasive, it is important to recognize that we still know far too little about why the Neanderthals became extinct while our own species survived. We should also not be too dismissive of Neanderthals' qualities, despite the common view of them as unintelligent brutes. Recent studies have suggested that they had some form of language, advanced skills in tool-making and use, and even a sense of the medicinal properties of herbs.[51] More controversially, there is evidence that Neanderthals may have buried their dead, indicating a possible sense of an afterlife—a subject that I will explore in more detail later—and sufficient creative ability to practise art. But it may still have been the superior ability of humans to negotiate complex social interactions, or a greater capacity for reasoning or thinking ahead, that in the end gave our species its advantage.

There is a limit to what we can learn about human brain evolution from fossil evidence alone. While fossil brain casts are helpful for determining overall brain size, they offer little insight about internal brain anatomy and organization. One way to investigate this question involves studying gross human brain structure and assessing how it differs from that of a chimpanzee, our closest living relative. Such a comparison, carried out by Suzana Herculano-Houzel at the Federal University of Rio de Janeiro, has revealed that most of the increase in the size of the human brain is due to an expansion of the cerebral cortex and its underlying nerve fibres.[52] The parts of the cortex that have grown most are those that integrate information from other brain regions and are involved in higher cognitive functions like planning and abstract thinking. Areas primarily devoted to one function, like the motor cortex or visual cortex, have grown less. Herculano-Houzel thinks that such an expansion may have had important functional consequences, for she believes that it helped to create new patterns of connectivity and new functions in the expanded brain areas.[53]

Another important feature of the human brain is that a larger fraction of its growth occurs outside the womb. While a newborn chimpanzee brain is 36 per cent that of its adult size, a newborn human brain is only 27 per cent that of an adult.[54] Subsequently, from birth until the age of five to six years, the human brain grows six times as fast as that of the chimp. For Chet Sherwood, 'it's as though our newborns and infants still have fetal brains growing in the outside world.'[55] Although humans reach adult brain size in childhood, brain development continues for decades afterwards. And while the human brain at birth has less than 2 per cent of adult levels of myelin—which increases the speed of electrical impulses travelling down neuronal axons—the chimp brain has 20 per cent of adult myelin levels at birth. This disparity continues into adulthood, with chimp brains being fully myelinated at puberty while human brains continue to add myelin until we are much older.

Such differences are important because myelination is guided by stimulation and learning. When neurons are more active, myelin is added to the fibres between them, strengthening the connections. With a more protracted period of myelination, humans have a longer window of time to strengthen these connections, which provides more opportunities for our brains to be shaped by culture, socialization, and environment. This would fit with the fact that compared to other primates, humans are more dependent on culture, interaction, and group identity. According to Sherwood, 'the idea is that the more that human ancestors were dependent on culture and the processes of social learning, the more that success and fitness required having the kind of brain machinery that could sustain ever more complex culture.'[56]

In humans, myelination moves in a wave, beginning in the cerebral cortex nearest the back of the neck and gradually moving forward, finally affecting the frontal lobes in our mid to late twenties.[57] Since the frontal lobes are important for planning, reasoning, and judgement, some scientists have suggested that the limited myelination of these areas in youth might account for the commonly held view that teenagers often make relatively hasty and uninformed decisions.[58] Of course, teenagers might consider such a viewpoint a typical example of older people showing a lack of understanding of young people and their thought processes! But the myelin difference is real, and therefore it warrants further exploration.

Certain human disorders involve defects in myelination. In the autoimmune disease multiple sclerosis (MS), the immune system turns on a sufferer's own neurons and attacks myelin, causing a gradually worsening weakness that can end in paralysis, and eventually death.[59] And while MS is primarily associated with abnormalities in the operation of the muscles that control our movements as well as keep us alive, some people with this disorder develop defects in decision making, showing an impact on the brain.[60] In addition, defects in myelination have recently been more specifically linked to a number of psychiatric conditions. For instance, a study led by Allen Fienberg of the Novartis Research Foundation in San Diego investigated differences in genes switched on or off in the prefrontal cortex of brains of schizophrenic individuals and found that 89 genes were abnormally regulated, with 35 of those being involved in myelination.[61] As schizophrenia tends to develop during adolescence, a time when the human forebrain is becoming myelinated, this provides one explanation for the fact that the disorder tends to manifest itself at this particular time in life.

Identifying differences in the size, structure, and development of the human brain compared to other species still leaves the question of which molecular mechanisms underlie such changes. Here though, we have a valuable new resource, namely increasing information about the human genome, the genomes of apes like chimps, and even those of extinct species like Neanderthals. Such new information will be one of the topics discussed in the next chapter, which focuses on the role that genes play in the formation of human consciousness.

CHAPTER 5

GENOME AND EPIGENOME

To understand how the human brain works we must also understand
how genes regulate the functions of each neuron and glial cell in that
brain, for ultimately all cellular activity is governed by changes in the patterns of 'expression' of the genes that are turned on or off in a cell. Here
though we face a problem, for while the genome has generally been
defined as the sum of the genes in an organism, the more we study real
genomes, the more we realize that viewing them in this way is a major
oversimplification of their true complexity.

Perhaps the most surprising outcome of the Human Genome Project—
which determined the sequence of the approximately three billion 'letters'
in the human genome—was the realization that genes only represent a
small fraction—less than 2 per cent—of the total DNA sequence. The
remaining 98 per cent was initially largely written off as 'junk' DNA, but
more recently evidence has been accumulating to show that a significant
proportion of this 'non-coding' DNA plays key roles in gene regulation—
that is, determining which genes get switched on or off and when.[1] In
addition, instead of DNA being seen as the sole controller of cellular function, there is now growing recognition that RNA—DNA's chemical
cousin—also plays a key role.[2] Moreover, the genome is increasingly
viewed as a complex 3D entity, with dynamic interactions between genes
and their control elements.[3] Meanwhile, the new science of 'epigenetics' is
revealing that the DNA 'recipe' for each organism appears far more
responsive to the environment than previously thought, to both changes
in the cellular environment and those outside the organism itself.[4] These
new ways of looking at the genome have important implications for our

understanding of how the human brain works, and of some of the factors that might make it unique compared to those of other species. But first, let us review what a gene is, and how it is regulated.

Traditionally, a gene has been defined as a stretch of DNA that codes for a specific protein. Proteins are the cell's building blocks, but also perform many other functions, such as transporting substances in and out of the cell, turning food into energy, and producing the chemical signals that regulate cell function. DNA acts as a linear code of four letters, the chemical 'bases' A, C, G, and T,[5] but proteins too can be viewed as a linear code, with their letters being amino acids. The connection between the two molecules is that a triplet of DNA bases codes for each amino acid. In fact the DNA information in each gene is copied into an intermediate form—a type of RNA called messenger RNA (Figure 13). This copying process, known as 'transcription', is carried out by a catalytic protein, or enzyme, called RNA polymerase, and it occurs in the cell nucleus. The messenger RNA is then transported out of the nucleus and into the cell's cytoplasm. Here, the information within it is 'translated' by a cellular machine called a ribosome into a linear chain of amino acids, which then folds into a 3D structure—a protein. Because each of the 20 amino acids has a specific

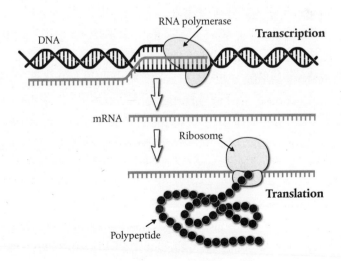

Figure 13 DNA to mRNA to protein.

size, shape, and chemical property, each type of protein will have a unique set of properties. Haemoglobin, for instance, is a globular protein exquisitely suited for its role carrying oxygen around the body, while collagen is a long, highly tensile protein that provides support for bones and tendons. Yet both are produced by information provided by the linear DNA code.

Genes are turned on or off by proteins called transcription factors, which attach themselves to regulatory DNA regions associated with a gene. They then interact with RNA polymerase, enhancing or suppressing its ability to transcribe the DNA information into messenger RNA.[6] Such is the situation in all organisms, from bacteria to humans. But the human genome is very different from the compact genetic entity in a bacterium, in that our genes make up less than 2 per cent of our genome. The proportion of the remaining 98 per cent of DNA that has some function is currently uncertain: it has been estimated at between 8 and 80 per cent of the genome, with continuing debate about the exact figure.[7]

A major source of such new insights is the Encyclopaedia of DNA Elements (ENCODE) project, a follow-up to the Human Genome Project and the subject of my book *The Deeper Genome*.[8] For geneticist Bing Ren of the University of California, San Diego, 'it's not overstating to say that ENCODE is as significant for our understanding of the human genome as the original DNA sequencing of the human genome.'[9] ENCODE sought to map all the functional activity in the human genome. One of its revelations was that the regulatory regions that switch genes on or off may number as many as four million, in contrast to the approximately 20,000 genes in our genome. Rather than being on/off regulators these regions are more like the many switches a sound engineer uses to control a band's live performance, since they control the expression of the messenger RNA in a graded fashion.

Another revelation was that such switches are often far away—at least in linear terms on the chromosome—from the genes they control. At first, this discovery made little sense; however, recent studies have shown that each chromosome in the living cell has an intricate 3D structure.[10] Consequently, switches that appear distant if we think of a chromosome as a linear entity can be very close to their target genes in this 3D conformation.

We have already noted that regulatory DNA switches work by interacting with protein transcription factors, which enhance or inhibit the activity of the RNA polymerase that generates the messenger RNA from each protein-coding gene. However, a surprise in recent years has been the discovery that various non-protein-coding RNAs also play important roles in the regulation of gene expression. Such RNAs are expressed over a large proportion of the genome and exist in various forms, some of which are still being identified. As well as regulating gene expression, they play important roles in the 3D architecture of the genome.[11]

Gene expression is now known to be far more sensitive to changes outside the cell, and indeed the organism, than previously suspected. For many years, the only way that information in DNA was thought to change was through mutations in the bases that make up the four letters of this coding molecule. Such mutations can be detrimental by disabling the functions of important proteins. But mutations are also important for evolution: they provide the variation in a species on which natural selection can act.[12] Mutations generally only drive evolution at a very slow rate. However, more recently, other so-called epigenetic changes have been shown to occur in the DNA, and also in proteins associated with it.

One such change involves the addition of methyl ($-CH_3$) groups to the DNA. Depending on where such changes occur in the genome this can either inhibit or enhance gene expression. Other important epigenetic changes affect the histones—proteins that wrap around the DNA to package and protect it.[13] The addition of acetyl ($-COCH_3$) groups to histones makes them less tightly bound to the DNA, enabling regulatory proteins to access the gene more easily.[14] Other chemical modifications to the histones allow them to 'recruit' regulatory proteins to a gene, just as an army or political party tries to recruit individuals into its ranks. The diversity in how histones can be modified, and thereby influence gene expression in different ways, has led to the idea of a 'histone code', which works at the epigenetic level alongside the genetic DNA code.[15]

The sensitivity, and rapid and reversible nature, of epigenetic changes means that the environment of both the cell and organism can greatly influence the type of changes that occur in the genome, and therefore its pattern of gene expression. This has led to recognition that stress, diet, and many other changes that occur during an organism's lifetime have a far more profound influence on the genomes of its cells than suspected. Most controversially, there is increasing evidence that epigenetic changes may be passed down through the generations, via changes in the sperm and eggs, although the impact of this on evolution of a species in the long term remains far from clear.[16]

While the epigenetic changes mentioned above involve chemical modifications to either the DNA or the histone proteins associated with it, another change to the genome, whose importance is increasingly being recognized, involves movement of DNA sequences around the genome.[17] Such 'transposition' of mobile DNA elements known as transposons challenges the view of the genome as an inherently stable entity. New evidence indicates that transposition has played an important role during evolutionary events in the past, such as the development of vertebrates, and perhaps in our own human evolution, as well as occurring in brain cells in the lifetime of an individual. Transposition may also play important functional roles in the brain, and be implicated in certain kinds of brain dysfunction. With that in mind, let us now look in more detail at how all these different facets of gene expression relate to brain function, and differences between the brains of humans and those of other species.

One important step forward since the completion of the Human Genome Project in 2003 has been the dramatic growth in our ability to study genomes and gene expression in both humans and other species. Although it took over a decade and around $3 billion to sequence the first human genome, thanks to 'next-generation' sequencing methods an individual's genome can now be sequenced in a matter of days for less than $1,000.[18] It is also possible to rapidly sequence the various forms of RNA in a cell, which provides a snapshot of the genes that are switched on or off, and of the expression of regulatory RNAs.[19] At the same time, epigenetic modifications like methyl groups on the DNA, or acetyl and other additions to histone proteins, can all be analyzed across the whole genome. Such 'global' analysis of gene expression and epigenetic information is important for the study of a structure as complex as the brain, for it can be applied to its different cell types, and also to brains at different stages of embryonic or post-natal development.

The most obvious way to try to identify the genetic basis of the unique characteristics that distinguish humans from other species is to compare the genome of our species with those of our closest animal relatives— apes such as chimps and gorillas. Here though, we face an apparent conundrum. For when the protein-coding sequences of humans and these apes were first compared, they showed a remarkable similarity between the two. There is a 99 per cent sequence similarity in the protein-coding genes between humans and chimpanzees, and a 98 per cent similarity between our species and gorillas.[20] This poses the question—does our uniqueness lie in that tiny 1 or 2 per cent difference and, if so, what can we learn about the relevance of this difference for the function of an organ like the brain?

One possibility is that certain genetic differences have a disproportionate effect on the brain. For instance, Marta Florio and Wieland Huttner at the Max Planck Institute in Dresden recently identified a gene that may have played a crucial role in the development of the brain during human evolution.[21] The researchers made the discovery while studying embryonic development of the neocortex, which, as already mentioned, is involved in higher-order brain functions. A particular feature of this brain region is that it is densely packed with neurons. To better understand the genetic basis of this dense neuronal packing, Florio and Huttner analyzed gene expression when this region is developing—in 14-day mouse embryos, and 13-week human embryos obtained from consenting women at abortion clinics. The researchers identified a gene named ARHGAP11B that was highly expressed in human neural stem cells—cells that give rise to new neurons. Intriguingly, this gene is not present in the genomes of any other mammals, including chimpanzees or other primates, but it *was* present in Neanderthals and another extinct proto-human species, the Denisovans.[22]

When Florio and Huttner artificially expressed the ARHGAP11B gene in mice, the brains of such mice grew larger neocortices and some even began forming the characteristic folds, or convolutions, found in the human brain, a geometry that packs a lot of dense brain tissue into a small amount of space, and which is not normally seen with the mouse brain.[23] Similar differences were found when this gene was expressed in ferrets; moreover, this latter study showed that the enlargement of the neocortex was due to enhanced action of a specific type of glial cells called basal radial glial cells.[24] This suggests a potentially important role for ARHGAP11B in human brain evolution, and provides a specific mechanism for its action. However, the expansion of the brain during this process, and importantly also the structural and functional changes that occurred during evolution of the human brain, are likely to have involved many other genetic changes besides this one, and these remain to be identified.

In fact, novel genes only found in humans, or in closely related extinct proto-human species, are currently thought to be rare. Instead a more general cause of the differences between humans and a closely related primate like the chimpanzee is likely to be not novel genes but rather subtle differences between the levels of expression of a gene in humans compared to other species. For instance, a study led by Debra Silver of Duke University in North Carolina has shown how changes in the regulatory regions that control gene expression can play an important role in this respect.[25] Silver's team focused on a regulatory region called HARE5, because it was close to a cluster of genes involved in neural stem cell function and because it differs in sequence between mice and humans.[26] Genetically engineered mice that expressed the human HARE5 regulatory sequence had enlarged brains, and the brain growth in such mice was 12 per cent greater than ones in which chimpanzee HARE5 was substituted. This finding may help explain why human and chimp brains grow at the same rate for the first part of foetal development, but then after 22 weeks, chimp brain growth levels off while human development continues. Meanwhile, another recent study has shown that specific differences in a gene called NOTCH2 increase the human brain's production of neural stem cells and delay their maturation into cortical neurons, which may also help to explain why our brains keep growing far longer than those of other primates.[27]

Size alone, however, is unlikely to explain the unique characteristics of the human brain. Scientists are also trying to identify the genetic basis of structural and functional differences between our brains and those of other mammalian species, particularly primates. Studies such as those above have identified novel human genes, as well as human-specific patterns of expression of genes present in both humans and chimpanzees, that may regulate unique human characteristics such as conscious awareness. Yet, other findings suggest that a key difference between humans and chimps is that our brains are *less* rigidly controlled by our genes, allowing the environment to play a greater role in our brain development.

One such study led by Aida Gómez-Robles at George Washington University in Washington, DC, compared hundreds of human and chimpanzee brains using magnetic resonance imaging (MRI), which provides a detailed image of the living brain.[28] The researchers obtained family trees for both humans and chimps, which meant they could measure similarity in the brains of genetically related individuals, including identical twins. The study showed that related humans vary far more in the shape and location of their neocortical folds than chimpanzees, with two chimp brothers being far more similar than two human brothers in the shape and location of these folds.[29] This suggests that chimps are more limited in the ways their brains can develop and their ability to learn new behaviours or skills than humans. And with such a loosening of the genetic ties in humans, our brains are probably more susceptible to external influences; in other words, they are more 'plastic'. This allows environment, experience, and social interactions with other individuals to play a more dramatic role in organizing the cerebral cortex. Mary Jane West-Eberhard of the Smithsonian Tropical Research Institute in Panama believes these findings 'illustrate and reinforce, using comparisons of closely related species, something...doubted by some evolutionary biologists in the past, namely the fact that plasticity itself can evolve.'[30]

If human brains are more loosely determined by genes than those of chimpanzees, epigenetic mechanisms may play a heightened role in different aspects of human consciousness, both at a species level and in individual human beings. And, indeed, new evidence is emerging for how this could work. I mentioned earlier that epigenetic modifications to the histone proteins that wrap around the DNA in the nucleus can affect the expression of genes with which they are associated. In one recent study, David Allis and his colleagues at the Rockefeller University in New York showed that histone modifications may regulate genes involved in structural alterations to synapses in response to neural signals—a key cellular mechanism underlying learning.[31] Further investigation showed that such histone modifications also play a key role in neuroglia. Allis believes that this 'may have implications beyond neurons, as a means for controlling plasticity in adult cells that have a set identity, but still must respond to their environment.'[32] With our heightened capacity for learning, such mechanisms could be particularly important in the human brain.

We met earlier DNA sequences called transposons, which can move about the genome. Presumably because of the disruption this might cause, there are mechanisms to supress transposon activity in the sperm and eggs that give rise to the next generation.[33] It was initially assumed that this would also be true of other cell types, for although such activity would not jeopardize the next generation as in the case of the sex cells, it could still disable vital genes or cause uncontrolled growth leading to cancer, if left unchecked. It came as a surprise then, when in 2002 Fred 'Rusty' Gage and colleagues at the Salk Institute in La Jolla, California, discovered high levels of transposon activity in the brain.[34] Gage and his team were studying a process called 'neurogenesis'—the development of neurons from stem cells in the adult brain. To understand this process better, they surveyed the genomes of these stem cells and discovered to their surprise that the most active elements were the transposons. This was a highly unexpected finding, for if transposon activity can interfere with the structural integrity of the genomes of each new neuron, then each nerve cell in the brain might be genetically different, conflicting with over a century of scientific dogma that has generally viewed each cell in the human body as having the same genome.

To investigate this possibility, Gage's team obtained brains from deceased people who had donated their bodies to medical research, and sequenced the genomes of hundreds of individual neurons.[35] Remarkably, they found that cells in the same brain are indeed genetically distinct. So what purpose might such genomic instability serve? Gage believes that transposons may generate novel responses in neurons in ways that the 20,000 or so genes in the human genome alone cannot. One interesting implication of his team's discovery is that identical twins might differ far more in brain function than suspected, since transposon events in their neurons will increasingly diverge as they develop, and so may their behaviour.

A puzzling aspect of the findings is how transposon movement could play a positive role given that such activity is thought to be mainly disruptive. One possible answer comes from a new type of transposon identified by Gage's team that has the potential to produce hundreds or even thousands of previously unknown proteins.[36] This transposon, called ORF0 and found in humans and other primates, can blend with the protein-coding sequences of existing genes, and as a consequence these can then form completely new proteins, with novel cellular roles. For Ahmet Denli, a researcher on the study, 'this discovery redraws the blueprint of an important piece of genetic machinery in primates, adding a completely new gear.'[37] Yet whether ORF0 plays specific, unique roles in humans remains to be shown.

Transposon activity is just one way that alteration of the genomes of different neurons can occur. There is also increasing evidence that brain cells can be altered genetically in a more subtle manner, via changes in the individual letters of the DNA code. Such changes, which occur in brain cells but also in cells in other organs, are termed 'somatic mutations' to distinguish them from the 'germline mutations' that affect the 'germ cells', that is the eggs and sperm.[38] Somatic mutations are caused by errors during DNA replication, chemical damage to the DNA, or inefficient DNA repair mechanisms. Recent studies indicate that such events occur in brain cells much more frequently than previously thought, which means that individual neurons may take on different properties than their neighbours as a consequence.

Mutations in neurons may be a factor in a variety of brain disorders. A study led by Christopher Walsh of Mount Sinai University, New York, found somatic mutations in the brains of children with hemimegalencephaly, a disorder in which one hemisphere is enlarged, causing epilepsy and intellectual disability.[39] These mutations cause specific cells in the brain cortex to proliferate abnormally, which raises the question of whether somatic mutations play a role in other brain disorders, including psychiatric conditions. An even more intriguing question is whether somatic mutations contribute to differences between individuals. Neuroscientist Alysson Muotri of the University of California, San Diego, believes this 'might explain why everybody's different—it's not all about the environment or genome. There's something else. As we understand more about [the role of somatic mutations], I think the contribution to individuality...will become clear.'[40]

Another revelation of recent years is the discovery that RNA can travel around the brain, transferring information. We have already seen how genes work by being transcribed into a messenger RNA copy; this then moves out of the cell nucleus and is translated into a protein in the cytoplasm. Initially it was assumed that both DNA and RNA remain firmly within the cell. However, recent studies suggest that certain types of RNA can leave one cell and end up in a different one. Such transport involves structures called exosomes, which are particles secreted from many types of cells (Figure 14).[41] In the brain and peripheral nervous system, there is growing recognition that exosomes can guide the direction of nerve growth, control nerve connections, and help regenerate nerves. According to Eva-Maria Krämer-Albers of the Johannes Gutenberg University in

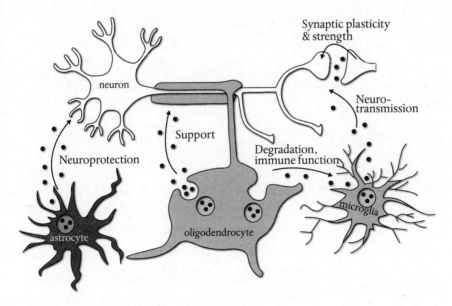

Figure 14 Communication in the brain via exosomes.

Mainz, Germany, 'the neuroscience community is just catching on to this relatively novel form of communication.'[42]

Felipe Court, of the Pontifica Universidad Católica, in Santiago, Chile, has recently discovered that glial cells use exosomes to stimulate the regrowth of axons from a damaged neuron.[43] His team found that exosomes secreted by glial cells are rich in messenger RNAs that code for proteins involved in the assembly, organization, and regeneration of cells, as well as in axon growth and regeneration. For Court, 'this means that the [glial] cells are making transcripts, putting them in exosomes, and exporting them just for neurons to use.'[44] As well as messenger RNAs, the exosomes also contain regulatory RNAs that influence gene expression. Court and his colleagues found that if they added exosomes from glial cells to damaged neurons, their axons began to regenerate immediately.

The importance of such novel mechanisms of gene expression is that they are starting to cast light upon key aspects of how the human brain functions and its plasticity and responsiveness to the environment. Two such aspects to consider are the linked processes of growth and development, subjects that we will look at in more detail in Chapter 6.

CHAPTER 6

GROWTH AND DEVELOPMENT

E ach human life begins with the union of a sperm and an egg. Embryogenesis is the process by which the fertilized egg divides repeatedly to produce the 37 trillion cells that make up a person.[1] But embryo development is about far more than just this dramatic increase in cell number. It also involves the formation of all the specialized cell types of the body, and their organization into tissues and organs. Initially, the embryo is just a ball of cells called a blastocyst. But then a dramatic transformation takes place called gastrulation. The biologist Lewis Wolpert, of University College London, has said that 'it is not birth, marriage, or death, but gastrulation which is truly the most important time in your life.'[2] This may be overstating things, but gastrulation is certainly a key event in the formation of a human being, for at this stage the embryo acquires the distinct cell layers that later give rise to different organs. During gastrulation, three cell layers form—the ectoderm, mesoderm, and endoderm.

From the ectoderm comes the skin and nervous system, including the brain. The guts develop from the endoderm, and the other organs from the mesoderm. The nerves and brain arise from a structure within the ectoderm called the neural tube that forms three weeks after conception.[3] The lower part of this becomes the spinal cord, while the brain develops from the upper part (Figure 15). At four weeks, the part of the neural tube that forms the brain is already divided into three regions which will become the hindbrain, midbrain, and forebrain.

While these are the gross structural changes underlying brain development, equally important are the cellular changes. The development of the

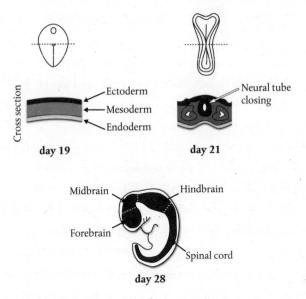

Figure 15 Development of the embryonic human brain.

human brain occurs at such a rapid rate that, at times, 250,000 neurons are added every minute![4] A newborn baby's brain has almost the same number of neurons as that of an adult. However, brain growth continues after birth due to the creation of new glial cells and connections between neurons.

Both neurons and glial cells originate from neural stem cells.[5] The process by which a stem cell gives rise to a more specialized cell type is called differentiation, and is driven by chemical messengers such as hormones and growth factors attaching themselves to receptors on the surfaces of stem cells. These receptors then transmit signals inside the cell that turn particular genes on or off, leading to changes in cellular proteins. Differentiation is a multistep process, with neural stem cells first giving rise to the prototypes of either neurons or glial cells, which then act as precursors to more specialized cell types within these different classes, to eventually form the hundreds of different cell types in the brain.[6]

A key feature of the brain is that incoming signals from the rest of the body also shape its development. This was graphically shown by experiments in the 1950s by David Hubel and Torsten Wiesel of Johns Hopkins

University, Baltimore, USA.[7] Hubel and Wiesel decided to study cats because they have well-developed visual centres, a reflection of their being nocturnal hunters. Not surprisingly many people feel uncomfortable about the idea of scientists experimenting on a species kept as a household pet. Yet these studies were central to shaping our understanding of how the brain 'sees' objects outside the body.

Hubel and Wiesel investigated how neurons in the brain's visual cortex respond to different images.[8] The experiments resembled something akin to the film *Clockwork Orange*—the cats, anesthetized to feel no pain, were strapped in chairs with their eyelids propped open, in front of a slide projector showing different images, while the electrical activity in neurons in their visual cortex was recorded. Initially, Hubel and Wiesel tried stimulating the cats' neurons with visual stimuli such as dots, but with no success. But then a chance accident took place of the sort that can sometimes lead to great scientific discoveries—a slide jammed in the projector, and one neuron began firing like a machine gun. Initially baffled, Hubel and Wiesel realized that the neuron was responding to the shadow that the jammed slide was making on the screen; in other words, it was responding to a vertical line.

Further studies showed that cells in the visual cortex responded not to spots but lines and edges, each cell having its own preferred orientation.[9] Such studies showed that the cortical neurons fire only when a line is in a particular place on the retina; neuronal activity changes depending on the orientation of the line; and sometimes neurons fire only when the line is moving in a particular direction. Indeed Hubel and Wiesel's studies showed that cats—and humans—see moving images much more easily than still ones.[10] This makes sense from an evolutionary perspective because it is probably more critical to detect moving objects—in a cat's case, dogs, mice, falling branches—than stationary ones, which can await attention.

From a brain development perspective, however, the most important finding by Hubel and Wiesel was made when they sewed shut one eye of a newborn kitten and only reopened it after a certain period.[11] Remarkably, kittens with one eye deprived of vision for only the first three months remain blind in that eye for their whole life. This showed that visual inputs at this stage of life are critical to development of this part of the brain. This feature of brain development was further explored by Colin Blakemore and Grahame Cooper of the University of Oxford in the 1970s.[12] They placed kittens in enclosures whose walls were painted with only vertical, or horizontal, lines. Kittens only exposed to horizontal lines could see the seats of chairs and would jump on to them to sleep, but could not see the chair's legs, and were constantly banging into them. In contrast, kittens only exposed to vertical lines were fully aware of the chair legs but could never identify a spot to snooze above these legs.

Not surprisingly, such experiments are controversial. Indeed, Blakemore endured years of attacks by animal rights supporters, despite, or perhaps because of, his willingness to discuss what might seem like barbaric experiments with the general public. According to him, 'there were times I was shocked by what happened to me—razor blades in envelopes, bombs, threats against my kids—but I never doubted the principle of public engagement.'[13] His justification for the experiments was that they revealed fundamental aspects of brain development and had potential relevance for treatment of human visual disorders.

Such studies show that the environment can shape the development of the brain, but do not explain how this occurs. However, further investigations have begun to reveal the neural basis for this environmental influence. For instance, recently Blakemore and his colleagues investigated how mice develop sensitivity in their whiskers.[14] Such whiskers are connected to clumps of neurons in the brain called 'barrel structures'— so-called because of their barrel-like shape. The researchers showed that if a clump of whiskers is removed at an early age, the brain region linked to that area never develops the barrel structure.[15] Such findings might suggest that the brain is merely a blank slate upon which the environment imposes itself, but Blakemore believes that a more accurate way of understanding the interaction between brain and environment is to view it as a complex feedback loop.

Intriguingly, studies of human beings with deficits in different senses indicate that the brain is highly plastic in responding to such deficits. For instance, it has been known for some time that people born blind or who become blind early in life often have a more nuanced sense of hearing, especially when it comes to musical abilities and tracking moving objects in space. Ione Fine and colleagues at the University of Washington in Seattle used functional MRI (fMRI) to compare the brains of blind versus seeing individuals, and found that that the auditory cortex of blind people showed narrower neural 'tuning' than sighted individuals in discerning small differences in sound frequency. According to Fine, 'this is important because this is an area of the brain that receives very similar auditory information in blind and sighted individuals. But in blind individuals, more information needs to be extracted from sound—and this region seems to develop enhanced capacities as a result.'[16] Such findings show how development of abilities within infant human brains is influenced by the environment in which they grow up.

Conversely, an fMRI study led by Christina Karns at the University of Oregon found that the brains of deaf individuals have enhanced capacity in other senses.[17] Given that we have seen how deaf individuals can develop an inner dialogue that relies not on sounds, but on visual symbols, it would be interesting in future studies to investigate how this is reflected in molecular and cellular changes in the developing human brain. However, here a major limitation is that scientists obviously cannot do experiments on living human foetuses. Instead, as in other areas of science, a variety of animal models are used to study the molecular and cellular processes that underlie nerve and brain development.

One species used to investigate brain development is the fruit fly. An advantage of studying the fruit fly brain is that it is much simpler than that of a person, containing only 200,000 neurons compared to the 100 billion neurons in a human brain.[18] And unlike mammals, fruit fly embryos develop outside the mother's body, and so can be studied continuously during development. The fruit fly also has a short lifespan, speeding up the analysis of its development, and gene mutants in this species can be generated using X-rays, which mutate the DNA. By studying fly embryos produced in this way, it is possible to identify mutants in which nerve and brain development is disrupted. Studies of the genetic basis of such mutants can then identify the genes involved.

Investigations of such fly mutants have identified genes that control how neurons connect with each other.[19] Recall that neurons possess a cell body, but also dendrites and an axon. They receive incoming signals from other neurons via their dendrites and in turn transmit messages to the dendrites of other neurons through their axon. In the embryonic brain, neurons must identify their correct partners in the brain's wiring circuit. Studies of flies have shown that neurons do this by expressing genes that code for proteins on the surface of the developing axon which recognize cues in the brain environment and guide neurons along molecular pathways—like train tracks—to their correct connections.[20] Other genes control the formation of synapses. Such findings can increase our understanding of human brain development since related genes often exist in the human genome and can play similar roles in the development of connections between human neurons cultured outside the body.

Because of this overlap in functional roles, genes that are linked to brain disorders in humans and are also present in the fly genome have been identified. In such cases, the fly can be used to study the functional role of such genes in brain development. Such an approach was used to study the gene associated with a condition called Fragile X, which is the most common cause of inherited mental disability in humans and is associated with a mutation in the FMR1 gene.[21] Studies of the brains of people affected by this disorder had indicated that they have slight differences in the structure of the dendrites of their neurons, but the functional significance of this difference was not clear. But investigations in flies showed that this gene is involved in the formation of the neuron's internal cellular 'skeleton', the cytoskeleton, and that defects in FMR1 affect the formation of synapses, which may be one reason for the learning disabilities of people with the disorder.[22] Further studies may help identify drugs that can offset some of the mental problems caused by this gene defect.

As well as providing a way to study the cellular architecture of the brain, flies offer a way to study the development of the nerve circuits underlying behaviour.[23] By generating fly mutants and then subjecting them to behavioural tests, it has been possible to identify the genes that regulate the formation of the neural circuits that underlie the fly's response to different odours, its sexual behaviour, and the body clock that regulates the sleep/wake cycle.

Although fruit fly studies have greatly expanded our understanding of the molecular mechanisms underlying nerve and brain development, there is clearly a big difference in complexity between the fly brain and that of a human being. So researchers are interested in studying brain development in more complex organisms. One organism that has proved very popular in this respect is the zebrafish.[24] Like humans, zebrafish are vertebrates. But zebrafish embryos develop outside the mother, so, like flies, they can be studied continuously as they develop. And because these embryos are also transparent, this makes it possible to use fluorescent probes to monitor the activities of the neurons in the developing embryo.[25]

Recently, ways have been developed to simultaneously record the activity of every single neuron in a zebrafish larva's brain. The approach, pioneered by Philipp Keller and Misha Ahrens at the Howard Hughes Medical Institute in Ashburn, Virginia, uses genetic engineering to modify zebrafish neurons so that they express a protein that fluoresces when the neuron becomes activated.[26] A sophisticated imaging system allows the changes in fluorescence of the whole larval brain to be recorded. Studies so far have recorded the global changes in activity that occur as a larva moves or is exposed to a stimulus. Future studies could explore how mutations in specific genes affect the activity of the whole brain.

Fluorescent probes can also be used to assess changes in the chemical messengers that regulate gene expression in developing neurons. One such messenger is calcium ions, changes in the concentration of which can act as a 'signal' that modulates the activity of cellular proteins. Yusuke Tsuruwaka and colleagues at the Agency for Marine-Earth Science and Technology in Yokosuka, Japan, used this approach to study changes in calcium signals in the embryonic zebrafish brain.[27] This identified a dramatic increase in such signals in the developing mid- and hind-brain around the time that neural circuits are forming in these brain regions. Such calcium signals may help form these circuits by activating the calcium-sensitive protein CaMKII, which is already known to play important roles in regulating many processes in the brain.[28] To explore this question further, the gene coding for CaMKII could be mutated to see what effect this has on zebrafish brain development. Such a strategy has been greatly facilitated by the development of new 'gene editing' technologies, which make it possible to generate subtle genetic changes in the zebrafish rapidly and economically.[29]

Despite the importance of zebrafish as a model of vertebrate embryo development, there are important differences in brain development between fish and mammals. For some years it has been possible to use genetic engineering to delete, or 'knockout', the expression of a gene in mice.[30] An alternative approach introduces a more subtle 'knockin' change, for instance a single letter mutation, into a gene. New gene editing methods now make it possible to produce knockout and knockin mice more rapidly, and also to edit genes in other mammalian species, including primates, humans' closest biological relatives.[31] Gene editing in primates has opened up the study of the role of genes in the development of brain regions like the prefrontal cortex, a region involved in planning and social behaviour, which is particularly prominent in humans but also in other primates.

Recently Yong Zhang and colleagues at the Chinese Academy of Sciences in Beijing used gene editing to knockout SHANK3—a gene linked to autism in humans—in monkeys.[32] SHANK3 protein is located at synapses, where it anchors other proteins in place. The study showed that monkeys lacking SHANK3 have fewer neurons in the prefrontal cortex, thereby highlighting a new role for SHANK3 in neurogenesis. Intriguingly, mice lacking this gene do not show this defect. The difference may reflect the fact that in newborn monkeys, the gene is mainly expressed in the prefrontal cortex, while in mice SHANK3 expression is highest in the striatum, a brain region governing movement and reward. Zhang believes this shows that monkeys 'have more prominent molecular and cellular changes compared with mice [and also] the value of pursuing non-human primate models further.'[33]

One major obstacle to studying brain development in mammals is that the embryo develops inside the mother's body; while embryos at different stages of development can be removed from the mother and studied, this can only occur at the expense of the embryo's life. This makes it impossible to observe embryonic development as a dynamic process, as is possible with fruit flies and zebrafish. The situation may soon change though, as Magdalena Zernicka-Goetz and colleagues at the University of Cambridge have recently succeeded in nurturing mouse embryos outside the mother's body for the first phase of embryogenesis.[34]

The mouse embryos survived for seven days, which represents a third of the normal 21-day pregnancy period in mice. Perhaps, then, such embryos could be cultured for a longer period of time outside the womb. There are undoubtedly many practical problems to overcome to achieve such a goal. An embryo growing inside the mother can only develop with the support of a yolk sac, which provides nourishment for the embryo and a network of blood vessels. However, such a support system might be generated in the future with other types of stem cells. As a consequence, it may be possible soon to study the development of the mouse brain and nervous system at later stages of development, using methods similar to those pioneered with fruit flies and zebrafish.

Zernicka-Goetz's findings also raise the question of whether human embryos might be nurtured outside the mother in a similar fashion. Indeed the Cambridge scientists are already planning to pursue such a line of investigation.[35] And although Zernicka-Goetz's team have said they are only interested in studying the very initial stages of human embryo development using this approach, it is hard to imagine that at some point scientists will not want to push the boundaries of this new technology. Currently, human embryo research in Britain is limited by a ruling by the UK Human Fertilization and Embryo Authority that says that such embryos should not be cultured for more than 14 days.[36] But this is a fairly arbitrary limit and it could be argued that to better understand the biological factors that lead to miscarriage and developmental disorders, the limit should be extended. However, whether such an approach could ever be used to study human embryonic brain and nervous system development would clearly be a highly controversial ethical topic.[37]

Another area of stem cell research that poses both ethical and technological challenges is the study of human brain organoids, or 'mini-brains'. These can be generated from embryonic stem cells but also from induced pluripotent stem (iPS) cells—ordinary skin cells transformed into a stem-cell-like state. By growing such cells in a 3D matrix, it is possible to create embryonic brain-like structures. An unexpected discovery of such studies has been the complex nature of such organoids. For instance, Paola Arlotta at the Massachusetts Institute of Technology has shown that mini-brains grown from iPS cells derived from healthy volunteers and cultured for nine months contain many types of neurons ranging from those that normally make up the cerebral cortex to ones that link the right and left hemispheres of the brain.[38] Glial cells were present as well as neurons. The neurons created in this way even formed electrically active neuronal networks. According to Arlotta, the neurons 'connect with each other, forming circuits, and once they're connected, they can synchronize their activity', thereby potentially mimicking higher-order functions of the human brain.[39] Mini-brain studies have also revealed important differences between those derived from individuals with specific brain disorders and those from unaffected people.[40]

Some scientists doubt the extent to which such mini-brains can ever mimic the complexity of brain structure in a living human. But neuroscientists are working hard to improve the power of this approach via a series of innovations. One involves growing relatively simple organoids representative of different brain regions and allowing them to connect with one another. For instance, In-hyun Park and colleagues at Yale University grew two separate organoids, one based on the brain cortex, the other on a structure called the medial ganglionic eminence (MGE) that in the embryonic brain produces inhibitory neurons which play a role in normal human brain development.[41] When placed together inhibitory neurons from the MGE structure migrated into the cortical organoid and began to integrate into the neural networks there, exactly as they do in the embryonic brain.

Park hopes such studies will provide insights into the developmental roots of conditions like autism and schizophrenia, for according to him, one feature of these disorders is 'an imbalance between excitatory and inhibitory neural activity'.[42] More generally this approach could allow researchers to study the development of specific brain subregions, and how cells from such regions interact once they start migrating and encountering each other.[43]

Another innovation involves culturing brain organoids for greater lengths of time. Sergiu Paşca and colleagues at Stanford University have grown such organoids for nearly two years.[44] This proved important in allowing the maturation of astrocytes, the class of glial cells we came across earlier. Astrocytes help neurons form synapses and prune them when they are no longer required; they also make connections with blood vessels and sense damage in the brain. But astrocytes also play a role in neurodegenerative conditions like amyotrophic lateral sclerosis, and may contribute to other brain disorders. However, a problem with brain organoids has been the limited development of astrocytes within them.

Paşca's team found that during the first five months of organoid growth, the astrocytes multiplied rapidly, eating away at the synaptic connections between neurons, much as they do in the foetus. But after nine months—mimicking what occurs in the brains of newborn babies—they began expressing a different set of genes, and switched to more supportive functions, like boosting calcium signals in nearby neurons to facilitate their maturation. By growing organoids for extended periods, Paşca believes 'we can really start to ask questions about what could go awry later in foetal development to lead to psychiatric disorder.'[45]

Studies like these are stimulating important new insights into the biological basis of brain development in the embryo and neonate. But our concern in this book is why human consciousness is qualitatively different from that of other animals, so it would be odd in a chapter about brain growth and development to solely discuss what happens in the womb or even just after birth. For a key feature of being human is the extent to which—unlike other mammals, including even our closest primate cousins—a major part of our development continues to take place after birth, in the growing infant and child.

Ironically, this feature of our species may have first arisen as a lucky accident due to a contradiction between two other important human features—our upright stance and our oversized brains. These two features allowed our proto-human ancestors to first begin to distinguish themselves from apes, but they also created a problem, because to walk upright efficiently the pelvis needs to shrink, and with it the birth canal in females.[46] Yet to develop a bigger brain, the skull of the developing baby must also increase in size, which makes it harder to exit the birth canal. This contradiction is one reason why giving birth as a human female is excruciatingly painful and in past times could often result in the death of the mother; indeed this remains a risk today despite advances in modern gynaecological practice.[47]

Yet even with such extra risks there are limits to how large a human baby's brain can grow without making birth impossible through a pelvis evolved for bipedalism. And here a fortuitous aspect of human evolution occurred—confirming that virtue can come from necessity—for natural selection led to the birth of babies that are 'underdeveloped' compared to other mammals. While solving the problem of how to give birth to a large-brained baby without dying in the process, the birth of such premature infants may have been key to the further development of humankind, by virtue of the fact that a baby born with an underdeveloped brain also possesses a plasticity amenable to another unique aspect of humanity—our particular facility for learning, and in the process sometimes transcending our teachers.[48]

The central argument of this book is that the human mind is the product of an interplay between natural, biologically determined development and the cultural impact on such development of the interaction of a growing individual with other people. According to this view, the process by which a child's mind becomes structured by society involves more than the simple acquisition of the values, expectations, and competencies of a specific culture. Rather, the entire system of naturally determined or 'lower' mental functions becomes restructured to produce higher mental functions. Or as Lev Vygotsky put it, 'when the child enters into culture, he not only takes something from culture, assimilates something, takes something from outside, but culture itself profoundly refines the natural state of behaviour of the child and alters completely anew the whole course of his development.'[49]

Vygotsky defined higher mental functions as behaviours mediated by language and other cultural 'tools', and their development as a transition from 'inter-individual'—that is, shared between individuals—to becoming a specific property of the individual. For young children, most higher mental functions exist only in an inter-individual form, as they share these functions with adults or other children. However, as a child develops, they begin internalizing these functions into their individual psyche. As Vygotsky put it, 'every function in the cultural development of the child appears on the stage twice, in two planes, first, the social, then the psychological, first between people as an "inter" mental category, then within the child as an "intra" mental category. This pertains equally to voluntary attention, to logical memory, to the formation of concepts, and to the development of will.'[50]

This process can be observed during the development of memory. Vygotsky distinguished between two types of memory in human beings. On the one hand, there is the natural retention of stimuli that we share with many other living organisms. On the other, there are the mediated forms of memory unique to our species. According to Vygotsky the first type of memory is 'very close to perception, because it arises out of the direct influence of external stimuli upon human beings.'[51] However, humans also have another, mediated form of memory, shown by the 'use of notched sticks and knots, the beginnings of writing and simple memory aids [which] all demonstrate that even at early stages of historical development humans went beyond the limits of the psychological functions given to them by nature and proceeded to a new culturally elaborated organization of their behaviour.'[52]

During a child's development, such mediated forms of memory come to dominate over natural forms of remembering, as Vygotsky demonstrated in a study carried out with children and adults of different ages.[53] Participants were presented with a list of fifteen words to be memorized. In the first stage of the study, remembering had to be done without any help, but in the second stage, participants were also offered fifteen picture cards that could be used as memory aids. The major finding of the study was that the degree to which the picture cards helped memorization depended on a participant's age.

In preschool children the use of memory aids did not lead to significant differences in remembering. However, the situation changed dramatically with seven- to eight-year-old children, who recalled about 75 per cent of the words with the help of the cards, but only 30 per cent without them. Interestingly, adults recalled 93 per cent of the words with the help of the cards compared to 60 per cent without them, showing that although the cards were useful in memorization, their overall influence was less. Vygotsky interpreted the findings as showing that external memory aids can help children at a certain stage of development, and also adults, to remember things, even in 'unmediated' conditions.[54] However, adults can additionally draw on memorization techniques that are internalized in a way that young children cannot. What does such a transformation of memory involve in terms of neuronal processes? To explore this question, it is time to look at what modern research is uncovering about the process of learning and memory in the animal, and human, brain.

PART III

THE DYNAMIC MIND

CHAPTER 7

LEARNING AND MEMORY

Without learning, there would be no human civilization. Learning is central to the development of a child into an adult, but also to the progression of human culture over time. But the ability to learn is also bound up with the capacity to remember. Speculation about the nature of human memory stretches back at least 2,400 years to Plato's attempts to understand this process. He believed that humans are born free of any knowledge and compared memory to a scribe making impressions on a wax tablet.[1] The eighteenth-century English philosopher David Hartley was the first to link memories to the nervous system. He proposed that life experiences set up 'neural vibrations' in the brain which, by continuing to resonate, persist as memories.[2] It was an interesting idea, but not one based on any actual observed process.

In 1949, Donald Hebb of McGill University in Montreal, Canada, suggested that 'neurons that fire together, wire together', implying that the encoding of memories occurs as connections between neurons are established through repeated use.[3] This was an intriguing suggestion with a plausible mechanistic basis in the known properties of neurons, but at the time there was no experimental evidence to support Hebb's hypothesis and it fell to others to demonstrate how such activity might lead to memory encoding at the molecular level.

A major pioneer in this work was Eric Kandel of Columbia University in New York. He chose to study a marine organism—the sea slug *Aplysia*—because its relatively simple nervous system and huge neurons made it an ideal tool for studying the biochemical changes that occur during a learned response.[4] In the late 1960s and early 1970s Kandel and his team showed that the process by which *Aplysia* learns to withdraw its sensitive gills in response to a stimulus involves production of a chemical messenger called cyclic AMP. This then activates the enzyme PKA, and PKA acts to modify the synapses between connecting neurons. Longer-term changes are regulated by the gene regulatory protein CREB, which is activated by PKA. The switching on of specific genes that trigger the production of new proteins leads to cellular changes such as an increase in synaptic connections.[5] This suggested that learning involves a physical change in nerve connections.

Another important step forward in our understanding of the cellular mechanism of memory formation was made in 1973 by Tim Bliss and Terje Lømo, at the University of Oslo in Norway. They showed that repeated stimulation of connected neurons in slices of the hippocampus region of a rabbit brain increased the efficiency of signalling between the cells by strengthening the synapses at which they communicate with each other. They called this process long-term potentiation (LTP) (Figure 16).[6] LTP leads to the long-term strengthening of the synapses between two neurons activated simultaneously. When axons that connect with neurons of the hippocampus are exposed to a high-frequency stimulus, the neurons become super-sensitized for up to several weeks. Subsequent work has shown that a strong input leads to more receptors for neurotransmitters being directed to the surface of the receiving neuron beyond the synapse, amplifying the signal it receives.[7] And it turns out that this movement of receptors to the cell surface is catalyzed by the calcium-sensitive enzyme CaMKII, which we have met before. It is activated by an increase in calcium ions in the receiving, or postsynaptic, neuron which occurs when the neuron is stimulated repeatedly.

One potential problem with the idea that LTP is the sole basis of memory is that synaptic components are short-lived and yet memories can last lifetimes. This suggests that synaptic information is encoded at a deeper, finer-grained molecular scale. And indeed, in support of this idea, Ryohei Yasuda and colleagues at Duke University in North Carolina have shown that such hard-wiring of memory may involve changes to the neuron's

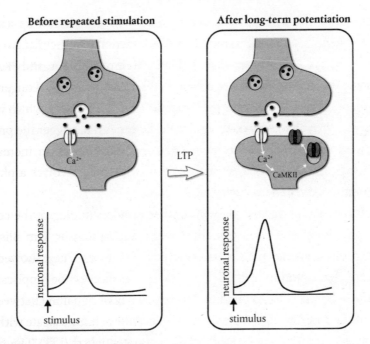

Figure 16 Long-term potentiation.

cytoskeleton that gives structure to synapses.[8] Their findings suggest that signalling molecules help to rearrange this framework, and give more volume and strength to the synapses.[9] Yasuda believes that a long-lasting memory could possibly come from changes in the building block assemblies. His team found that certain proteins that regulate the cytoskeleton are activated by CaMKII, and trigger long-term changes in the synapse.

Changes such as those identified by Yasuda and colleagues are in the postsynaptic neuron. However, recently evidence has emerged that changes in the presynaptic neuron may also be important. In particular, a study led by Troy Littleton of the Massachusetts Institute of Technology has identified an important role for spontaneous release of neurotransmitter by the presynaptic neuron in the absence of action potentials.[10] These 'mini' neurotransmitter releases have previously been thought to represent noise occurring in the brain. But Littleton's team found that mini events can drive synaptic structural plasticity, for when the researchers stimulated neurons, these events, which are normally of very low frequency, suddenly ramped up and stayed elevated for several minutes before going down. The enhancement of minis stimulates the postsynaptic neuron to release a factor, still unidentified, that causes the presynaptic cell to stimulate the enzyme PKA.[11] This enzyme in turn stimulates growth of new connections between presynaptic and postsynaptic neurons, making the latter even more responsive to future communication from the presynaptic neuron.

I mentioned previously that there is increasing recognition of the important roles played by glial cells in the brain. Until recently such cells were not thought capable of being involved in processes such as LTP, because glial cells lack the electrical activity of neurons. However, there is increasing recognition of a role for these cells in memory formation. For instance, Dmitri Rusakov and colleagues at University College London recently found that astrocytes release the amino acid D-serine, which activates a protein on the surface of the neuron called the NMDA receptor, which plays an important role in inducing LTP. Andrea Volterra of the University of Lausanne in Switzerland believes this shows that while neurotransmitter release and electrical changes at the synapse are important for synaptic memory formation, also required is a 'burst from the astrocyte to complete the process'.[12]

While changes to the synaptic structure of a neuron clearly play an important role in the initial formation of memories, there is increasing evidence that altered expression of genes underpins long-term memory formation. I mentioned previously that gene expression can be regulated at a number of different levels. A central way is by gene regulatory proteins switching genes on or off by altering their production of messenger RNA transcripts.[13] But recent studies suggest that gene expression can also be controlled by regulatory RNAs. These are found in far greater amounts and with much increased diversity in cells of the human brain compared to other organs in the body.[14] So it is worth considering how this alternative control mechanism affects gene expression in neurons, and how it relates to memory.

One class of regulatory RNAs called micro RNAs regulate the translation of messenger RNAs into proteins.[15] This is important because unlike transcription, which occurs in the central cell nucleus, translation can happen locally in different parts of a neuron, not only in the main cell body but also in other cellular regions. And there is evidence that such local translational activity is important in memory formation: the ribosomes on which messenger RNAs are translated accumulate at synapses following strong neuronal activity.[16] Other studies have shown that specific micro RNAs are generated at postsynaptic regions on dendrites in response to signals at the synapse. Slices of hippocampus taken from a mouse brain can undergo long-term potentiation in response to stimulation, and slices that undergo such LTP also show an increase in the amount of neuronal micro RNAs.[17]

Changes in gene expression mediated by regulatory RNAs in neurons persist for weeks, suggesting a potential role in long-term memory formation. An important feature of regulatory RNAs is that they can interact with other such RNAs. This interaction affects their function and means that a given regulatory RNA can either inhibit, have no effect upon, or even enhance the activity of a messenger RNA, depending on which other regulatory RNAs are present in its vicinity. This means that gene expression at individual synapses may be very sensitive to subtle changes in the chemical signals at such synapses.[18]

While regulatory RNAs can be generated at synapses and act to control the production of particular proteins at these sites, there is also emerging evidence that some proteins produced at synapses are gene regulatory proteins that can conversely travel to the cell nucleus and influence expression of genes there.[19] This could be a way in which local activity at different synapses has an impact on the more long-term genetic behaviour of a neuron, which could be one basis for lasting memory formation.

As I mentioned previously, a surprising recent discovery about regulatory RNAs is that they can be secreted from the cell in membrane-bound structures called 'exosomes' and then carry information to other cells. There is evidence that such exosomes can travel from the postsynaptic neuron to the presynaptic one, and help enhance the connection between the two as part of the learning process.[20] Exosomes are also able to travel to other cells in the brain besides their synaptic partner neurons. They may provide a way for neurons to influence the activity of not only other neurons, but also of glial cells in their vicinity.

Neurons in the brain cortex are organized into functionally related minicolumns, which tend to show highly correlated input and output activity.[21] Neil Smalheiser, a neuroscientist at the University of Illinois, believes that 'if neighbouring neurons that are activated together release exosomes together, this may provide a means to "synchronize" their gene expression.'[22] This, in turn, may help establish or reinforce a circuit-level memory representation that is retained by the minicolumn as a whole. Communication via exosomes could allow cells to activate their neighbours, but also inhibit them. This could be important in fine-tuning a learned response by enhancing valuable synaptic connections while eliminating inactive ones that have ceased to contribute to the learning process.

If studies like these provide insights into the molecular and cellular mechanisms by which memories can be stored, where in the brain does such activity take place? One brain region that plays a particularly important role is the hippocampus.[23] The first recognition that the hippocampus has a role in memory came in the 1950s because of the unfortunate consequences of an operation on a man called Henry Molaison, of Hartford, Connecticut.[24] He suffered from a severe form of epilepsy—the brain disorder in which seizures occur that are caused by intense bursts of electrical activity across the brain. Molaison's seizures were so frequent and violent that as a teenager he had to leave school for several years, and when he finally finished his studies in his early twenties, the only job that he could manage with his disability was winding copper coils on electric motors at a local factory.[25]

In 1953, a neurosurgeon called William Scoville proposed to drill into Molaison's skull, insert a metal tube, and suck out the hippocampal region from which his seizures seemed to originate.[26] At that time, almost nothing was known about the functional role of the hippocampus, apart from a suspicion that it was involved in mediating our sense of smell. Losing such a sense seemed a small price to pay if Molaison's violent seizures could be stopped, and the operation was given the go-ahead. Initially, the operation was judged a success since the seizures were greatly reduced. But it soon became clear that Molaison—then aged 27—had been left without the ability to store or recall any experiences after his operation. From that point on, for the remaining 55 years of his life until his death in 2008, he was never able to remember anything for more than half a minute after it had occurred.

While a personal tragedy for Molaison, his condition was a revelation for scientists seeking to understand the biological basis of memory.[27] Until this point, it was assumed that memory was a property of the whole brain. But only Molaison's capacity to store and recall new memories appeared defective. In contrast, his memories of life before the operation seemed to have been retained, but only in a very general fashion. So Molaison could remember that his father was born in Louisiana and that he had accompanied his parents on family holidays in Massachusetts. But if asked about specific episodes in such holidays, he had no idea. This led to the proposal that different parts of the brain play distinct roles in memory formation, with the hippocampus being the place where new memories are formed, while long-term memory storage takes place in other parts of the brain, particularly the prefrontal cortex, the region of the brain involved in planning and problem-solving.

Intriguingly, Molaison seemed perfectly capable of learning new practical skills. For instance, in one study he was told to trace the image of a star on a piece of paper while looking only at the image of the hand doing the drawing in a mirror. Such a task is difficult to perform at first, but gets easier with practice. And Molaison improved with each time that he carried out the task, even though he had no conscious memory of ever having tried it before.

Molaison's memory defects suggested that the hippocampus acts to knit together our experiences—which can encompass sights, sounds, smells, emotions, and every other aspect of a given moment, each processed in a different part of the brain—into a single remembered event that can be recalled in the future.[28] This type of memory is called 'episodic', to distinguish it from 'semantic' memory, which is concerned with general facts about the world or ourselves, such as where we used to go on holiday as a child, or that Adolf Hitler ruled Germany during World War II. However, there is also another type of remembering called 'procedural' memory. A classic example of this is learning to ride a bike—once learned, you do it without thinking. The fact that Molaison could learn new skills initially suggested that procedural memory did not involve the hippocampus. But studies of another person with no hippocampal function—Lonni Sue Johnson—have challenged this straightforward scenario.

In Johnson's case, it was not surgery that caused the problem, but a viral infection.[29] Normally herpes simplex only results in cold sores, but in Johnson's case, the virus burrowed into her brain and completely destroyed her hippocampus. Unlike Molaison, Johnson was already in her late 50s when she lost the use of her hippocampus.[30] As well as being an accomplished artist whose commissions included drawing covers for the *New Yorker* magazine, she was also a talented viola player and a pilot who owned and flew two small planes. In other words, she had a lifetime's worth of experiences and skills and so provided an important subject for tests to see how much the loss of her hippocampus had affected her abilities. One test involved Johnson sitting in a flight simulator and attempting to fly a 'virtual' plane.[31] The test was complicated by the runway in the simulation appearing in a different place each time, and the joystick of the plane being programmed to resist Johnson's movements; consequently she could not rely on her previous experience of flying, but had to adapt to the new conditions. In fact, Johnson did learn to compensate for the changed situation, but only partially. This suggested that procedural learning also involves the hippocampus, although to a much lesser degree than for episodic memory.[32]

While studies of individuals with such a tragic loss of brain function have revealed the importance of the hippocampus for human memory, it is through animal studies that we have begun to learn how specifically this brain region helps form memories. One way it does this is to build up a map of a person's surroundings. This feature of the hippocampus was discovered by John O'Keefe of University College London in studies performed in 1971.[33] O'Keefe implanted microelectrodes in the hippocampal neurons of rats, and then monitored neuronal activity as the rats moved around a box. To his surprise, individual neurons fired when the rats moved to a particular place in the box. O'Keefe concluded that such 'place-cells' help create a memory of the environment, so 'that if you put them all together, you could have something like a map.'[34] These days we would liken the brain cells to the global positioning systems (GPS) now routinely used to make journeys by car, by bike, or on foot.

But a memory of place is only one of the features of neurons in the hippocampus. Other neurons in this brain region seem able to recognize memorable objects, or even people. Recognition of this fact came from a recent study by Rodrigo Quian Quiroga and colleagues at the California Institute of Technology.[35] The study involved human patients suffering from epilepsy whose brains had been implanted with microelectrodes in different regions in order to try and identify the sites from which seizures were originating, so that these could be removed surgically to prevent them affecting the rest of the brain. Quiroga used the opportunity to monitor the activity of different hippocampal neurons when the patients were presented with various pictures, of animals, objects, landmark buildings, and celebrities.

A surprising outcome of the study was just how specific certain neurons seemed to be in relation to recognizing a particular picture.[36] For instance, in one patient, a particular neuron only became activated when the patient was shown different photos of Bill Clinton, the US President at the time. In another patient, it was an image of the actor Jennifer Aniston who triggered a reaction in a particular neuron. However, this neuron was not activated by a photo showing Aniston with her former partner Brad Pitt! The actor Halle Berry also activated a specific neuron in the hippocampus of one patient; intriguingly, this neuron also responded to photos of Berry playing the DC comic character in the eponymous film *Catwoman*, released in 2004, in which her features were obscured by a mask, as well as to a caricature cartoon of the actor, or even a picture of the word 'Halle Berry'.

Specific neurons in other patients were activated by famous landmarks like the Eiffel Tower or Sydney Opera House, both as photos, caricatures, or simply their written name.[37] This suggests that a memorable person, or landmark, is registered by a particular neuron in a way that unites their visual image with more abstract representations. Moreover, neurons in the hippocampus can also form associations, for instance between a celebrity and a famous landmark. So when the patient with a specific neuron that responded to Jennifer Aniston was shown photos of the actor alongside the Eiffel Tower, it took very little time for that neuron to begin firing in response to images of the Eiffel Tower alone. According to Quiroga, this shows that a neuron will change its firing properties to encode a new memory.[38]

How does such neuronal activity in the hippocampus contribute to long-term memory formation? Based on studies of individuals such as Henry Molaison, the initial view was that the hippocampus only acted as a relay station, with more long-term memory formation being the responsibility of the prefrontal cortex. But this idea of a rigid division of labour in memory involving different brain regions has been challenged by a recent investigation of memory formation in mice by Susumu Tonegawa and colleagues at the Riken-MIT Center in Tokyo.[39] They used an approach that I have mentioned previously called optogenetics—in which neurons in a mouse brain are genetically engineered to respond to a particular wavelength of light—to activate or inhibit neurons in different brain regions. Previous studies by Tonegawa's team using this approach identified specific neural circuits involved in memory and showed that optogenetics can be used to artificially reactivate memories.

In the more recent study, Tonegawa and his colleagues used optogenetics to show that memories can be formed simultaneously in the hippocampus and prefrontal cortex.[40] However, the long-term memories remain 'silent' for several weeks, as shown by the fact that the neurons in the prefrontal cortex showed clear signs of remembrance of a painful stimulus when artificially activated by optogenetics, but they did not show any activation during natural memory recall. These findings suggest that the hippocampus and the cortex act in a complementary manner. What seems to be happening is that at first only a silent copy of a memory is made in the prefrontal cortex, and then gradually this becomes cemented as a long-term memory while the hippocampal one is erased. However, the nature of this long-term cement still remains to be determined. And the devastating effects of loss of hippocampal function in Henry Molaison and Lonni Sue Johnson show that, at least in human beings, the hippocampus can clearly not be bypassed for normal memory formation.

Other recent studies have suggested that the hippocampus acts as a 'convergence zone' that pulls together different bits of information stored in the cortex that relate to distinct elements of an event into a coherent whole. One such study was led by Ed Wu Xuekui of the University of Hong Kong.[41] His team used optogenetics to trigger low-frequency stimulation of neurons in a part of the hippocampus known as the dorsal dentate gyrus and found that this increased functional connectivity between different regions of the cortex.

Such a dynamic interaction of the hippocampus and the cortex may be key to the ability of the brain to form new associations between people, places, and objects. For instance, if you were to chance upon Halle Berry at the Eiffel Tower, your brain would form an association between the two different elements of this encounter, along the lines of the studies of the epileptic patients mentioned earlier.[42] Yet what still remains unclear is how the brain manages to create an association between two separate objects, like Halle Berry and the Eiffel Tower, yet still maintain a separation between the two individual objects, so that if you visited this Paris landmark on another occasion, but this time happened to bump into Brad Pitt, your memory could nevertheless create a new association. And it may be that the interaction between the hippocampus and cortex is somehow key to this.

One feature of the low-frequency stimulation used by Xuekui on hippocampal neurons, which triggered connectivity between regions of the cortex, is that this produces electrical activity that is a characteristic of slow-wave sleep, often referred to as deep sleep—a state we usually enter several times each night.[43] In fact there is increasing evidence that sleep plays a very important role in memory consolidation.[44]

During slow-wave sleep, neuroscientists have identified brain waves with three types of rhythms: slow oscillations, ripples, and spindles. Slow oscillations originate from neurons in the cerebral cortex; ripples are quick bursts of electrical energy produced within the hippocampus; and spindles, electrical impulses that occur around 7–15 times per second, are triggered by a brain structure called the thalamic reticular nucleus (TRN), which is part of the thalamus. A study by researchers at the Institute for Basic Science in Daejeon, Korea, has identified a central role for these three types of brain waves, and a particular coordinating role for spindles, in memory consolidation during sleep (Figure 17). According to Charles-Francois Latchoumane, a researcher on the study, 'often during the night a regular pattern is manifested, where a slow oscillation from the cortex is immediately followed by a thalamic spindle and while this happens, a hippocampal ripple appears in parallel.'[45] The correct timing of these three rhythms appears to act like a communication channel between different parts of the brain and thereby facilitates memory consolidation. The

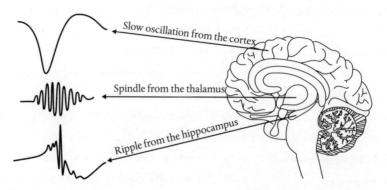

Slow oscillation from the cortex

Spindle from the thalamus

Ripple from the hippocampus

Figure 17 Role of brain waves during sleep in memory consolidation.

researchers focused on spindles because their number has been shown to increase following a day full of learning experiences, and to show a decline in the elderly and in some patients with schizophrenia.

In the study, the researchers used optogenetics to trigger artificial spindles in the TRN region in sleeping mice that had taken part in a memory task while awake. Such stimulation led to the coordination of the three groups of brain waves, and enhanced memory. In contrast, suppressing spindle number had a negative impact on remembering. The implication is that the thalamus mediates information exchange between the hippocampus and cortex as a vital part of long-term memory consolidation. Latchoumane believes that 'memorization during deep sleep has to do with time coordination.'[46] If the hippocampus tries to exchange information when the cortex neurons are not ready to receive it, the information could be wasted. Slow oscillations might be the signal used by the cortex to flag that it is ready to accept information. Then the thalamus would alert the hippocampus via the spindles. In other words, different brain regions seem to actively coordinate with each other during sleep.

As well as these findings by Latchoumane's team relating to sleep, other recent studies have confirmed the general importance of brain waves for the memory process. I previously mentioned that Earl Miller of the Massachusetts Institute of Technology has highlighted the importance of brain waves as general coordinators of brain function. Such waves can be categorized by their frequency (Figure 18). Alpha waves (8 to 12 Hz) are dominant during quiet thoughts, and in some meditative states; beta waves (13 to 32 Hz) dominate our normal waking state of consciousness when attention is directed towards cognitive tasks and the outside world; and gamma waves (33 to 100 Hz) are the fastest brain waves and relate to simultaneous processing of information from different brain areas. In addition, there are two types of slower frequency brain waves: delta waves (up to 4 Hz) and theta waves (4 to 7 Hz), which are characteristic of deep sleep and light sleep, respectively.

One topic that Miller has investigated is the involvement of brain waves in working memory.[47] This form of memory allows us to hold multiple

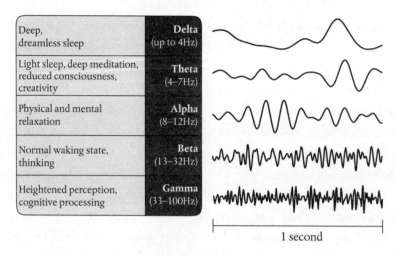

Deep, dreamless sleep	**Delta** (up to 4Hz)	
Light sleep, deep meditation, reduced consciousness, creativity	**Theta** (4–7Hz)	
Physical and mental relaxation	**Alpha** (8–12Hz)	
Normal waking state, thinking	**Beta** (13–32Hz)	
Heightened perception, cognitive processing	**Gamma** (33–100Hz)	

1 second

Figure 18 Different frequencies of electrical waves in the human brain.

pieces of information in mind—a telephone number, the time of an appointment that day, or a grocery list for our evening meal—from a few minutes to several hours. Miller notes that 'working memory allows you to choose what to pay attention to, choose what you hold in mind, and choose when to make decisions and take action. It's all about wresting control from the environment to your own self. Once you have something like working memory, you go from being a simple creature that's buffeted by the environment to a creature that can control the environment.'[48]

In one recent study, Miller and his colleagues found that the primate brain uses beta waves to consciously switch between different pieces of information. The findings add to insights that emerged from another study by Miller's team, which found that gamma waves are associated with encoding and retrieving sensory information. They also found that when gamma waves increased in intensity, beta rhythms decreased, and vice versa. Previous work by these researchers has indicated that beta waves are associated with 'top-down' information such as a current goal, how to achieve this, and the rules of a task. Such findings have led to the suggestion that beta waves act as a control mechanism that determines which pieces of information are allowed to be read out from working memory. Miller has proposed that beta waves act like a signal that gates access to working memory. They clear out working memory, and can act as a switch from one thought or item to another.[49]

So far in our consideration of the biological mechanisms underpinning memory, we have been focusing on the individual brain. But the main focus of this book is the importance of social interaction in the development of the human mind. An indication of the importance of social interaction in the formation of human memory capacity came from a study in the late 1980s led by Hallam Hurt, a specialist in child development at the University of Pennsylvania in Philadelphia.[50] She was concerned about the effect that the crack cocaine epidemic devastating US inner cities at that time was having upon the development of children of addicted mothers. To investigate this, Hurt and her colleagues studied four-year-old children from low-income families and compared the IQ of infants who had been exposed to the drug with those who had not. Somewhat to Hurt's surprise, the study found no significant difference between the two sets of children. Instead, the IQ of both sets of children was much lower than average. According to Hurt, 'these little children were coming in cute as buttons, and yet their IQs were like 82 and 83. Average IQ is 100. It was shocking.'[51]

This discovery led Hurt and her colleagues to investigate how a factor that the two sets of children had in common—their poverty—might have affected their intellectual development; for instance was it possible that a lack of time, resources, and education, which could be characteristic of parents of children in such a state of poverty, might have had an impact? The researchers began to monitor how frequently the parents of children in such families spoke to them affectionately, spent time answering their questions, and hugged, kissed, and praised them.[52] They also asked whether the parents had at least ten books at home for the children, a record player with songs for them, and toys to help them learn numbers.

They found that children who had received more attention and nurturing at home tended to have a higher IQ. And the same children also did much better at memory tasks. These included being asked to remember people's faces or a list of words, and to find a token hidden in one of a number of boxes. The latter task measured working memory, as finding the token quickly was helped by being able to remember the contents of each box that had already been opened. Years later, when the children had entered their teens, the researchers took MRI images of their brains and then matched these with the records of how well nurtured the children had been at both four and eight years old. They found a strong link between nurturing at age four and the size of the hippocampus but no correlation between nurturing at age eight and the hippocampus. A nurturing social environment at a very young age is evidently important in the development of human memory capability.[53]

So how might social interaction have enhanced the memory process? To address this question, let us return to the way that language helps human beings to group, distinguish, and differentiate between things in the world around them. In other words, what is the basis of conceptual thought, and how does this relate to language capacity as a whole?

CHAPTER 8

THOUGHT AND LANGUAGE

Conscious awareness, the ability to think through a problem in a rational way, the capacity to communicate with other human beings through language—these are all key features of humans that make our species unique. But what is the connection between thought and language? And what can science tell us about the biological basis for our species' ability to think and talk, and the relationship between these two characteristics?

Language may be defined as the ability to generate a limitless range of expressions with a limited set of elements and rules. Other animals communicate, but human language involves more than just communication. A key feature of language is that it allows humans to describe objects, places, events, and people in the present, but also in the past and future.

I have previously mentioned two opposing viewpoints about language. On the one hand, behaviourists like B. F. Skinner have argued that human beings start life as a 'blank slate', and our language capacity is something that we learn by exposure to the language of others.[1] On the other, Noam Chomsky proposed that humans are born with an innate capacity for language, as shown by the ease with which children rapidly learn a wide vocabulary, but also by their remarkable capacity for linking words together in complex grammatical structures. So what have we learned subsequently, if anything, about the molecular and cellular processes that underlie this capacity? And is the ability of children to absorb language and use it to structure the world around them really largely innate—a 'language instinct' as it has been called—or is there an important role for environmental influence, not just in learning a specific tongue, but in general language acquisition?

An intriguing development in the search for a biological basis for human language was the discovery of a British family with a language disorder clearly inherited across three generations.[2] Studies in the 1990s by researchers at London's Institute for Child Health showed that family members with the disorder could not select and produce the fine tongue and lips movements required to speak clearly. For Faraneh Vargha-Khadem, who studied the family, the most obvious feature of affected individuals was that they were 'unintelligible both to naive listeners and to other...family members without the disorder.' Affected individuals also had difficulty processing sentences, as well as poor spelling and grammar.[3]

Subsequently, in 2001, geneticists at the University of Oxford led by Anthony Monaco and Simon Fisher pinpointed the source of this language defect to a subtle difference in the FOXP2 gene. According to Fisher, the findings 'opened a molecular window on the neural basis of speech and language.' But Monaco was keen to stress that 'this is *a* gene associated with language, not *the* gene.'[4] And this turned out to be wise advice.

One factor arguing against the idea that FOXP2 represents the sole genetic basis of human language is that this gene is found not just in humans, but in a range of other species. But there are subtle differences in the gene between humans, chimps, and mice that may have functional significance.[5] So the human FOXP2 protein coded by the gene differs from the chimp protein by two amino acids, and by three amino acids from the mouse protein.

To further investigate the function of FOXP2, recently a team led by Svante Pääbo at the Max Planck Institute for Evolutionary Anthropology in Liepzig have created mice whose FOXP2 protein was altered to be identical to the human version.[6] Studies of such mice showed that they have subtle changes in the frequency of their high-pitched vocalizations, and distinctive changes to wiring in certain brain regions. Subsequent studies of the mice have suggested that FOXP2 is involved in the brain's ability to learn sequences of movements. In humans this could be important for the complex muscle movements needed to produce the sounds for speech, but in other species it may have a different role, coordinating other movements. And, indeed, while FOXP2 may well have a particularly important role in language in humans, even in our species the gene also appears to play a role in other parts of the body besides the brain, in processes ranging from reproduction to immunity.

In fact, the evidence points to FOXP2 having an influence on human language capacity, but also suggests that it is far from being the only factor; so Gary Marcus, who studies language acquisition at New York University, believes that FOXP2 is almost certainly 'part of a broader cascade of genes'.[7] This would fit with the FOXP2 protein's most basic function being to switch other genes on and off. One gene it controls is SRPX2, which helps some of the brain's neurons enhance their connections to other neurons. Richard Huganir, of Johns Hopkins University in Baltimore, USA, first identified this connection.[8] Huganir and his team were trying to identify genes that play a role in forming synapses and identified SRPX2 as such a gene. Further investigation showed that SRPX2 enhanced connections in a part of the mouse brain equivalent to a region particularly associated with language capacity in humans. This discovery prompted Huganir and his colleagues to look for a connection with FOXP2. And, indeed, they found that the FOXP2 protein inhibits the action of SRPX2. When the researchers artificially inhibited SRPX2 in a living mouse, they found that this reduced the calls that baby mice make to attract the attention of their mothers.

Findings like these suggest that language in humans involves a complex interplay between a number of different genes which affect the connections between neurons in specific parts of the brain. However, it still remains unclear whether FOXP2 affects neurons involved in language processing per se or those that control muscles involved in speech, although it is worth remembering that human individuals with a defect in the gene had problems processing sentences, as well as a poor capacity for spelling and grammar.

Quite what we can learn from the mice studies mentioned above is also far from clear, since in this species both the FOXP2 and SRPX2 genes may be playing quite different roles to the situation in humans. Because of such potential limitations of using the mouse as a model, scientists have begun to look to other species to study the role of FOXP2 in language.

One interesting line of investigation is being pursued by Stephanie White of the University of California, Los Angeles. She has focused on songbirds, since these are one of the few animals besides humans that can produce a wide variety of complex vocal sounds.[9] In addition, much like the situation in humans, songbirds have a critical period in youth when they are best at learning vocal communication skills. In birds, this is when they learn a song they will use later in life as a courtship song. After this critical period ends, just as it is more difficult for human beings to learn languages, so it is for songbirds to learn their songs.

White and her colleagues have shown that changes in the level of FOXP2 protein in a part of the songbird brain involved in song led to changes in the activity of thousands of other genes.[10] White believes this shows that the activities of these genes are being manipulated by FOXP2, much like an orchestra is led by a conductor. According to her, 'it's not that all the genes (or instrumentalists) became loud or became quiet; it's that they change in a coordinated way. We refer to these as "suites of genes", and one of these suites of genes is highly correlated to learning in young birds.'[11] Intriguingly, preventing such changes in FOXP2 disrupted learning, which may mimic what happens in human beings with a FOXP2 defect. And, indeed, White hopes that studying the relationship between FOXP2 and the genes that it controls in birds may lead to new insights about FOXP2's role in humans.

In another line of investigation, scientists are seeking to study the role of FOXP2 in non-human primates. Although such primates cannot speak, we saw previously how apes can be taught sign language; however, they also seem to have natural limits to this capacity, with no indication that it is comparable to the complex sophistication of human language. Yet given that it is becoming increasingly possible, through the new technology of gene editing, to create genetically modified versions of practically any species, one potential line of investigation would be to create a monkey or ape with the human version of FOXP2 instead of its normal form, and see how this affects their capacity for learning sign language.

Indeed, Su Bing, a geneticist at the Kunming Institute of Zoology in China, is already planning to use gene editing to make such a change to FOXP2 in macaque monkeys. He remarks, 'I don't think the monkey will all of a sudden start speaking, but will have some behavioural change.'[12] However, such a move would raise an obvious ethical issue, if it turned out that the genetic change altered a monkey's state of conscious awareness.

Another approach for studying the link between genes and language is to continue to search for human beings with genetically based language defects. Since the discovery of FOXP2, other genes with potential roles in human language have been identified based on studies of individuals with various abnormalities in their ability to express themselves verbally, but also read and write.[13] These genes have been shown to play important roles in a variety of brain processes, including migration of neurons, connections between brain cells, and cell signalling pathways. This further confirms that it may be a mistake to imagine that genetic links to language will necessarily reveal specific 'language genes'. Instead the evidence from recent studies suggests that those genes that, when defective, can give rise to specific language abnormalities will also play multiple roles in brain function.

What can we learn about the biological basis of language by moving from genetic analysis to brain structure? One view is that specific regions of the human brain are associated with our species' language ability. This view has drawn support from studies of individuals with a condition called aphasia, from the Greek for 'without speech'. In 1861, Paul Broca studied a patient who could only say one word—'tan'. When he died, Broca carried out an autopsy on this patient and found that he had damage to a region on the outer cortex on the left side of his brain. This region subsequently became known as 'Broca's area' (Figure 19).[14] Some 15 years

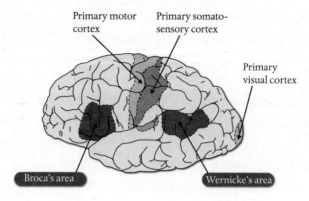

Figure 19 Broca's and Wernicke's language areas in the human brain.

later, Karl Wernicke studied other patients with language defects and identified another region on the left side of the brain associated with language, now known as 'Wernicke's area'.[15] People with damage to Broca's area generally understand language but cannot generate intelligible speech, whereas those with damage to Wernicke's area cannot understand language but can produce it. Findings such as these led to the idea that discrete brain regions play defined roles in the language process.[16]

However, more recent studies of brain-damaged patients, experimental treatments that stimulate or inhibit particular regions of the brain in patients, and imaging studies of the human brain have all undermined the idea of such a simple division of language production and comprehension.[17] Rather, they suggest that language is a dynamic process distributed across much wider areas of the brain than had been suspected. Moreover, while there is no doubt that regions such as Broca's and Wernicke's play important roles in language, there is increasing evidence that they are also involved in other aspects of brain function.[18]

Indeed, the idea that either a particular gene or a specific region of the brain is responsible for human language capacity betrays an overly simplistic view of how evolution works. One feature of evolutionary change is its conservative character. Driven by trial and error, rather than a preformed plan, evolution tends to co-opt existing processes and structures, rather than generate completely novel ones.[19] And there is good reason to believe that a similar process may have occurred during the evolution of human language.

One interesting recent finding is that Broca's region—and its equivalent in an ape brain—seems to be the main location of the 'mirror neurons' that I mentioned previously. Mirror neurons, recall, become activated when an individual performs an action, but also when the individual observes someone else making the same movement. Their association with Broca's region has led to the suggestion that these neurons may play a role in language in humans.[20] This would fit with the fact that imitation plays an important role in the way that children learn language. More controversial is the possibility that human language may have its origins in the gestures that other primates use to communicate with other.

While an attractive idea, allowing for a way for language to emerge from an existing biological mechanism in our ape ancestors, one objection, voiced by linguist Robbins Burling of the University of Michigan, is that such a transition would have required a 'move from a visual language to an audible one.'[21] Yet since in monkeys mirror neurons can be activated by sounds—for instance, the tearing of paper, or cracking of a nut—the transition from a gestural to a sound-based language system might still be feasible. Another factor in favour of such a transition is that deaf people have developed sign language systems as sophisticated as any other form of human language, without using sound. Of course there is a world of difference between the language systems used by our species—whether sign-based or verbal—and the gestures used by apes, which as we have seen seem incapable of conveying abstract symbolic information as human language does. But by providing an initial biological framework in the brain for language to develop, it is easier to imagine how our amazing language capacity might have gradually evolved by commandeering existing neural circuits, rather than appearing, fully formed, out of nowhere.

Another interesting recent discovery about Broca's region has come from a brain imaging study of individuals with mutations in the FOXP2 gene. Brain activity in such people was compared to related family members without the mutation when the two sets of individuals were asked to perform different language tasks, such as thinking of verbs that go with nouns.[22] The study found that neural activity in Broca's region was substantially reduced in the individuals with the FOXP2 mutation. Interesting questions to now explore would be whether FOXP2 plays a specific role in mirror neurons in Broca's area in humans, and how this compares to its role in equivalent areas of the brain in ape species like chimps.

More generally, brain imaging has provided a new way to study the neural mechanisms underlying human language ability. One recent study, led by Jack Gallant, a neuroscientist at the University of California, Berkeley, has aimed 'to build a giant atlas that shows how one specific aspect of language is represented in the brain, in this case semantics, or the meanings of words.'[23] Gallant's team imaged brain activity in human volunteers listening to stories read out on The Moth Radio Hour, a US radio show. They then studied how words triggered neural responses in specific regions across the brain cortex and found that conceptually related words affect the same regions.[24] For instance, on the left-hand side of the brain, one region was stimulated by the word 'victim' but also by 'killed' and 'confessed'. On the right-hand side, another region responded to terms relating to the family like 'wife', 'husband', and 'children'. The regions activated were similar in all volunteers, suggesting that their brains organized the meanings of words similarly.[25] Although indicating that words of different meaning can trigger activity in localized brain regions, the findings also suggest that as a whole, language capacity is distributed across our mental organ.

Brain imaging is having an impact on our understanding not only of language processes in adult humans, but also of language acquisition in children. Various approaches now make it possible to investigate this question in children, in newborn babies, and even during foetal development. One approach involves studying international adoptees—babies adopted soon after birth who grew up hearing a different language than they had heard in the womb. For instance, Anne Cutler and colleagues at the Max Planck Institute for Psycholinguistics in Nijmegen in the Netherlands investigated children who had been born in Korea, but then adopted by Dutch couples.[26] They found that such individuals could pronounce Korean sounds significantly better than those born and raised in the Netherlands. This suggests that the ability to recognize but also generate the different types of sounds in a particular language begins before birth. Cutler believes this shows that babies are 'picking up useful knowledge about language even though they're not actually learning words.'[27]

Another indirect way of investigating language acquisition during foetal development is by imaging the brains of babies born prematurely. Studies of full-term babies—those born after the normal 37 weeks in the womb—show that such infants can recognize their mother's voice, tell apart two languages heard before birth, and remember short stories read to them while in the womb. To investigate whether such language capacity develops in the womb or immediately after birth, a team led by Fabrice Wallois of the University of Picardy Jules Verne in Amiens, France, carried out brain imaging in babies born 2–3 months before the normal gestation period.[28] The researchers played soft voices to them while they were asleep in their incubators a few days after birth, then monitored their brain activity. This analysis showed that the babies could distinguish between male and female voices, and similar sounds such as 'ga' and 'ba'. Moreover, the regions of the brain cortex that became activated were the same as those used by adults for sophisticated understanding of speech and language. Wallois believes the findings show that the neural connections underlying our species' language capacity are already present and functional at this foetal stage of development.[29]

Of course it is possible that the brain of a premature baby might differ from that of a foetus of the same age. To address this concern, Moriah Thomason and colleagues at Wayne State University in Detroit used fMRI to study foetuses in the womb directly.[30] This type of study was impossible until recently, as a moving foetus is hard to track. But new computational tools now make it possible to take into account such movement in the analysis.

Thomason's team have used such an approach to monitor the activity of brain regions known to be involved in language and their connections with other regions of the brain.[31] For instance, they have shown that foetuses that would later be born prematurely showed weaker connections between a region that develops into a language processing centre and other brain regions compared to foetuses carried to full term. This is interesting given that babies born prematurely often suffer from language problems later in life. Are the differences in brain connectivity a consequence of underlying problems that lead to an early birth, or might they actually contribute to a baby being born prematurely? In other words, could abnormal linguistic connectivity between mother and foetus somehow contribute to a premature birth? This is just one of the interesting questions raised by such imaging analysis.

While imaging of language acquisition in foetal brains is still at an early stage, the use of such approaches to study language development in babies is already yielding some very interesting findings. Young children have an amazing ability to learn new languages, and there is a cut-off for this ability; it becomes much more difficult to learn a new language after the age of ten.[32] This feature of language development is so widely recognized that it is easy to overlook how surprising such a state of affairs is from the standpoint of human learning in general, in which the capacity to learn tends to improve as we get past early childhood.

One reason for this difference is that language learning involves a 'critical' or 'sensitive' period.[33] At birth, a baby can perceive the 800 or so sounds—technically known as phonemes—that can be linked together to form all the words in every language. The neonatal brain is most open to learning the sounds of its native language from six months old for vowels and nine months for consonants.[34] This period lasts for only a few months but is extended for children exposed to sounds of a second language. But it is also becoming clear that the child's social environment has a major role to play in language development.

The role of social interaction is evident from recent studies which have demonstrated that babies showing more 'gaze shifting'—when a baby makes eye contact with a parent or teacher and then follows their gaze to an object—have enhanced brain responses to the sounds of a new language. One such study by Rechele Brooks and colleagues at the University of Washington in Seattle has led Brooks to conclude that 'young babies' social engagement contributes to their own language learning—they're not just passive listeners of language. They're paying attention, and showing parents they're ready to learn when they're looking back and forth. That's when the most learning happens.'[35]

A curious finding of such imaging studies is that babies first exposed to their native language show activation not only in the brain areas involved in hearing, but also in the superior temporal gyrus, Broca's area, and the cerebellum—'motor' regions associated with speech production. This suggests that an infant's brain starts laying down the neural connections necessary for the formation of words long before it actually begins to speak. According to Patricia Kuhl, who led the above study at the University of Washington, 'most babies babble by 7 months, but don't utter their first words until after their first birthdays. Finding activation in motor areas of the brain when infants are simply listening is significant, because it means the baby brain is engaged in trying to talk back right from the start and suggests that 7-month-olds' brains are already trying to figure out how to make the right movements that will produce words.'[36] The findings confirm the importance of adults speaking to babies during social interactions even if the latter cannot yet respond in kind.

The consequences for language development of not being exposed to such social interaction in the first years of life were tragically shown by the example of children in orphanages in Romania.[37] From the 1960s onwards, the regime led by dictator Nicolae Ceausescu implemented a drastic process of forced industrialization. To increase the working population, contraception and abortion were banned and childless couples taxed, while people from the countryside were pressurized to find work in the cities. During these drives many newborn children were abandoned and subsequently ended up in state-run orphanages. Only when a revolution overthrew Ceausescu in 1989 did it become apparent how appalling conditions were in these orphanages. Babies were left in their cots for hours, with their only human contact being when a carer—typically assigned to 15–20 children—came to feed or bathe them. As toddlers, the orphanage children were virtually ignored. Unfortunately, even following Ceausescu's overthrow it took a long time for any positive change to occur in these institutions, and by the end of the 1990s, children were still being largely neglected.[38]

Three US researchers—Charles Zeanah, a child psychiatrist at Tulane University; Nathan Fox, a developmental psychologist at the University of Maryland; and Charles Nelson, a neuroscientist at Harvard—studied the effects of such neglect on the orphanage children's development.[39] They found that most children showed no attachment at all to their carers. According to Fox, 'they showed these almost feral behaviours...aimlessly wandering around, hitting their heads against the floor, twirling and freezing in one place.'[40] Imaging analysis showed that the children's brain activity was weaker than in children of a similar age in the general Romanian population. For Fox, 'it was as if a dimmer switch had been used to turn their brain activity down.'[41] Such differences were reflected in a significantly lower IQ in the orphanage children, and an even greater effect on language development. Nelson remarked that while 'IQ was deleteriously affected, language was clobbered.'[42]

In fact, increasing evidence points to the importance of language in general intellectual development. This might not seem so surprising given that a key way in which we humans express ourselves is through words. However, as I have already mentioned, Lev Vygotsky and Valentin Voloshinov, both of whom worked in Russia in the 1920s and early 1930s, argued that there is a more fundamental relationship between thought and language. In particular, Vygotsky believed that 'inner speech' is a crucial motor of thought,[43] while Voloshinov stressed the dialogical character of both spoken and inner speech.[44]

Vygotsky viewed inner speech as a product of the internalization of outer speech, a process that takes place during childhood. And as we noted, he believed that an important intermediary stage in this internalization process is 'egocentric speech', the phenomenon observed in young children of talking to themselves as they play. Vygotsky argued that far from egocentric speech reflecting a young child's self-centredness, it remained connected to the sphere of social communication. To test this idea, Vygotsky placed children in situations that disrupted their sense that their egocentric utterances could be heard and understood by others.[45] So children were placed in a room with children who were deaf-mute or spoke a foreign language, physically separated from their peers, or loud music was played in an adjacent room. In these circumstances, egocentric speech was much reduced. Later studies revealed that children's use of egocentric speech is sensitive not only to the possibility of being heard and understood, but also to the presumed attitudes of others.[46] For instance, children produce more egocentric utterances in the presence of an adult perceived as willing to help them with a problem-solving task than with one who is not.

The role of egocentric speech in assisting with problem-solving activities in young children raises the question of why we do not generally talk to ourselves in adulthood. Charles Fernyhough, of the University of Durham, believes one reason why adult humans tend not to think out loud is that 'when we acquired language and when we started to use language in out-loud private speech, we'd have learned pretty quickly that it's not a good idea to talk to yourself out loud when you're in a difficult dangerous situation.... [T]hen there's a sort of social and cultural pressure as well. If you're going around saying what you think, your competitors, your rivals, the other people around you will know what you're thinking and then it's hard to fulfil your plans. So there are some good reasons for doing it silently.'[47]

Another, more fundamental reason may be that inner speech has some very important characteristics compared to external speech that suit its role in thought and which cannot be replicated just by speaking out loud. Fernyhough believes that inner speech can occur in two main forms: expanded inner speech, which retains many characteristics of external speech, and condensed inner speech, with characteristics like those mentioned earlier.[48] In fact, there is probably an organic connection between these two forms, with condensed inner speech being closer to what might be called pure thought, and expanded inner speech that which takes place when we introspect our thoughts, or try and explain them to others. The peculiar qualities of condensed inner speech—and how difficult it can be to express it verbally in external speech—was recognized by Voloshinov when he characterized inner speech as 'the still undifferentiated impression of the totality of the object—the aroma of its totality, as it were, which precedes and underlies knowing the object distinctly.'[49] Unfortunately, this 'fuzzy' character of condensed inner speech poses major challenges in trying to use introspection to report on its qualities. William James, one of psychology's original pioneers, recognized this when he described reflecting on one's own thoughts as 'trying to turn up the gas quickly enough to see how the darkness looks.'[50] The very act of trying to express our inner thoughts in external speech may therefore misrepresent their true character.

Despite such problems, recent studies suggest that condensed inner speech is highly fluid, with word meanings being dynamic rather than static formations. Also, in line with Vygotsky's view that 'higher mental functions appear on the inter-psychological plane before they appear on the intra-psychological plane',[51] or Voloshinov's belief that consciousness is 'a social entity that penetrates inside the organism of the individual person,'[52] inner speech appears not so much a monologue but closer to a dialogue in character. As Voloshinov noted, 'the units of which inner speech is constituted are certain whole entities somewhat resembling a passage of monologic speech or whole utterances. But most of all, they resemble the alternating lines of a dialogue.'[53] Extending this idea, Fernyhough believes that 'a solitary mind is actually a chorus. We can go so far as to say that minds are riddled with different voices because they are never really solitary. They emerge in the context of social relationships and they are shaped by the dynamics of those relationships.'[54]

Coupled with Voloshinov's claim that different speech genres represent the interests of different groups within society, this dialogic aspect of inner speech suggests that an individual consciousness also contains perspectives drawn from different parts of society. As an individual develops from childhood to adulthood, such inner 'voices' may represent the views of parents, siblings, friends, schoolteachers, or any assortment of people who have had an influence over the years. Importantly, this dialogue may become argumentative and can also be influenced by new voices as an individual's personal circumstances change.

In fact, as I will explore later, the conversation going on in our heads may often be far from straightforward, because of the role of the unconscious. And if the dialogue becomes unbalanced this may lead to what are characterized as disorders of the mind. In this case, according to Fernyhough, 'when somebody hears a voice, what's happening is that they're actually producing some inner speech but for some reason they don't recognize that speech as having been produced by themselves. It's experienced as something that doesn't belong to the self, that comes from outside.'[55] Why this might happen is something I will return to later when we consider mental disorders in more detail. But for now, let us turn to another crucial aspect of human uniqueness—the links between consciousness, creativity, and imagination.

CHAPTER 9

CREATIVITY AND IMAGINATION

The ability to think rationally is an essential feature of being human, but it is hard to imagine how our species could have gone from living in caves to sending rovers to Mars in the space of 40,000 years without another crucial element—our creative impulse. Both Einstein and Picasso believed that their respective genius in science and art was based upon an ability to view the world as would a child.[1] But clearly there is a difference between an adult with a childlike ability to think outside the box, and actually being a child. So what is the basis of human creativity and imagination, and how does it differ between adults and children?

Human interest in imagination is at least as old as the ancient Greek philosophers who pondered the link between the sacred statues—known as 'images'—in their public squares and the mental processes of representation.[2] William James also believed that imagination was the ability to form images or ideas in the absence of sensations: 'fantasy, or imagination, are the names given to the faculty of reproducing copies of originals once felt.'[3] Such a view suggests that mental images are essentially copies of objects in the real world, but reproduced later when the object is no longer there. Yet there is increasing evidence that remembered images involve a transformation of objects in the real world, and that the process of perception is a highly mediated mental action and therefore itself an act of imagination.

Recent studies of vision challenge the view of perception as an activity that produces images of real-life objects inside the brain. They show that our eyes are in a constant state of motion, even if looking at a stationary object.[4] Such findings have led Etienne Pelaprat and Michael Cole of the

University of California, San Diego, to claim that constant discoordination with the world is a normal aspect of human perception.[5] Pelaprat and Cole's views have been influenced by the ideas of Lev Vygotsky, who, as you will recall, believed that our consciousness is transformed by social interactions, particularly through language, in a fashion similar to how tools have allowed us to transform the natural world.

Vygotsky saw imagination as no less influenced by this social input than other aspects of consciousness, stating that it is 'the means by which a person's experience is broadened, because he can imagine what he has not seen [and] can conceptualize something from another person's narration and description of what he himself has never directly experienced.'[6] A feature of the imagination is its ability to sometimes come up with the unexpected. This makes evolutionary sense, because if the brain were limited to retaining previous experience, a human being could adapt only to familiar, stable conditions of the environment; new or unexpected changes in the environment not encountered previously would fail to induce the appropriate adaptive reactions.[7] To deal with a changing environment, Vygotsky argued that the human brain 'combines and creatively reworks elements of this past experience and uses them to generate new propositions and new behaviour [making] the human being a creature oriented toward the future, creating the future and thus altering his own present.'[8]

Vygotsky thought that imagination was particularly stimulated in the developing child through the process of play, which for him has three key features:[9] children create an imaginary situation; they take on and act out specific roles; and they follow rules determined by those roles. Children create imaginary situations initially by external actions, and then increasingly by internal ones. Importantly, the emergence of the internal actions signals the beginning of a child's transition from earlier forms of thought processes—sensory motor and visual representational—to more advanced and symbolic 'higher' forms of thought.

Play allows the child to achieve mastery of the object and furthers symbolic ability. Vygotsky believed that 'play is a transitional stage in this direction. At the moment when a stick—i.e. an object—becomes a pivot for severing the meaning of horse from a real horse, one of the basic psychological structures determining the child's relationship to reality is radically altered.'[10] Play thereby acts as a transitional stage from a child's thinking constrained by a situation's concrete properties to being free of these constraints. Typically, young children at this stage of development feel that a play prop must have properties similar to the real object which it represents. So only things that can be rolled—for instance a pencil—were considered by children in Vygotsky's studies to be acceptable substitutes for a train. At this stage, the child cannot yet use symbols arbitrarily but they are mastering the prerequisites of symbolic thinking.

In contrast to the common view that imagination precedes play and is necessary for its emergence, Vygotsky believed that play in humans instead acts as a transitional stage in the development of imagination, which he described as 'a new formation that is not present in the consciousness of the very young child, is totally absent in animals, and represents a specifically human form of conscious activity.'[11] A key aspect of play is that it allows the child to reach further than their current stage of development. As Vygotsky put it, 'in play a child is always above his average age...it is as though he were a head taller than himself.'[12]

If play is part of the process whereby the powers of imagination are formed during childhood, how does this relate to creativity in the adult and what we have said about the socially mediated nature of human consciousness? Here it is worth noting that some aspects of adult human culture have attributes in common with play, including the link with imagination. For instance, when as adults we read a novel or watch a film, we can also imagine ourselves and others in the roles of the characters experiencing situations within it.[13] By identifying with a character in an imaginary world created by the book or movie, we can become a 'head taller' than we typically are—just as in childhood. The potentially greater role for the imagination when reading the text of a novel as opposed to seeing a movie may explain why filmed dramatizations of novels may sometimes fall short of expectations.[14] In such cases, the filmmaker has failed to match the power of the reader's imagination. But of course it takes great skill in both novel writing and filmmaking to really get the reader or viewer to truly empathize with what are after all imaginary characters. And it was in reference to such a skill that the novelist Albert Camus noted: 'The first thing for a writer to learn is the art of transposing what he feels to what he wants to make others feel.'[15]

If play is central to the development of imagination during childhood, but also has similarities to imagination in adults, what can modern neuroscience tell us about the imaginative process? The areas of the brain involved in this process in adults are now beginning to be explored using sophisticated brain imaging methods.

The potential of imaging approaches for investigating the neural mechanisms of imagination and creativity has been demonstrated by Charles Limb, of the University of California, San Francisco, and Allen Braun, at the US National Institutes of Health. They used fMRI to study jazz musicians performing musical exercises that ranged from a memorized scale to a fully improvised piece of music.[16] They found that during the improvised pieces, a brain region called the dorsolateral prefrontal cortex became less active. This region plays roles in planning, and self-censorship, and is inhibited during daydreaming, meditation, and REM sleep. So does a person's mindset during the creative process just become less rigid because of loosening control of this region compared to the rest of the brain? In fact, the situation seems rather more complex, as shown by a related study carried out by Charles Limb and Malinda McPherson.

This study also involved imaging the brains of jazz musicians playing improvised melodies on the piano.[17] However, in this case, the musicians were given a photograph of a woman with either a positive or negative expression, and then asked to match their musical compositions to the photo's mood. Based on Limb and Braun's previous study, it might be expected that the dorsolateral prefrontal cortex would be inhibited in all cases. In fact, the inhibition was pronounced only in the musicians who had seen the positive face. In contrast, those who had seen a negative face showed greater activity in brain regions linked to control and reward.

Such findings suggest that imagination is not a product of one specific brain region or neural process and that emotion has a major impact on the creative process.[18] We will look in more detail at the role of emotion in human consciousness later in this book.

Imaging can identify interesting changes in the whole brain during the creative process, yet tells us little about what is happening at the cellular and molecular levels. But another study explored the more granular picture by returning to the idea of imagination as the remembrance of images no longer directly in the field of vision. Julio Martinez-Trujillo of Western University in Ontario and Diego Mendoza-Halliday of the Massachusetts Institute of Technology recorded neuronal activity in monkeys trained to use a keyboard to report the direction of movement of a cloud of dots on a computer screen, or the cloud's direction a few seconds after it had disappeared, based on a memory of the image.[19] The findings of the study were surprising. According to Mendoza-Halliday, 'we might have expected that the neurons that are active when we perceive a visual object are the same ones that memorize it; or, on the contrary, that one group of neurons perceives the object and a completely different group memorizes it; but instead, we found that all of the above are true to a certain extent. We have perception neurons, memory neurons, and also neurons that do both things.'[20] This suggests that seeing and remembering are closely linked, but in a complex fashion.

Yet, as already mentioned, true imagination is likely to involve more than just remembering a past image, and therefore its neural basis may also be different. Take for instance a scenario in which you are asked to imagine a pineapple. This could involve remembering a pineapple you once saw. But now try and imagine a dolphin balancing a pineapple on its snout. Unless you have been to a very unusual show at the zoo, it is likely that you will have never actually seen such a phenomenon, yet this should not prevent you picturing it. This raises the question of how your brain can do this, in the absence of any past images to draw upon. One possibility is that the same neuronal mechanisms I mentioned previously that allow us to combine a memory of meeting Halle Berry with the place where we met her, the Eiffel Tower, might also allow us to create completely imaginary visions. In other words, by employing such mechanisms of memory, the imagination allows us to juxtapose images that we have seen separately in other contexts in completely novel ways.

In fact, most of us are far more likely to be able to conjure up an imaginary vision of Halle Berry at the Eiffel Tower than actually remember meeting her there. Yet from a neuronal point of view, it is potentially much harder to conjure up the imaginary vision, for the simple reason that here we have two juxtaposed images, like seeing a dolphin balancing a pineapple on its snout, that are highly unlikely to have occurred at the same time. And while an original memory of two juxtaposed objects could rely on their physical association in the real world, a purely imaginative vision has no such help from reality. Instead, it is left to the brain to make this association. So how does it do this, and which brain regions are involved?

Traditionally the area of the brain assumed to be involved in mediating a 'higher' function such as imagination has been the cerebral cortex, particularly the prefrontal cortex, the section of the cortex that lies at the very front of the brain, just under the forehead.[21] This brain region is often described as responsible for the 'executive' functions of human consciousness, with its role being to manage complex processes like reason, logic, problem solving, planning, and memory, just as the chief executive of a company is meant to guide and control the operations of the organization. To do so, the prefrontal cortex is believed to direct attention, develop and pursue goals, and inhibit counterproductive impulses.

A historical example often used to support this idea of the dominant coordinating role played by the prefrontal cortex in human consciousness is the famous case of Phineas Gage, who was involved in a tragic accident in the mid-nineteenth-century.[22] In September 1848, Gage, a 25-year-old construction foreman, was in charge of men building a railway in Vermont, USA. The team employed explosives to clear a route for the railway, and at one point Gage was using an iron rod to compact gunpowder in a borehole. Unfortunately, the iron produced a spark that prematurely ignited the explosive, and the resulting blast propelled the rod—which measured 3.5 feet long and 1.25 inches wide—straight through his head. Remarkably Gage survived, and made what initially appeared to be a full recovery. However, subsequently, he was said to have undergone dramatic changes in personality. Previously described as efficient, well-balanced, shrewd, and energetic, Gage was now apparently 'fitful, irreverent, indulging at times in the grossest profanity, impatient of restraint or advice when it conflicts with his desires, obstinate, devising many plans of operation which are no sooner arranged than they are abandoned in turn for others appearing more feasible.'[23]

In 1860, Gage began having epileptic convulsions of increasing severity, and he died that year. A post-mortem examination of his brain showed that the rod had destroyed much of Gage's left frontal lobe and prefrontal cortex. This, and the reported changes in his personality after the accident, has been used as evidence that the prefrontal cortex plays an organizing role in the 'higher' brain functions that distinguish us from 'lower' species.[24]

Yet recent reconsideration of this case and new evidence about how the human brain functions have challenged this established view. One problem with the standard account of Gage's accident and subsequent life is that he maintained several jobs in the years after his injury, working first in a stables in Hanover, New Hampshire, later as a long-distance coach driver on the Valparaiso–Santiago route in Chile, and finally on a small farm in Santa Clara, California.[25] This raises questions about whether the effect of his accident on Gage's personality was as far-reaching as has been claimed. Second, the claims made about his personality suggest possible effects of the injury on Gage's ability to handle his emotions, but they do not necessarily indicate an all-important executive function for the prefrontal cortex.

Most importantly, recent studies have challenged the master coordinator view of the prefrontal cortex, pointing instead at a more distributed role for many different brain regions in mediating higher mental functions. Imagination and creativity may similarly be more a product of connections between different brain regions than a function of one specific region. For instance, a recent study led by Alex Schlegel of Dartmouth College, New Hampshire, has provided evidence that this is the case. His team used fMRI to investigate changes in the activity of different brain regions in people asked to use their imagination.[26] Participants were asked to imagine specific shapes and then mentally combine them into more complex figures, or dismantle them into their separate parts. The study indicated that imagination requires a highly distributed neural network, with 11 areas across the brain being activated when an individual is asked to take part in creative acts. Schlegel believes that this network is equivalent to a 'mental workspace'—that is the connected neurons provide the basis for our ability to alter and manipulate images, symbols, and ideas and give us the intense mental focus that allows us to formulate new ideas and solutions to complex problems.[27]

One way of reconciling the original view of imagination and creativity as being located in the prefrontal cortex with the possibility that such higher mental functions are at the same time distributed across many different brain regions is to return to something already mentioned, namely recent evidence that suggests brain waves play a key role in coordinating functions across the brain. As we saw, Earl Miller's team at the Massachusetts Institute of Technology have recently identified an important role for such brain waves as mediators of working memory—the process that allows us to focus on several pieces of information in the mind simultaneously. Interestingly, one brain region that Miller's team have been focusing on is the prefrontal cortex.[28] Their studies indicate that this brain region seems to help construct an internal model of the world, sending 'top-down' signals that convey this model to lower-level brain areas. Meanwhile, other regions of the brain send raw sensory input to areas of the prefrontal cortex, in the form of 'bottom-up' signals.

Differences between the top-down model and the bottom-up sensory information allow the brain to figure out what it is experiencing, and to tweak its internal model accordingly. His team's findings have led Miller to compare the prefrontal cortex—somewhat ironically given the nature of Phineas Gage's accident—to a railway switch operator and the rest of the brain to railway tracks, with the switch operator activating some parts of the track and taking others offline.[29] This model would explain how attention works—for instance, a person can focus on a particular picture while ignoring a noise in the background. It may also explain how different pieces of information might be juggled around in the brain while we try to solve a problem creatively. It suggests a directing role for the prefrontal cortex in this process, but argues against the idea that this region is all-powerful, for the fact that this brain region requires incoming signals from these other regions shows it is also directed by them.

Another important aspect of Alex Schlegel's study mentioned earlier is that its findings contradict a common idea about creativity: that it is primarily concentrated on one side of the brain. According to this idea, the right brain is the site of creative thinking and the left brain that of logical thinking. Yet Schlegel's team found evidence that both the right and left parts of the cerebral cortex are involved in the imaginative process.[30] But perhaps the most intriguing aspect of this study was that it unexpectedly highlighted an important role for a region of the brain called the cerebellum in imagination and the creative process.

Previously the cerebellum has been more linked to control of balance and coordinated movement of the body's muscles.[31] Yet recognition of a role for this brain region in higher mental functions should maybe not be a surprise, given emerging facts about its contribution to brain function, particularly in humans. For although it occupies only 20 per cent of human brain volume, the cerebellum contains 70 per cent of its neurons. And a sign that the cerebellum may have specific roles to play in our species is that it has increased three- to fourfold in size compared to that in other primate species in just the past million years of human evolution.[32] So how might this brain region influence imagination and creativity? Surprisingly, it seems to do so in ways that are similar to its role in coordinating movement.

Research indicates that the human cerebellum can enhance the functions of the cerebral cortex in four different ways.[33] First, it increases the speed, efficiency, and appropriateness of processes in the cortex by sending it signals. Second, it develops neural routines in preparation for both expected and unexpected circumstances—for example, playing the piano without sheet music, memorizing multiplication tables, or throwing a basketball into a hoop. Third, by encoding serial events and then reconstructing the sequence, the cerebellum can alert the cortex to what may happen before it actually does, thereby predicting future events and helping the cortex prepare for them. Fourth, the cerebellum aids in correction of errors through extended experience and practice.

Evidence for a role for the cerebellum in the creative process, and important insights into how it plays such a role, have come from a study by Allan Reiss and colleagues at Stanford University.[34] Volunteers were placed in an fMRI scanner and asked to draw either 'action words'—such as 'vote', 'levitate', 'snore', and 'salute'—or, as a control, a simple zigzag line. Drawing a zigzag line engaged the fine-movement and attentional-focus areas of the brain, but did not require creative processing. Later, the drawings were rated by experts for creativity, accuracy, and other parameters, and then compared to the brain scans.

The study produced two unexpected findings. First, greater creativity displayed by individuals carrying out the tasks was associated with higher levels of activity in the cerebellum.[35] Second, although more difficult drawing tasks increased activity in the cerebral cortex, higher creativity scores were associated with reduced activity. Reiss believes the findings show that the cerebellum is an important coordination centre for the rest of the brain, allowing other regions to be more efficient.[36] But he also thinks they illustrate why a deliberate attempt to be creative may not be the best way to optimize a person's creativity. While greater effort to produce creative outcomes involves more activity of executive-control regions, it may actually be necessary to reduce activity in those regions in order to achieve creative outcomes.[37] In other words, trying too hard may prove counterproductive when it comes to creativity. This has obvious implications for education, for it suggests that teaching approaches that are based on making the student consciously 'work' at learning may be counterproductive. Indeed, Michel Thomas, who pioneered a unique approach to language learning that I, like millions of other people around the world, have used successfully to learn a variety of foreign languages, specifically implored his students not to 'try and remember'.[38] The reason was that he believed that the active decision to do so could actually impede learning ability. And since the cerebellum is particularly adapted to unconscious, repetitive acts, this provides an explanation for how it can apparently play an important role not only in regulating movement, but also in creativity and imagination.

Interestingly, given what I have said previously about the importance of tool use in human evolution, Larry Vandervert, of American Nonlinear Systems, has hypothesized that the great expansion of the human cerebellum in the past million years coincided with a blending of neural mechanisms of vocalization with visual and spatial working memory, and that this blending was based on progressive tool use.[39] Drawing on Vygotsky's arguments about the importance of play in the development of imagination, he has also proposed that play in both animals and humans is based on the same mechanisms of repetition in the cerebellum, but play in humans has evolved from simple training for the unexpected to long-term survival by predicting, mitigating, and preventing the unexpected. As animal play evolved toward specifically human play, a cerebellum region called the dentate nucleus became a river of nerve tracts running from the cerebellum to different regions of the cortex. And as it increased the number of nerve tracts interconnecting with higher mental functions, it may also have supported imagination in play and the development of creativity in the adult.

If imagination and creativity are a characteristic of all humanity, what about especially gifted people? 'The genius', noted nineteenth-century German philosopher Arthur Schopenhauer, 'lights on his age like a comet into the paths of the planets.'[40] With no tools other than his own imagination, Albert Einstein predicted that massive accelerating objects—which at that time meant primarily regular stars and planets, but nowadays could also encompass such phenomena as black holes and pulsars— would create ripples in the fabric of space-time. It took one hundred years, highly powerful computers, and massively sophisticated technology to definitively prove him right, with the recent physical detection of such gravitational waves.[41] In the arts, Michelangelo is often held up as an example of a genius, based on his creation of some of the most famous artistic works of all time—in a range of different mediums—which include the statue of David and the frescoes in the Sistine Chapel.[42] So what makes a genius and how does this relate to brain function?

Einstein believed that a key part of his own creative process involved unconscious intuition. As he put it, 'a new idea comes suddenly and in a rather intuitive way. That means it is not reached by conscious logical conclusions. But thinking it through afterwards you can always discover the reasons which have led you unconsciously to your guess and you will find a logical way to justify it. Intuition is nothing but the outcome of accumulated earlier intellectual experience.'[43] Indeed, studies such as the one above that implicate the cerebellum as well as the cerebral cortex in the creative process (with the cerebellum dealing with unconscious aspects, the cortex with the conscious) fit with the fact that a flash of inspiration can arise at unexpected times—in a dream, in the shower, on a walk, but also after a period of conscious contemplation of a problem. According to this model, information arrives consciously, but the problem is processed unconsciously, the resulting solution leaping out when the mind least expects it.

All humans have a faculty for imaginative thought, but why do some individuals have this faculty to such a degree that they see that much further than the rest of us? What, in other words, is it about the mind of a genius that makes them different? The study of Einstein's brain following his death in 1955 has revealed some potentially interesting features. One was the fact, mentioned previously, that Einstein seemed to have an unusually high proportion of glial cells compared to neurons. Yet this difference, observed by Marian Diamond at the University of California, Berkeley, was only found in one of four of the samples of Einstein's brain that she studied, casting doubt on its general relevance.[44]

Further analysis by Britt Anderson at the University of Alabama, Birmingham, indicated that the number of neurons in Einstein's prefrontal cortex was equivalent to brains in a control group, but they were more tightly packed, suggesting that this might allow faster processing of information.[45] Meanwhile Sandra Witelson of McMaster University in Hamilton, Canada, has claimed that the inferior parietal lobule—the brain region responsible for spatial cognition and mathematical thought—is wider than normal, and better integrated. Witelson has suggested that this might relate to Einstein's own descriptions of his thinking: 'words do not seem to play any role', but there is an 'associative play' of 'more or less clear images'.[46] Most recently, Dean Falk at Florida State University has studied Einstein's corpus callosum—the bundle of fibres connecting left and right hemispheres of the brain—and claimed to have found evidence that it was thicker than normal, which he thought might mean that the physicist had enhanced cooperation between his brain hemispheres.[47]

As interesting as these findings might be, this type of analysis, and resulting speculation about their relevance to Einstein's genius, suffers from several potential flaws.[48] For a start, in the absence of the brains of other 'geniuses' with which to compare, the relevance of findings relating to one individual is unclear. Moreover, the differences identified above were selected as being potentially interesting, yet those that did not fit the picture of a genius—like Einstein's slightly smaller-than-average brain—were largely ignored. And finally, even if the differences identified are unusual compared to the average, it remains to be shown that they have relevance to Einstein's scientific genius. Einstein was an amateur violin player, bilingual, and, it has been suggested, autistic.[49] So any differences found might actually relate to these other aspects of his character. This latter point is important because another unusual feature of Einstein's brain, his dramatically asymmetric parietal lobes, has also been identified in musicians who use their left hands, and so could merely be a consequence of the physicist's lifelong playing of the violin.[50]

There is another reason why we should not expect the brain of a person of exceptional creativity to have unique features that stand out like a

Eureka moment. This is the emerging evidence that there may be a thin line between genius and mental disorder, which provides a demonstration of the potential problems in trying to identify a clear biological source for genius. In fact, the belief that creativity and madness are closely linked goes back at least to ancient Greek times, when Plato drew attention to the eccentricities of poets and playwrights, and Aristotle noted that 'those who have been eminent in philosophy, politics, poetry, and the arts have all had tendencies toward melancholia.'[51] The seventeenth-century English poet and critic John Dryden observed that 'great wits are sure to madness near allied, and thin partitions do their bounds divide.'[52] Recent studies have shown that such a link between creativity and mental disorder may be based on actual neural mechanisms.

For instance, Andreas Fink and colleagues at the University of Graz in Austria used fMRI of the brain to study a variety of human individuals, including some that were prone to schizophrenia.[53] While inside the brain scanner, individuals were asked to come up with original uses for everyday objects—a common assessment of creativity. The researchers found important similarities in the brain images of people who showed evidence of high creativity and those prone to schizophrenia, with both showing higher activity in the right precuneus, an area of the brain that helps us gather information. This suggests that creative individuals and people prone to schizophrenia may both share an inability to filter out irrelevant material. And the findings support a proposal by Shelley Carson, a Harvard University psychologist, who believes that creativity and mental illness are both characterized by a process that she calls 'cognitive disinhibition' (Figure 20).[54] This involves a failure to keep off-the-wall images or ideas out of conscious awareness. This failure may make people with a tendency towards mental disorders like schizophrenia or bipolar disorder more prone to delusional thoughts or mental confusion, but it could also be a fertile source for the creative mind. According to Carson, 'you have more information in conscious awareness that could be combined and recombined in novel and original ways to come up with creative ideas.'[55] An individual with such tendencies

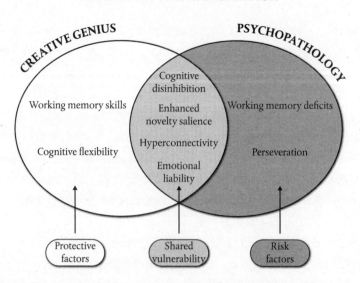

Figure 20 Overlaps between creativity and schizophrenia.

could become either a highly productive creative person or a victim of mental disorder, depending on the particular circumstances of their life and the potential to express themselves in a positive fashion. I will look at such possibilities in more detail when I examine the subject of mental disorder. But before I do so, I want to explore an aspect relevant to discussion of both highly functioning and disordered minds: the relationship between emotion and rational thought.

CHAPTER 10

EMOTION AND REASON

Until now I have been focusing on the human mind as a rational entity, guided by rules. And indeed, a common, current view of the mind is that it can be compared to a highly sophisticated computer.[1] But it does not take much familiarity with human behaviour to recognize that much of it can also appear very irrational. To some extent such behaviour can be regarded as abnormal and filed under the category of mental disorder. But this raises the question of what is normal or abnormal, for it would be hard to find anyone who has not acted in a reckless, thoughtless, or 'out-of-character' manner at various points in their life or even on a regular basis. Binge drinking, road rage, a lovers' tiff, more serious 'crimes of passion', all are signs that even 'normal' human beings sometimes behave in decidedly irrational ways. One important reason is that far from being simply a computer, the human brain is also an organic structure with an evolutionary history of hundreds of millions of years.[2] As a consequence, our brains are guided not only by reason, but also by what have come to be known as the emotions. In this chapter, I will look at the physiological basis of emotions, and the role they play both in distinct brain regions and in the brain as a whole.

The word 'emotion' comes from the French verb *émouvoir*, meaning to move or stir up.[3] This gives an idea of the strong sense of physical disturbance that can underlie an emotional response. But while emotions can clearly affect the body as a whole, they also have an important effect on the mind. So how are these two actions connected? Emotions have been found to be linked to chemical changes that can affect both mind and body. The brain is connected with the rest of the body through direct

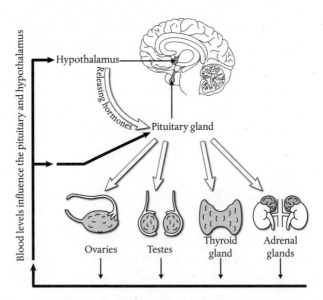

Figure 21 Control of hormone production by the hypothalamus.

nerve connections to the organs, but also by brain regions like the hypothalamus, and glands linked to it such as the pituitary that secrete hormones into the blood (Figure 21).[4] The pituitary gland in turn can influence the activity of the adrenal gland, the thryoid gland, the gonads, and various other organs.

A revelation of recent years has been that many bodily regions both receive signals from the brain and send reciprocal messages back to it.[5] Research shows that the gonads, stomach, intestines, liver, kidney, heart, and even fat tissue not only receive signals from the brain but also secrete hormones that can influence it and, as a consequence, our mental state. So when someone says that they have a 'gut feeling' about something, it may have a real biological basis. In addition, we are beginning to realize that these different parts of the body can communicate with each other in a far more interactive fashion than had been realized.

For this reason, some scientists now argue that even human consciousness cannot be thought of as a product solely of the brain, but also of the body. In what has become known as the '4Es' theory of consciousness, the human mind is said to be 'embodied', because of the reciprocal interaction of the brain and the other bodily organs.[6] For instance, Antonio Damasio, a neuroscientist at the University of Southern California, has argued that a crucial area of investigation for psychology should be 'why and how we emote, feel, use feelings to construct our selves...and how brains interact with the body to support such functions.'[7] The three other E's stand for 'embedded', meaning that the human brain/body also has an interaction with the world around it; 'enacted', because this interaction is an active process; and 'extended', because humans employ tools as part of that interaction with the world.[8] I will return later to this theory of consciousness, and how it relates to the theme of this book.

To complicate the picture further, a chemical defined as a hormone because it is released into the bloodstream by a gland and subsequently fulfils particular functions in the body may also be secreted directly into the brain, where it can play quite different roles. One example of such a hormone is oxytocin. For some years this hormone has been known to play an important role both before and after a baby is born, by inducing labour, and stimulating milk production by the mother's breasts. But more recently, oxytocin has also been shown to be released by the hypothalamus directly into the brain.[9] There it influences behaviours including social recognition, bonding, and parental behaviour.[10] In addition, this chemical plays an important role in orgasm, and its concentration in the brain increases when a person falls in love, and decreases as a couple become accustomed to each other. For this reason oxytocin is often referred to as the 'love hormone'. But this term is misleading, since the chemical can also intensify memories of bonding gone wrong, such as in men who have poor relationships with their mothers. It can also make people less accepting of individuals seen as outsiders. In other words, whether oxytocin makes you feel warm and trusting towards other human beings or suspicious of them depends on the particular context of the interaction.

One intriguing possibility recently put forward by Erich Jarvis of the Rockefeller University in New York is that oxytocin may facilitate the acquisition of language.[11] I have mentioned that views about how children acquire the capacity for language have tended to be split between behaviourists who see this capacity as a learned response and those, like Noam Chomsky, who believe it is an innate 'instinct' in humans. However, Jarvis believes the real situation is somewhere in between and that oxytocin may play a crucial role.[12] He points to evidence that oxytocin is active not only in the brains of mammals, but also in those of songbirds, which, as we have seen, have been used as a model for the study of the genetic basis of language in humans. Noting that oxytocin is particularly active in the brain regions involved in song acquisition in songbirds, Jarvis proposes that oxytocin might control the social mechanism of vocal learning in both songbirds and human beings. As he puts it, 'when a child says "Daddy" and receives a pat on the back or a smile, that gives the child a feeling of reward. That feeling may release oxytocin into the vocal-learning circuits, to strengthen the memory in the vocal-learning pathway of how to say that sound.'[13]

Another chemical that plays dual roles in the brain and the rest of the body is dopamine.[14] Like the more commonly known adrenaline, this chemical is a modified amino acid, the molecules that also serve as the building blocks for proteins. Dopamine is known to be involved in regulating the activities of the kidney, pancreas, and immune system. However, it is in the brain that this chemical has been shown to play its most important roles to date.

Dopamine is vital for our control of movement, as shown by its depletion in people with Parkinson's disease, a disorder associated with mobility problems, tremors, and speech impairments.[15] But it also regulates many higher mental functions in humans, such as learning, concentration, planning ahead, and pleasure-seeking behaviour. The role of this chemical in the pleasure response was demonstrated in the 1980s by Wolfram Schultz of the University of Fribourg in Switzerland.[16] Studying monkeys, he developed methods for recording activity from neurons that use dopamine to transmit information to other neurons.

Schultz found that in an untrained monkey, the dopamine neurons responded whenever a reward was given to the animal, but if it was shown various visual patterns to which it had to respond to secure the reward, the pattern of response changed as the monkey learned.[17] The dopamine neurons now responded when the correct visual pattern appeared, and the response to the reward itself disappeared. If no reward was given, the activity of dopamine neurons actually decreased at the expected time after the visual signal; but if the reward was delivered at an unexpected time, the neurons responded to it. According to Schultz, 'this is the biological process that makes us want to buy a bigger car or house, or be promoted at work. Every time we get the reward, our dopamine neurons affect our behaviour. They are like little devils in our brain that drive us towards more rewards.'[18]

One indication that dopamine has a particularly distinctive role to play in our brains has come from a recent study led by Nenad Sestan of Yale University.[19] This compared gene expression in 16 different regions of adult human brains with expression in these brain regions in other primates. A striking discovery was that humans—but not other primates—express a gene called TH that is involved in the production of dopamine in the brain neocortex, as well as in the striatum, which is a region associated with movement.

The significance of this discovery remains unclear, but Sestan has speculated that since humans, but not chimpanzees, bonobos, or gorillas, have a rare population of interneurons in the neocortex that is also capable of producing dopamine, it is possible that these are able to produce dopamine locally and therefore have the capacity to modulate the local circuitry in a more precise way.[20] These findings show that although humans have the same basic neurochemicals as other species, how these chemicals modulate brain function may be qualitatively different in humans, compared to even our closest animal relatives.

Such findings are worth considering in light of the central theme of this book, which is that a crucial distinguishing feature of human consciousness is its mediation by 'cultural tools', a primary one being language.[21] This idea can also be applied to our understanding of emotions in humans. It has been common in psychology to distinguish two classes of emotions:

those with analogues in animals and human infants, such as joy or fear; and those with no such analogies, such as guilt or pride. Yet for the culturally mediated view of human consciousness I have been proposing in this book, this is a false division; rather, all emotions have a social dimension and develop as an individual human's mind matures into adulthood.

A feeling so apparently basic as hunger can clearly be very different in an adult human compared to an animal or for that matter a human child. Imagine also the difference between a person lost in the wilderness desperate for food compared to the hunger of someone about to eat at an expensive restaurant with a partner or close friend. Clearly here context is everything. This is also true of other feelings, like sexual desire. Far from sex being merely a physical act, in humans it is closely entwined with the idea of love, traditionally seen as a higher emotion. At the same time sex in humans can take on a dazzling—and sometimes quite bizarre—number of different forms. In both these cases, a biological desire becomes transformed by the social context, but also by the intellect. And it seems therefore possible that the chemistry of the brain may be subtly different in humans.

While emotions are radically altered in humans because of the influence of higher mental functions, equally higher functions cannot be separated from emotional responses. Lev Vygotsky made this point in arguing that 'when associated with a task that is important to the individual [and] that somehow has its roots in the centre of the individual's personality, realistic thinking calls to life much more significant emotional experience than imagination or daydreaming.'[22] This may seem counterintuitive, as rational thought is often counterposed to emotional responses, but this ignores the importance of emotion for rational thought.

One way of examining how rational thought and emotional responses are intertwined is to study how these two factors develop from childhood to adulthood. Giovana Reis Mesquita of the University of Bahia in Brazil has recently argued that, 'in the field of emotions, one can draw a parallel reasoning between thought and language, as Vygotsky proposed...the child, when born, is endowed with thought, but a practical thought, very focused on a concrete relationship with the environment. Similarly...the child's innate emotions [are] basic, somewhat elaborated, common to all human beings, composed of standardized facial expressions and, therefore, [of] a distinctly biological determination.'[23] However, when the emotional system meets language, there is a qualitative leap in the former. This not only makes us unique in being able to name the emotions that we feel, but also means that human emotions are highly influenced by the society in which they arise.[24]

One interesting aspect of Vygotsky's view of the interaction between rational thought and emotional responses is its implications for education. I noted earlier that Vygotsky's ideas have had particular resonance among developmental psychologists, educationalists, and teachers, because of their relevance for understanding the process by which children learn, and how this might be aided in the classroom. One important concept in this regard is what Vygotsky termed the 'zone of proximal development' (ZPD), defined as 'the distance between the actual developmental level as determined by independent problem-solving and the level of potential development as determined through problem-solving under adult guidance, or in collaboration with more capable peers' (Figure 22).[25] Vygotsky developed this concept to counter the view that a child's intellectual potential is fixed, like the amount of water in a bucket. In contrast,

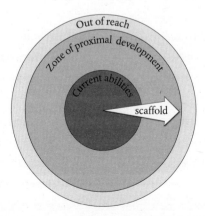

Figure 22 Vygotsky's zone of proximal development in learning.

he believed that learning is most effective when the child gets assistance from a teacher, carer, or older child in such a way that the child can accomplish tasks that would be outside their level of competence, if left unaided. The ZPD is sometimes described as a kind of 'scaffolding' in aiding a child's learning.[26]

The ZPD is often discussed purely in terms of intellectual support. However, Holbrook Mahn and Vera John-Steiner of the University of New Mexico have argued that Vygotsky also saw emotional support of the child as critical to successful learning.[27] Such support includes encouraging confidence and the sharing of risk in presenting new ideas, constructive criticism, and creating a safety zone.

To facilitate such interactions, in studies carried out in schools in New Mexico in the mid-1990s, Mahn and John-Steiner encouraged teachers to use 'dialogue journals'.[28] These involved students writing for 10 to 15 minutes at the beginning of class on whatever topic they chose. The teacher then made their own comments in the journals. Students were encouraged to focus on communication, not to worry about mistakes, and were also free to jump topic and draw on their own interests and experiences. As students became less anxious about their writing, they became more fluent. A recurring theme in the students' reflections about such journals was that the responses they received from the teacher played an important role in motivating them and giving them the confidence to take risks with their writing.[29]

The example above shows that a caring school environment and emphasis on meaningful communication between teacher and student can help create positive emotional responses that facilitate learning. But in some circumstances, emotional responses may become negative and even dangerous to health and well-being. This is the situation with addiction.[30] Earlier in this chapter, we saw how different chemicals play important roles in brain function. These chemicals not only are crucial to life, but also modulate key emotional responses. However, such responses are also central to how recreational drugs work. While these drugs are an important feature of human society, they also carry the risk of addiction, which can lead to much misery, and even death. It is therefore worth looking at how such drugs elicit the responses they do, and how individuals can become addicted to them.

Human beings have probably used naturally occurring chemicals for both medicinal and recreational purposes since long before our species developed writing and could thereby record our experiences. Indeed, Elisa Guerra-Doce of the University of Valladolid in Spain has recently compiled evidence from around the world which indicates that ancient human societies were using recreational drugs at least as early as 13,000 years ago.[31] For instance, it seems that human societies were cultivating opium poppies by at least 6000 BC. Discussing the cultivation of such poppies, Guerra-Doce notes that 'apart from its use as a food plant, there is also uncontested evidence for the exploitation of its narcotic properties.'[32]

While opium dulls the senses, other drugs stimulate them, leading to psychedelic effects. Guerra-Doce found evidence that between 8600 BC and 5600 BC, inhabitants of caves in Peru's Callejon de Huaylas valley were using *Echinopsis pachanoi*—a cactus that contains the psychedelic substance mescaline. The most ancient signs that Guerra-Doce uncovered of recreational drug use were reddish stains on 13,000-year-old human teeth found in a burial pit on Palawan Island in the Philippines, thought to be caused by chewing the leaves of the betel plant, which is still used in much of Asia as a stimulant. Other drugs that ancient societies appear to have used include tobacco, cannabis, and hallucinogenic 'magic' mushrooms, alongside the more commonplace alcohol.[33]

Such drugs are well known, in an often notorious way, in modern society. However, it is important that we do not assume that modern use of narcotic and psychedelic drugs can necessarily be equated to the way these drugs were used in the past. For instance, Elisa Guerra-Doce points out that most traces of the drugs identified by this study were in tombs and sites with ritual or religious significance.[34] Because of this, she thinks the substances were used either during mortuary rites, to provide sustenance for the deceased in their journey into the afterlife, or as a tribute to underworld deities, with use of such substances perhaps being restricted due to their sacred role.[35] However, given the often illicit way drugs are used in contemporary society, it is also important not to disregard the possibility that in ancient times, recreational drugs were taken more generally for pleasure, even if religious rules meant they were only supposed to be consumed as part of ritual ceremonies.

Recently, scientists have begun to realize that the brain mechanisms that underlie the pleasure we can get from recreational drugs have roots in the same ones that give pleasure in the rest of our lives.[36] Essentially, the brain's reward system learns the actions that produce positive outcomes, such as obtaining food or sex. It then reinforces the desire to initiate those behaviours by inducing pleasure in anticipation of the relevant action. But in some circumstances this system becomes oversensitized to pleasurable but harmful behaviours, producing pathological impulses like addiction, a term derived from the Latin for 'enslaved by' or 'bound to'.[37] Addiction exerts a long and powerful influence on the brain in three distinct ways: craving the object of addiction; loss of control over its use; and continuing involvement despite adverse consequences.[38] In the 1930s, when researchers first began investigating addictive behaviour, they believed that people with addictions were morally flawed or lacking in willpower. However, today addiction is recognized as a chronic condition that changes both brain structure and function. Just as cardiovascular disease damages the heart and diabetes impairs the pancreas, addiction hijacks the brain. Recovery from addiction involves willpower, but it is often not enough to 'just say no'. Instead, multiple strategies—including psychotherapy, medication, and self-care—may be required to treat addiction.[39]

Addiction can prove fatal. In the case of tobacco, an addict's life can be shortened by the increased likelihood of getting lung cancer or other respiratory problems.[40] A chronic alcoholic may die of liver cirrhosis or cancer of the liver, although drinking far too much alcohol in one go—as in binge drinking—can also prove fatal.[41] In the case of an opioid like heroin, although the drug is taken because of the pleasurable sensations it produces, it has other effects, one being to supress the brain regions that control breathing.[42] Opioids additionally depress the brain's ability to monitor and respond to carbon dioxide when it builds up to dangerous levels in the blood. According to Bertha Madras, a professor of psycho-biology at Harvard Medical School, 'it's just the most diabolical way to die, because all the reflexes you have to rescue yourself have been suppressed by the opioid.'[43]

The potentially devastating effects of drug addiction were brought home to me as a teenager when my 21-year-old cousin Thomas died of a heroin overdose. He had been off the drug for months but some addict friends persuaded him to try it again for old time's sake. In typical fashion, after being off the drug for so long, the dose he had been used to proved fatal, his body having become less tolerant compared to someone taking heroin on a regular basis.

Although traditionally addiction has been associated with drugs, it has become clear that excessive gambling, eating, sex, and even shopping can also affect the brain in ways that can be viewed as addictive. Recently, concerns have been raised about the possibility of individuals becoming addicted to the Internet, with overuse of computer games of particular concern, but also excessive use of social media. So how valid are such concerns and to what extent can Internet addiction be explained by the model of the mind proposed in this book?

That social media was designed to be addictive was acknowledged by Sean Parker—the founding president of Facebook, although retired from this role since 2006—when he admitted that the social network was deliberately founded to distract its audience. Noting that 'the thought process was "How do we consume as much of your time and conscious attention as possible?"',[44] to achieve this goal Parker explains that Facebook exploits a 'vulnerability in human psychology. Whenever someone likes or comments on a post or photograph, we...give you a little dopamine hit.' According to New York Times columnist David Brooks, 'tech companies understand what causes dopamine surges in the brain and they lace their products with "hijacking techniques" that lure us in and create "compulsion loops".'[45] Social media sites create irregularly timed rewards, a technique also employed in slot machines, based on the work of behaviourist B. F. Skinner who showed that the strongest way to reinforce a learned behaviour in rats is to reward it on a random schedule. This is the key to social media's success—we compulsively check such sites because we never know when the enticing stimulus of social affirmation may next occur.[46]

Of course it could be argued that excessive time spent on social media is quite different to drug addiction in that, while it may fritter away hours that could be spent on more productive activities, this type of addiction is not going to prove fatal, like a heroin overdose. Such a distinction is an important one to make; however, it is also important to recognize that addiction to Internet sites may have more profound effects than might be supposed.

For instance, Earl Miller, whose studies of working memory I have previously discussed, has pointed to the potential adverse effects of an individual being continually bombarded with information from the Internet, whether social media updates, e-mail messages, or a 24-hour news network.[47] Miller has criticized the view that being able to respond to all these different inputs simultaneously demonstrates that an individual is a good 'multitasker'. Instead, he believes that our brains are 'not wired to multitask well…When people think they're multitasking, they're actually just switching from one task to another very rapidly. And every time they do, there's a cognitive cost in doing so.'[48] One reason for this inefficiency is that such multitasking can increase the production of the stress hormone cortisol as well as adrenaline, which then overstimulate the brain and cause mental fog or scrambled thinking.[49] To make matters worse, the prefrontal cortex has a novelty bias, meaning its attention can be easily hijacked by something new. Ironically, the very brain region we need to rely on for staying focused is in fact very easily distracted![50]

Of course, the Internet has also changed our lives in many positive ways—allowing friends and family to keep in touch, facilitating exchanges between work colleagues, and even playing an important role in political campaigns.[51] There is also a world of difference between someone so hooked on social media that it begins to dominate their life and a person who uses it occasionally. In addition, a major theme of this book is that there is a danger of reducing human behaviour to simple biology, and this applies to Internet addiction as much as any other form of human behaviour. Instead, we should be sensitive to the negative social influences that can fuel addiction in its many forms—whether to alcohol, heroin, or the latest computer game or social media site. In particular, we need to consider whether excess use of a drug or excessive time spent in front of a computer screen is a cause of social dysfunction, or rather a consequence of estrangement from society.

Moreover, it is important that drugs themselves are not only considered from a purely negative point of view. This is particularly true in a book devoted to exploring the unique aspects of human consciousness compared to that of other species, given interesting new ideas about how drugs may have played a role in the development of that consciousness. Until recently it was assumed that consumption of alcohol by our species on a regular basis only occurred with the agricultural revolution that took place about 12,000 years ago, because only then would it have been possible to stay sufficiently long in a place to grow and ferment the raw materials for making an alcoholic beverage. But evidence has been emerging that production and consumption of alcohol not only may have predated the agricultural revolution, but also may even have been a stimulus for this key event in human history.[52]

Brian Hayden, an archaeologist at Simon Fraser University in Canada, has recently found evidence of beer-making activities among the Natufians—sedentary hunter-gatherers who inhabited an area of the Middle East that is now part of Syria, Jordan, and Lebanon, and who are thought to have acted as forerunners of the agricultural revolution—as early as 13,000 years ago.[53] Archaeological remains found in this region include stones for grinding barley and brewing vessels that could have potentially been used to make beer. Hayden believes that cultural factors, not economic ones, stimulated the domestication of grains like barley, in that once people understood the effects of alcohol on the mind, it became a central part of feasts and other social gatherings that forged bonds between people and inspired creativity. According to Hayden, 'it's not that drinking and brewing by itself helped start cultivation, it's this context of feasts that links beer and the emergence of complex societies.'[54] Such consumption of alcohol in a social setting could have been important in the development of human consciousness by allowing our ancestors to become more expansive in their thinking, as well as more collaborative and creative. A night of drinking may have ushered in these feelings of freedom—although the morning after, instincts to conform and submit may have kicked back in to restore the social order.[55]

According to this view, alcohol played an important role in the development of human consciousness, but only late in our evolution, at the dawn of civilization. However, there have also been claims that drugs helped shape our species' consciousness much earlier in prehistory. For instance, the 'stoned ape' hypothesis, put forward by Terence McKenna in his 1992 book *Food of the Gods*, proposes that the proto-human species *Homo erectus* came across the 'magic mushrooms' that contain the psychedelic drug psilocybin because of the tendency of this fungus to grow in the dung of animals that they tracked for food, and that this chemical induced the *Homo erectus* brain to reorganize and take on new mental properties; this in turn kick-started an evolution of cognition that eventually led to the development of the sophisticated art, culture, and technology of *Homo sapiens*.[56]

Like all highly speculative theories of human evolution, a major problem with the 'stoned ape' theory is the lack of direct supporting evidence. It is also far from clear how such a behavioural change in a proto-human species like *Homo erectus* would then manifest itself in the evolution of our own much later species. Nevertheless, the evidence mentioned previously that has been emerging about the important role of different frequency brain waves in coordinating different regions of the human brain means that it would be unwise to be totally dismissive about the idea that a drug might have had radical effects on brain function—including such important attributes as creativity and imagination—and therefore might have played an important role in the evolution of human consciousness.

Meanwhile, on surer scientific footing, modern methods of brain imaging have recently been used to reveal the effects of psychedelic drugs on human brain activity. The study led by Anil Seth of the University of Sussex in Brighton examined brain scans of individuals exposed to the psychedelic substances LSD, psilocybin, and ketamine—at drug doses described by Seth and his team as having 'profound and widespread effects on conscious experiences of self and world'.[57] Such drugs were described by those taking them as 'broadening' the scope of their conscious mind's content, with accompanying vivid changes in imagination during consciousness. The study found that administration of these drugs led to an increase in neural activity in certain specific regions of the brain compared to the situation with individuals who had not taken the drugs. These are brain regions known to be important for perception, rather than involvement in language and movement.

The increase in brain activity in this study accompanied a host of peculiar sensations that participants said ranged from floating and finding inner peace, to distortions in time and a conviction that the self was disintegrating. The researchers have speculated that the increased neural activity in particular regions of the brain could explain the dreamlike hallucinations some people experience when under the influence of psychedelic substances. Robin Carhart-Harris of Imperial College London, who was involved in the study, believes the sudden increase in randomness in brain activity may reflect a deeper and richer conscious state. Other studies by Carhart-Harris's team have shown that certain types of meditation can have similar effects on the brain.[58] He has noted that 'people tend to associate phrases like "a higher state of consciousness" with hippy speak and mystical nonsense. This is potentially the beginning of the demystification, showing its physiological and biological underpinnings.'[59]

One hope is that such research could help scientists understand what neural activity corresponds to different levels of consciousness in humans. Another is that by understanding how people respond to the drugs, doctors might more accurately predict which patients could benefit from taking psychedelic drugs as therapeutic agents. This is important because recent studies have shown that ketamine may have potential in treatment of certain types of depression, in ways that I will discuss later. For Anil Seth, 'the evidence is becoming clear that there is a clinical efficacy with these drugs. We might be able to measure the effects of LSD in an individual way to predict how someone might respond to it as treatment.'[60]

Perhaps most intriguingly, study of the effects of psychedelic drugs on the brain might provide insights into the relationship between the conscious and unconscious aspects of brain function. And it is to a discussion of the role of the unconscious that we will now turn.

CHAPTER 11

CONSCIOUS AND UNCONSCIOUS

Until now this book has focused mainly on the conscious human mind. It is now time to delve deeper and look at the unconscious aspects of our minds. The unconscious consists of those mental processes that occur automatically and are not available to introspection. In fact, increasingly, studies of brain function are revealing that a surprising amount of that function is automatic and, therefore, unconscious.[1] Obvious instances are breathing and control of the rate of our heart beat and the functions of other organs, but as I have said, not only sporting ability but also a certain amount of creativity and imagination may be a partially unconscious process. Recent studies in this area are throwing up some interesting, and unexpected, findings. For example, sophisticated new imaging methods have shown that the processes that underlie the decision to press a right or left button are, in fact, activated milliseconds to seconds before we are actually consciously aware of making the decision.[2]

Although such findings might appear to challenge the idea of 'free will', maybe we should not be surprised to find that so much of the function of the brain is automatic, given the many tasks that it must undertake in a typical day. Take, for instance, just a small portion of that day when someone is making their way to work. As much of the journey—whether on foot or by bike, car, or public transport—is part of a well-rehearsed routine, many aspects of it may be automatic. Even the thoughts we have on that journey, whether about the work we expect to do that day, or some argument we had the previous evening with a partner, friend, or colleague, will be on our mind, but perhaps at different levels of conscious experience.

In a sense then, there is nothing apparently too controversial about the idea that human consciousness is to a large degree unconscious, if we consider this to mean that it allows the brain to successfully combine a number of different tasks at the same time, in an efficient manner. One limitation of consciousness is its finite capacity, and here the unconscious plays a vital role. An individual can only keep a certain number of items or thoughts in consciousness, but the unconscious is potentially limitless. If we were consciously aware of everything that was going on in our unconscious, it would be overwhelming. The idea of a person completely aware of their unconscious is inconceivable.

However, there are more controversial aspects to discussions about the unconscious, particularly the idea that it may play a far more contradictory role in consciousness, even acting in opposition to conscious thoughts. Such a viewpoint sees the unconscious as a force that may supress unwelcome thoughts, but which might also make us carry out actions that do not appear to have been planned by our rational minds and may even appear wholly irrational. This view of the unconscious is most associated with Sigmund Freud. Opinions about Freud tend to be as polarized today as when he first put forward his controversial views about the mind. A recent BBC television programme hailed Freud as a 'genius of the modern world'.[3] Yet biologist Peter Medawar called psychoanalytic theory 'the most stupendous intellectual confidence trick of the twentieth century.'[4] When I first started working at the University of Oxford I asked a psychologist colleague what he thought of Freud's view of the mind. He replied that the Oxford psychology department dealt in science, not creative fiction!

We will look at Freud's ideas in more detail shortly and try to identify what, if anything, is valuable about such ideas for the view of human consciousness that I have been seeking to develop in this book. But ideas about the existence of an unconscious mind that might often be at odds with the rational, conscious one did not begin with Freud. To understand how these ideas arose, we need to return to another important thinker about the mind—John Locke, who put forward the first mechanistic view of the mind by comparing it to a blank slate.[5] We noted earlier that this view ignores the possibility that an individual's biological make-up may also have an important impact on their personality. But there is another problem with it, which was recognized by some thinkers in the eighteenth and nineteenth centuries: much of human behaviour is not immediately explainable in terms of experience alone. Rather, an individual's motives and reasons for acting in a particular way are often far from transparent. This led to the idea of an unconscious side to the human mind.

The most important exponents of such an idea in the early nineteenth century were the intellectuals of the Romantic Movement. Sometimes assumed to only consist of poets like William Wordsworth and painters such as Caspar David Friedrich, this Europe-wide movement also included scientists like Jean-Baptiste Lamarck—who proposed the first theory of evolution—and the polymath Alexander von Humboldt, as well as individuals like Johann Wolfgang von Goethe, who straddled both the arts and sciences.[6] A number of Romantic thinkers developed ideas about the unconscious mind. Thomas De Quincey studied at Worcester College, Oxford, where I now teach medicine, but left without graduating. This abrupt exit may have been connected to his experiments with opium, but the experience later brought him fame when he recounted it in his memoir *Confessions of an English Opium-Eater*, in which he also speculated about the possibility of an unconscious store of memories.[7] De Quincey was far from the only English Romantic interested in the unconscious. Samuel Taylor Coleridge's fascination with the creative sources of the imagination, and Wordsworth's with childhood recollection, also led them to speculate about an unconscious aspect to consciousness.[8]

It was, though, the German Romantic tradition that pioneered the most far-reaching notion of the unconscious as a formative principle underlying human life. In particular, the philosopher Friedrich Schelling saw the human mind as having two main aspects: one conscious, enquiring, rational; the other unconscious, buried, but intimately connected with natural forces.[9] Carl Gustav Carus extended these ideas in his 1846 book, *Psyche: On the Development of the Soul*, in which he said that 'the key to an understanding of the nature of the conscious life of the soul lies in the sphere of the unconscious'.[10] Carus's view fused—often rather haphazardly—ideas about unconscious memory with the question of unconscious biological and natural processes governing the development of life, enveloping both in an aura of mysticism. 'Can the free activity of the conscious soul', he asked, 'ever match the perfection and abundance of the creations of the unconscious soul?'[11]

Despite such important antecedents, it is Sigmund Freud who has become the principal figure associated with the notion of an unconscious mind. Freud was born in 1856, in Freiberg, Moravia—then part of the Austro-Hungarian Empire, now in the Czech Republic; he originally planned to study law at the University of Vienna, but opted instead for medicine, with a particular focus on the new field of psychiatry. Freud became interested in 'hysteria', a disorder commonly diagnosed in women at this time, characterized by anxiety and irritability. With his colleague Joseph Breuer, he began exploring the life histories of clients with hysteria, which led them to the view that talking to such patients was a 'cathartic' way of releasing 'pent up emotion'.[12] They published *Studies on Hysteria* in 1895, and began developing the ideas that led to psychoanalysis.[13]

Breuer and Freud viewed the unconscious as the epicentre of people's repressed thoughts, traumatic memories, and fundamental drives of sex and aggression. Freud saw it as a storage facility for hidden sexual desires, resulting in 'neuroses'—or what we would now call anxiety disorders. He considered the human mind to consist of the id, the ego, and the superego (Figure 23).[14] He believed that the id forms our unconscious drives and is not bound by morality but seeks only to satisfy pleasure; the ego is our conscious perceptions, memories, and thoughts that enable us to deal effectively with reality; and the superego attempts to mediate the drives of the id through socially acceptable behaviours.[15]

Freud believed that we can learn much about an individual through the interpretation of their dreams.[16] He argued that when we are awake our deepest desires are not acted upon because they are inhibited by

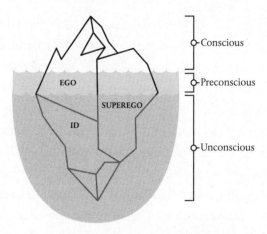

Conscious

Preconscious

Unconscious

Figure 23 Freud's model of the mind (the iceberg metaphor).

considerations of reality (through the ego) and morality (via the super-ego). But during sleep these restraining forces are weakened and we may experience our desires through our dreams. Freud also believed that our dreams contain repressed or anxiety-provoking thoughts that cannot be acknowledged directly even while sleeping for fear of anxiety and embarrassment, but which can enter our dreams in a disguised, symbolic form.

Although dreams may indicate supressed unconscious thoughts, there are limits to how much we can learn from them, mainly because of the difficulties in trying to recall dreams without at the same time supressing their content. So Freud developed a method of analysis that he called 'free association'.[17] The idea behind this was to get the patient to express anything they might be thinking, however odd. The psychoanalyst notes instances when the patient fails to free associate, because for Freudians, this represents the unconscious disrupting the capacity for an individual to express their feelings. And this can provide clues about the nature of unconscious desires in the individual.

So what are the unconscious desires that an individual is driven to supress? One of the most controversial aspects of Freud's view of the mind is the emphasis he put on sexuality. He felt that suppression of sexual desires was the biggest motivating force behind human behaviour, and particularly the cause of mental disorders. This is clear from his theories regarding psychosexual development in children, which draw on stories from ancient Greek literature.[18] In what he named the Oedipus Complex, Freud suggested that male children sexually desire their mothers and resent their fathers, whom they see as competing for the attentions of their mother. In the Electra Complex, female children desire their fathers and wish to supplant their mothers. Because of these desires, Freud believed that young boys fear that their fathers will destroy their penises in punishment for their feelings towards their mother. For female children, the realization that they do not have a penis and cannot have a relationship with their mother leads to 'penis envy'. Such feelings later progress towards sexual desire of a daughter for her father. Freud believed these anxieties will then be repressed and can lead to defence mechanisms that may result in psychological problems.[19]

A big problem with such ideas is that although they are intriguing and have undoubtedly affected the way we view family relations, there is very little empirical evidence to support the idea that boys generally lust after their mothers and hate their fathers, despite many investigations into the question.[20] As for Freud's proposal that women suffer from 'penis envy', this now seems a rather laughable and sexist notion. And there is no evidence from modern studies of the brain that there is any biological basis for an id, ego, or superego. There is also little evidence to support another proposal by Freud that normal sexual development during childhood requires progression through a defined series of stages— oral, anal, phallic, and genital.[21] Nor does it appear that problems in progressing through these stages are at the root of differences in sexuality between humans.

For instance, Freud proposed that homosexuality was a failure to progress beyond the anal phase, or overcome the Oedipal complex, although he also showed a progressive side in replying to one mother who wrote to him about her son's homosexuality that his sexual orientation was 'nothing to be ashamed of, no vice, no degradation.'[22] Freud also suggested that only 'mature' women could orgasm from vaginal sex, and that women who could only climax via clitoral stimulation were somehow stunted, stuck at a latent phase. Such ideas now seem a product of a much more backward era in terms of views of women and sexuality.

Criticisms like these have led to Freud's view of the mind being often disparaged in academic psychology and neuroscience. So does this mean that Freud's ideas about the mind—and particularly his proposals about the unconscious—have nothing to offer the view of human consciousness that I have been developing in this book? In fact, I believe we have much to learn from Freud, but only if we go beyond the superficial framework of his ideas, which were heavily influenced by the prejudices of his time, and instead focus on what I consider the most important aspect of his view of the mind: that it is a highly dynamic entity.

Previously I mentioned Valentin Voloshinov and his view that language restructures the developing mind, in a dialogue with society. Voloshinov acknowledged the importance of Freud's discoveries about the unconscious mind; however, he provided a very different explanation of the material basis of the unconscious.[23] Voloshinov believed that rather than representing our repressed animal instincts, the complex and subtle behaviour displayed by the unconscious suggests it is better understood as that portion of the conscious which is not yet articulate, and also as an ongoing struggle among various motives and voices within the conscious. And while Freud thought that the id and ego which emerge during psychoanalysis are repressed inner realities in the process of discharge, for Voloshinov they are rather reflections of real social tensions, including potentially those between analyst and patient.

If the unconscious represents the least articulate part of a continuous dialogue taking place within our minds, this is because the themes with which it is concerned are those where the 'official' ideology of society is most at variance with an 'unofficial' one based on a person's actual experience. For Voloshinov, the levels of consciousness 'corresponding to Freud's unconscious lie at a great distance from the stable system of the ruling ideology. They bespeak the disintegration of the unity and integrity of the system, the vulnerability of the usual ideological motivations.'[24] Viewed in this light, it is perhaps no surprise that Freud discovered the unconscious through his studies of sexuality. For it is hard to think of another area of people's lives where the gap between public pronouncements and private feelings and practice are more divergent than those concerning sexuality, especially in the late nineteenth and early twentieth centuries, when Freud was developing his ideas.

Freud's work is of importance in the context of this book in two further respects. First, some of the findings of psychoanalysis concerning the dynamics of the unconscious may help us better understand the unique features of inner speech. I have already mentioned Voloshinov's distinction between sense and meaning. This was a distinction shared by Lev Vygotsky, who noted that 'the senses of different words flow into one another...so that the earlier ones are contained in and modify the later ones.'[25] This is remarkably similar to what Freud observed about words as they manifest in dreams: 'The process may go so far that a single word, if it is specially suitable on account of its numerous connections, takes over the representation of a whole train of thought.'[26] And, secondly, Voloshinov's interpretation of Freud allows us to see how unconscious themes in our minds may be made conscious by a process of articulation. This occurs through a realization that the contradictory feelings and desires which exist within our thoughts have a basis in a contradictory social reality. Such articulation is powered by an individual's personal experience of the world, but may also require dialogue with a third party. This may occur in psychotherapy, indicating that descriptions of therapy as the 'talking cure' are not so far off the mark. However, such a possibility also suggests that social and political activities that challenge the basis of such contradictions in society may also prove therapeutic—not to speak of the effect such activities have on the real world, as I will explore later in this book.

We can also explore the extent to which Freud's insights can be related to findings emerging from psychology and neuroscience about the link between mind and brain. One of Freud's key claims was that memories of traumatic events can be suppressed by the mind. Recent research has provided evidence for such repressed events in the brain. For instance, a recent study led by Jelena Radulovic of Northwestern University in Illinois showed how a neurochemical called GABA can suppress traumatic memories.[27] Together with another neurochemical, glutamate, GABA plays an important role in regulating the emotional state of the brain. If we are hyper-aroused and vigilant, a surge of brain glutamate helps the storage of memories in our neuronal networks. In contrast, GABA calms us down, aiding sleep and inhibiting the excitable potential of glutamate. The most commonly used tranquilizing drug, benzodiazepine, activates GABA receptors in our brains. There are two kinds of GABA receptors. Synaptic ones work with glutamate receptors to balance the excitation of the brain in response to external events like stress. In contrast, GABA receptors on other parts of the neuron work independently of glutamate; being internally focused, they adjust brain waves and mental states according to the levels of internal chemicals like GABA, sex hormones, and micro RNAs—regulatory RNAs of the type we have already come across.

To explore the role of such 'extra-synaptic' GABA receptors in memory, Radulovic and her colleagues infused the hippocampuses of mice with gaboxadol, a drug that stimulates such receptors.[28] The mice were then given a brief, mild electric shock, and when they were returned to the same box the next day, they moved about freely and were not afraid, indicating that they did not recall the earlier shock in the space. However, when the mice were put back on the drug before returning them to the box, they froze, anticipating another shock. The experiment indicated that when the extra-synaptic GABA receptors were activated with the drug, they changed the way the stressful event was encoded. In the drug-induced state, the brain used completely different molecular pathways and neuronal circuits to store the memory. Vladimir Jovasevic, a researcher on the study, believes that this indicates an entirely different system even at the genetic and molecular level than the one that encodes normal memories.[29] This system is regulated by the micro RNA miR-33 and may be the brain's protective mechanism when an experience is overwhelmingly stressful.

Such findings suggest that in response to traumatic stress, in some individuals, rather than the glutamate system being activated to store memories, the extra-synaptic GABA system is triggered instead, and this leads to the formation of inaccessible traumatic memories, exactly as Freud's theories would predict. Radulovic believes that the brain functions in different states, much like a radio operates at AM and FM frequency bands. It is as if the brain is normally tuned to FM stations to access memories, but needs to be tuned to AM stations to access subconscious memories. If a traumatic event occurs when these extra-synaptic GABA receptors are activated, the memory of this event cannot be accessed unless these receptors are activated once again, essentially tuning the brain into the AM stations.[30] Radulovic hopes the findings may lead to new treatments for patients with psychiatric disorders for whom conscious access to their traumatic memories might help them to recover.[31] But of course, access to such memories would need to be handled carefully.

While findings such as these are revealing potential molecular and cellular mechanisms involved in the repression of unwelcome thoughts in the unconscious, other research has begun to cast light on the regions of the brain that are involved in such a process. Such research is revealing that far from the unconscious having a specific location in the brain, instead it seems to involve a dynamic interplay between its different parts.

Further insights about this interplay have emerged from studies of individuals with so-called 'split brains', an unfortunate consequence of attempts to treat epilepsy. We saw earlier that damage to the hippocampus during attempts to treat epilepsy in a patient called Henry Molaison in 1953 led to the discovery of the key role that this brain region plays in the process of memory. Another strategy used at this time to treat severe epilepsy involved cutting the 'corpus callosum', the bridge of nerve fibres that connects the right and left halves of the brain, to prevent the epileptic 'electrical storm' from spreading from one half of the brain to the other.[32] This treatment generally cured the epilepsy, and initially it seemed to make no real difference to the patients treated in this manner.

However, studies of such split-brain patients in the 1950s by Roger Sperry of the California Institute of Technology, working with his PhD student Michael Gazzaniga, began to indicate profound differences between such patients compared to untreated individuals.[33] An unintended consequence of cutting the corpus callosum was that it seemed to have left patients with two separate spheres of consciousness. For example, one patient tried to button up his flies with one hand, while the other hand tried to undo them. Another tried to embrace his wife with one hand, while his other hand pushed her violently away, as if his conscious love for his wife was being opposed by an unconscious dislike.[34]

Sperry and Gazzaniga made their most interesting discoveries while probing the visual properties of split-brain patients. A curious feature of the brain that humans share with other mammals and even more diverse animal groups is that the left side of the brain senses and controls movements on the right side of the body, while the right brain does the same for the left body. This is also the situation with incoming senses so that inputs from our right eye enter the left side of the brain, and vice versa. The researchers decided to see what would happen if split-brain patients were shown an object they could only see with one eye—the other eye being covered—and then asked to describe, or alternatively draw, what they had seen.[35] In one such experiment, if a patient viewed the word FACE with their right eye and was asked what they had just seen, they replied 'face'. However, if they viewed the same word with their left eye, they replied 'nothing'. Yet if they were then asked to draw what came into their mind, they drew a picture of a face.

Such findings confirmed previous suggestions that while the left side of the human brain primarily controls our verbal language capacity, the right side regulates more abstract human faculties, like reading, writing, and drawing. Other studies by Sperry and Gazzaniga showed that if a split-brain patient bumped into an object with the left side of their body, this being controlled by the right brain, the patient was unable to verbalize their complaint. Not only did the split-brain operation appear to have given the patient two separate minds it also seemed to restrict their identity—or ego—to the left side.[36] Moreover, if a split-brain patient was shown a picture of a nude person in such a way that it was only presented to the right, non-verbal side of their brain, they often grinned, giggled, or blushed. Yet when asked why they were responding in such a manner, the patient could not explain their behaviour.

Such findings are often cited in support of the idea that the left brain is the seat of logic and rationality, while the right side is the source of intuition and creativity. However, in reality the situation seems far more complex.[37] It is true that the functions of the two brain hemispheres appear different, but the differences lie in how each side processes specific kinds of information. For example, the left brain processes details of visible objects whereas the right processes overall shape. Meanwhile, the left hemisphere plays a major role in grammar and decoding literal meaning whereas the right plays a role in understanding verbal metaphors and decoding indirect or implied meaning. Yet this is far from providing backing for the idea that the two halves of the brain have rigidly defined roles. Indeed, Sperry himself warned that 'experimentally observed polarity in right–left cognitive style is an idea in general with which it is very easy to run wild…it is important to remember that the two hemispheres in the normal intact brain tend regularly to function closely together as a unit.'[38] In other words, the two hemispheres normally have distinct, but highly complementary, roles.

At the same time, the findings have implications for our understanding of the unconscious, for they suggest that an individual might have quite opposing points of view about a particular issue—and even different emotional responses—coexisting in their brain, all of which are important for providing a material basis for the idea of a dynamic unconscious.

Another way of viewing the unconscious from a neural perspective is to see the brain as governed by two competing systems. On the one hand, there are the subcortical areas that control basal drives for immediate pleasure and avoidance of pain.[39] These work outside of awareness. In contrast there is the prefrontal cortex, which works to weigh up the future consequences of a person's actions. Usually these two systems are in balance, which leads to adaptive behaviour balancing immediate rewards with the possible future outcomes of action. However, when these two systems are out of balance, people may have impulsivity problems where they act on their basic drive for immediate pleasure, or avoidance of pain, despite the consequences. This can occur when they have a lesion to the prefrontal cortex, overactivation of limbic areas, or poor connectivity between these two systems.[40]

When an individual is keeping memories at bay, unconscious processes and thoughts are allowed to operate in the brain without any self-reflection. But Heather Berlin, a neuroscientist at the Icahn School of Medicine at Mount Sinai, New York, believes that unconscious thoughts may become more accessible when the prefrontal cortex, which is normally suppressing such thoughts and keeping them at bay, is inhibited by the use of techniques like hypnosis, certain types of meditation, or during dreams.[41] The same may happen with certain drugs, including alcohol, which also lead to inhibition of this region.

Linking to our previous discussion of the biological basis of creativity and imagination, we can consider how different parts of the brain interrelate in people who are being spontaneously creative during improvisation. Recent neuroimaging studies have shown that when an individual is in an improvisational state, for instance through playing jazz or taking part in freestyle rap, there is decreased activation of the dorsal-lateral prefrontal cortex—the part that is active when people are suppressing information.[42] In addition, there is increased activation of the medial prefrontal cortex, which is involved with the generation of new ideas. It appears therefore that when the suppression of the prefrontal cortex is removed, unconscious processes can be brought to the surface.

Such findings are important from the point of view of understanding how the brain works. But they might also help in treating a distressed mind. One idea is that challenging the thought processes of the brain through the engagement that can occur between two individuals may allow the brain to form new connections. For instance, consider a repressed memory that is manifesting its effects behind the scenes and causing an individual to act out a specific maladaptive pattern of behaviour.[43] If that memory can somehow be brought to the surface and reinterpreted, associating it with a new positive or neutral emotion or feeling, that might break a pattern of unwanted behaviour by creating new neural pathways. However, this process can only occur if the repressed memory is brought into consciousness and re-encoded so it can connect directly with the prefrontal cortex.

I said earlier that approaches developed by Freud to identify repressed unconscious desires included the interpretation of dreams. Recent studies have begun to cast light on the brain processes that underlie dreams.[44] They indicate that in the dreaming brain, the prefrontal cortex is massively deactivated. In contrast, the limbic structures, which regulate instinctual emotional responses, are switched on. As an individual emerges from dreaming sleep, the frontal lobes come back online, and then there is an inhibition and a removal from consciousness—a forgetting—of the dream.

Freud hoped that interpretation of dreams, or use of the free association method that he introduced into psychoanalysis to try and mimic the dreamlike state, would make it possible to use the information obtained in this way about a person's repressed desires to treat disorders of the mind. But this raises the question of what success psychoanalysis—or for that matter other types of 'talking cure'—has achieved in treating mental disorders. To address this question properly, it is important that we take a detailed look at what is known about the basis of such disorders. And this will be the focus of the next part of this book.

PART IV

MIND IN TROUBLE

CHAPTER 12

SANITY AND MADNESS

———◦———

For a species with such a unique gift of conscious mental awareness, it is remarkable how often that gift can turn into a burden. Mental distress is recognized as a major problem in modern society, with around one in four people treated for a psychiatric disorder in Britain, and a similar pattern in North America and mainland Europe.[1] Because of this prevalence, and a perceived lack of clarity about both the underlying basis of conditions such as schizophrenia, bipolar disorder, and clinical depression and the drugs used to treat them, some people believe that the very designation of such conditions as 'illnesses' or 'disorders' is wrong.[2] Instead, they are seen as understandable psychological responses to trauma, stress, and other insults that an individual may encounter in their life,[3] or even as manifestations of normal differences within the human population.[4] In contrast, a dominant idea within mainstream psychiatry is that conditions like schizophrenia, bipolar disorder, and severe depression are due to defects in the genetic make-up of affected individuals.[5]

Yet whatever their origin, what is surely inescapable is the misery and suffering that results from such mental conditions. At the very least, such conditions can result in an individual being unable to find the fulfilment and happiness that should be accessible to everyone; at worst, they can result in the premature ending of lives, through both individual suicide[6] and acts such as mass shootings[7]—sadly increasingly common in the USA over recent years. If we are to tackle this major social problem, we badly need a better understanding of the causes of mental ill-health.

The central difficulty in diagnosing and treating mental disorders is reaching a level of agreement about their underlying basis. In this respect, proponents of the view that mental disorders are primarily biological in origin were particularly excited by the Human Genome Project, as they believed this would make it possible to truly get a grip on the nature of such conditions. The way ahead seemed quite straightforward—identify a large enough group of people with a particular mental disorder, and then use sophisticated new approaches for screening the genome to compare such individuals with those unaffected by the disorder. Such genome-wide association studies (GWAS) have now been used to study a range of mental disorders.[8] The problem has been that far from identifying a few clear associations with specific genes, GWAS findings have challenged all those who imagined that the link between mental disorders and the human genome would be a straightforward one, and even raised questions about the suitability of this approach to understand such conditions.

Take schizophrenia, for example. A recent major GWAS of this disorder studied over 36,000 patients and found association with 108 genomic regions.[9] The discovery of so many genomic 'hits' itself confounds previous expectations by some people that there would be just a few 'schizophrenia genes'. It also raises questions as to whether schizophrenia can be considered a distinct disorder if so many genomic regions are involved in its development.

On the positive side, some GWAS hits do make sense with regard to some of the previous scientific explanations for this disorder. For instance, one dominant theory in psychiatry is that schizophrenia represents an imbalance in the brain chemicals dopamine and glutamate,[10] whose roles in the brain we have already noted. This theory was originally proposed on the basis that certain drugs used to treat schizophrenia have an effect on these neurochemicals. And indeed some of the GWAS hits are in genes linked to both the production and sensing of glutamate or dopamine by brain cells.[11] Yet many other GWAS hits show no such connection. But these are only some of the problems raised by the findings. Another is that each association only appears to increase the risk of succumbing to the disorder by a small amount—less than 2 per cent.[12] Equally confusing at first glance is that the vast majority of GWAS hits—around 90 per cent—are not even in protein-coding genes, but in regions between genes.[13] So how are we to make sense of such a confusing situation?

That most associations are not in protein-coding genes is actually not as surprising as it might sound given recent findings by initiatives like the ENCODE project.[14] As already mentioned, and discussed at length in my book *The Deeper Genome*, this project showed that a significant proportion of the 98 per cent of the genome that does not code for proteins is far from the 'junk' DNA it was once thought to be. Instead, within the non-coding part of the genome, ENCODE scientists discovered about four million genetic 'switches' that regulate the activity of the 22,000 or so protein-coding genes.[15] Many genomic hits identified through GWAS in schizophrenics are in such regulatory regions.[16] This points to a more subtle influence of such GWAS hits on gene expression, and much more possibility of environmental influence, than for single-gene disorders, for which there is such a strong link between a gene and the disorder that the latter will always tend to manifest itself, no matter to what environment a particular individual is exposed.

Yet the problem remains that the individual genetic variants identified by GWAS seem to contribute so little to the disorder. In fact this is also proving the case with GWAS of other disorders, both those that affect the mind like depression or bipolar disorder and those that affect the body like diabetes and heart disease.[17] Consequently, some scientists have begun to question whether, despite the millions of dollars invested in this method of analysis, it has anything useful to tell us. So Jonathan Pritchard, a geneticist at Stanford University, has argued that many GWAS hits have no particular biological relevance to disease and would not make good drug targets. He notes that 'the implicit assumption of GWAS has been that when you find hits, they should be directly involved in the disease you're studying. When you start to think that all of the expressed genes in a tissue can matter, it becomes untenable that there's a simple biological story for each one.'[18] Instead, Pritchard believes that GWAS may be detecting 'peripheral' variants because they act through complex regulatory networks to influence the activity of 'core' genes that are more directly connected to an illness.

GWAS proponents have countered that by identifying the components of such networks, scientists can build up a clearer picture of how the cell normally works, and how it can go wrong during the development of a particular disease.[19] But this does not address the possibility that GWAS may be failing to identify the core genes that might be playing a much greater role in a disorder than those so far identified.

One potential flaw of GWAS in this respect is its focus on 'common' variations—that is, single-letter differences in the human DNA sequence found in at least 5 per cent of individuals.[20] When this type of analysis was first initiated, there was a good rationale for such a focus. For reasons of cost, the only method of analysis for studying a large number of individuals was not to sequence the whole genome but rather to identify the presence, or absence, of common variations in the genomes of those individuals that were being assessed. But what if the common variations being assessed have no role in the disorder? This then raises the question of why they would show any association in the first place. To understand this we need to look in a little more detail at how gene association studies work.

GWAS make use of the fact that genes are linked to other parts of the genome by virtue of belonging to the same chromosome, there being 23 pairs of chromosomes in a typical human cell. Each gene also comes as a pair, with subtle variations in sequence between the two members of such a pair, which is why scientists often refer to gene 'variants'. Now it might be expected that a gene on any particular chromosome would always be linked to other genes on that chromosome. However, a process called 'crossing over' that occurs during the formation of egg and sperm cells shuffles genetic information between any particular chromosome pair— that is, between the chromosome derived from the mother and that which comes from the father.[21] This exchange of information swaps sections of one gene variant with that of its pair, which is why no two eggs or sperm from a particular individual are genetically the same, and why siblings can differ significantly in looks, personality, and other characteristics. The process of crossing over means that genomic regions closest together on a chromosome are more likely to be linked together over subsequent generations than ones far apart.[22] This has provided a way for scientists to 'map' genes on a chromosome. But when GWAS identifies certain genomic regions as associated with a particular disorder, it does not necessarily mean that such regions have a direct role in that disorder; instead, they may merely be linked to other regions that do.[23]

Such a possibility could explain why genomic regions identified by GWAS seem to carry such a small risk of the individual succumbing to the disorder. Rather than representing an actual role in the condition, such associations may reflect the existence of rare variants only present in a few individuals, but whose presence greatly increases the risk of those individuals having the disorder; this in turn skews the apparent importance of linked common variants. Until recently, it was impossible to test such a possibility because of the prohibitive costs of identifying such rare variants by sequencing the whole genomes of large numbers of individuals. However, with 'next-generation' methods of DNA sequencing resulting in a dramatic decrease in the cost of analysing a whole genome,[24] it is now becoming feasible to carry out such whole genome sequence analysis for a variety of mental disorders.

In the meantime, other recent studies have indicated that genomic changes known as 'copy number variants' (CNVs) may have a much more significant impact on the risk of succumbing to schizophrenia than the GWAS hits mentioned earlier.[25] CNVs are defined as additions or losses from the genome larger than one thousand letters of the DNA sequence, but they can be much larger; they generally arise due to errors in the crossing-over process that occurs during the formation of the eggs and sperm that was mentioned earlier. Because of the relatively large size of CNVs, it was possible to detect these changes even before it became economical to sequence the whole genomes of many different individuals. Although loss of such large segments of DNA, which may contain multiple genes, can give rise to multiple abnormalities, interest has focused on the fact that particular CNVs are associated with a greatly enhanced risk of an individual succumbing to schizophrenia.

For instance, one CNV, in which approximately one and a half million letters is lost from a region of chromosome 3, leads to a 40-fold increase in the chance of an affected person becoming schizophrenic.[26] This region of the genome contains 22 protein-coding genes and there is now interest in investigating whether any of these are involved in the increased risk of schizophrenia. One particularly interesting discovery is that some of the genes eliminated by CNVs associated with this disorder seem to play important roles in the way that glial cells regulate inflammation—one of the ways the body responds to infection—in the brain. This has led to the idea that one factor that might trigger schizophrenia could be an inappropriate response to trauma, resulting in damage to regions of the brain.

Other recent findings have suggested that schizophrenia may be triggered by an inappropriate immune response. For instance Oliver Howes and colleagues at the London Institute of Medical Sciences have shown that the earliest stages of schizophrenia are associated with a surge in the number and activity of glial cells in the brain.[27] We observed earlier that glial cells help to fight infection, so a surge in their numbers could typically occur in response to an infective agent in the brain, but they also act in a 'gardening' role, pruning unwanted connections between neurons. But in schizophrenics, the pruning seems to become overly aggressive, leading to vital connections between neurons being lost.

The most extensive pruning in schizophrenics occurs in the prefrontal cortex, a region that as we have seen plays an important coordinating role in the brain, as well as in regions involved in hearing, which may explain why people suffering from this disorder often hear voices.[28] The prefrontal cortex also indirectly controls the brain's levels of dopamine, which could be important since, as we noted earlier, imbalances in dopamine may underlie some of the symptoms experienced by schizophrenics. Most drugs used to treat schizophrenia work by blocking dopamine release, which can bring psychotic symptoms under control, but probably does little to address the underlying problem in the brain.

According to Howes, 'the current drugs...all still work in exactly the same way. They are only able to target the delusion side of things. It's like getting a sledgehammer and squashing it down.'[29] In contrast, he hopes that studying how an inappropriate immune response might play a major role in the development of schizophrenia could lead to new drug treatments in the future.

Another emerging idea about why certain individuals succumb to schizophrenia is that while the symptoms of this disorder generally only become obvious in early adulthood, the condition itself may be set in train by changes in the brain at a much earlier stage of development. Evidence for such a view comes from studies of the link between schizophrenia and CNVs mentioned above. One intriguing aspect of such studies is not only the new insights that they are generating about the underlying mechanisms of this disorder, but also the revelation that quite different mental disorders—as defined by their symptoms—may be more closely related in terms of their developmental origins than had been suspected.

For instance, an interesting feature of many CNVs is that they can both increase the chance of an individual becoming schizophrenic and make them more likely to succumb to conditions like autism or attention deficit/hyperactivity disorder (AD/HD), which first show symptoms in children.[30] And this raises an important question: what might have caused such a change in the brain during its development—is it primarily due to a genetic difference, or is the environment also an important factor?

In addressing this question, it is useful to look at a very different way of explaining schizophrenia, one based on the idea that it is not genetics but the social environment that determines whether an individual succumbs to this disorder. One version of this hypothesis sees schizophrenia as primarily the result of defective family relations, and the disorder as a consequence of those distorted relations, but also as a response to wider pressures in society. Such a view of schizophrenia is particularly associated with British psychiatrist R. D. Laing.

Ronald David Laing was born in Glasgow in 1927. In the 1960s he began to define psychosis in terms of its relation to society, and psychotics as individuals who in their own way are making sense of their social circumstances.[31] One theory particularly associated with Laing is that of the 'double-bind'. Although first put forward by British anthropologist Gregory Bateson in the 1950s, it was Laing who popularized this theory.[32] It proposes that a key factor in the development of schizophrenia is major problems of communication within the patient's family. The basis of the theory is that the person who becomes mentally unwell is caught in a situation in which messages from other family members are deeply contradictory; however, the contradiction is never brought into the open, and the unwell person is unable to leave the field of interaction with such family members.[33]

The mechanics of the double-bind are best conveyed by real-life examples, as in Laing's 1964 book *Sanity, Madness and the Family*.[34] The book was based on studies by Laing and his colleague Aaron Esterson of patients, their parents, and siblings, individually, in pairs, and as groups. One case that shows some of the dysfunctional behaviour that can exist in a schizophrenic's family is that of a patient code-named 'Maya Abbott', who believed her parents were trying to influence her by telepathy and thought-control.[35] While that may sound like a classic case of paranoid schizophrenia, Laing and Esterson discovered on speaking to the girl's parents that for years they had indeed been trying to influence her thoughts, believing her to be telepathic. Long before she became ill, the parents had been sending each other signals which 'Maya' was not supposed to perceive, in order to test their hypothesis. Yet when confronted by 'Maya', the parents denied having made the gestures.[36] More generally, numerous studies carried out over the past few decades suggest that sexual abuse, physical abuse, emotional abuse, and a variety of other types of adverse childhood experiences may be important factors in the development of a schizophrenic.[37]

If schizophrenia can be a product of both differences in individual biology and a dysfunctional or harmful social environment, how does it actually manifest itself in the brain? One confusing issue is that two people can be diagnosed with schizophrenia with completely different sets of symptoms. The officially defined symptoms for this disorder include hallucinations, delusions, disorganized speech, disorganized or catatonic behaviour, and 'negative' symptoms, which can include emotional flatness, apathy, and lack of speech.[38] But for a person to be diagnosed with schizophrenia they only need show two of these five symptoms. This raises the question of whether this mental condition is a single disorder, or a collection of different ones. Nevertheless, that does not prevent us trying to find an underlying structure to the way this condition affects human consciousness, particularly from the view developed in this book, which sees our minds as mediated by language and culture.

Lev Vygotsky, who as we have seen was a key proponent of such a view of the mind, believed that since the development of human consciousness follows a specific pattern, so its descent into a disordered state like that in schizophrenia might also follow such a pattern, but in reverse.[39] He argued that compensation is an important process in the development of this disorder, and what are often perceived as impairments in schizophrenia may actually be a consequence of this process. If a higher mental function is 'switched off', the mind tries to compensate by resorting to a developmentally more archaic function, but finding itself without higher levels of control, this may produce abnormal forms of behaviour and thought.

Evidence for such a scenario comes from studies of conceptualization in schizophrenics. Recall that Vygotsky identified the transition from pre-conceptual to conceptual forms of thought as a key stage in adolescents. To test whether this process might occur in reverse in schizophrenia, in the late 1930s two US psychologists, Eugenia Hanfmann and Jakob Kasanin, used the same block sorting test that Vygotsky had used to study concept formation in childhood, to assess schizophrenic adults.[40] Similar to the way in which Vygotsky studied conceptualization in children, in Hanfmann and Kasanin's study subjects sorted blocks that differed in two ways: in their physical properties—size, height, form, and colour—and by their hidden 'names'—nonsense triplets of letters attached to the bases of the blocks that defined a particular concept; in this experimental test as in Vygotsky's original studies, these names always referred to a combination of the height and the size of the object.[41] Intriguingly, some schizophrenics refused to accept the triplets of letters as names or conceptual labels and were unable to use them as instruments to organize their behaviour.

As Hanfmann and Kasanin noted, for such individuals the word on the block was seen as mere 'lettering', one among many attributes that the blocks happened to have.[42] This reflected the fact that schizophrenics showed a regression from true conceptual thought in their sorting activities. Depending on the individual, this ranged from linking objects in an inconsistent manner—such as one block linked to another by similar colour, the second one to a third block by virtue of size, and so on—to what Vygotsky had described as pseudo-conceptual reasoning.[43] The latter type of reasoning was identified when the experimenter showed the erroneous nature of an individual's sorting selection by turning over a block and exposing its inaccuracy by the nonsense word describing it. An unaffected individual would respond that 'something has been changed' and begin searching for a new correct category. In contrast, a schizophrenic person tested in this way typically continued to persevere with the original method of grouping for a while, and then reverted to a trial-and-error approach.

While pseudo-conceptual grouping superficially resembles conceptual classification, it lacks hierarchical organization. If such organization is a feature of higher mental functions, lack of it may indicate a regressive trend. Such regression may split a functional system of verbal thought into communicative speech and nonverbal thinking. A schizophrenic may retain an ability to communicate with the same words as an unaffected person in terms of the objects they refer to; however, their words will exist in different systems of meanings.

Such a breakdown in shared meaning may explain why schizophrenics typically show peculiarities in their communication with others. An illustration of this was provided by a study by Heinz Werner and Bernard Kaplan of Clarke University, Massachusetts, in 1963. Schizophrenic and unaffected subjects were asked to describe visual stimuli under different conditions, such as a description for him or herself, one for a hypothetical person, and another for a real individual.[44] The schizophrenic subjects used almost as many idiosyncratic references—that is ones that could only be understood by themselves and not by others—while describing stimuli for another real or hypothetical person as they did in descriptions for themselves, and in all cases use of idiosyncratic references was more than twice that used by unaffected subjects. Werner and Kaplan concluded that in schizophrenics there is a lack of differentiation between idiosyncratic inner-speech-for-oneself and social-speech-for-others. And this may be one reason why people with severe schizophrenia can find it difficult to understand what other people are saying to them, and explain what they mean to others.

How do such findings relate to the typical symptoms of schizophrenia, and can they help attempts to both diagnose and treat the disorder? One point to consider here is Vygotsky's claim that while the onset of schizophrenia in specific individuals may have quite different biological and social roots, its development as a regression from higher forms of human consciousness may follow a specific pattern that can be studied.[45] This suggests that one way to treat schizophrenia might be to try and find a way to reassemble the higher functions that have regressed. When it comes to how such an endeavour could work in practice, something to consider is that schizophrenics do not always show signs of such regression of higher mental faculties, but then 'normal' individuals do not always think conceptually; instead, they can employ what Vygotsky called 'associative' forms of thinking.

As an illustration, consider a TV advert for yoghurt.[46] The advert first shows an attractive young woman playing the piano with skill and passion. The voice-over tells us that the woman's performance is 'without compromise'. Suddenly the image switches to a pot of yoghurt, and we are now told that the yoghurt is also made 'without compromise'. The aim is to make the viewer conflate two different activities—piano playing and yoghurt eating—to the extent that they might then subliminally imagine that by buying and eating this particular foodstuff they are also participating in the performance, and perhaps enjoying the attentions of the woman herself. In this example the viewer is being asked to take part in what Vygotsky called 'complex' thinking, not true conceptual thought. Yet they are doing so on the clear understanding that what they are watching is a commercial, and they are not literally either participating in the piano playing or engaging sexually with the attractive young woman.

In contrast, if a schizophrenic teenager says she is the Virgin Mary, it is assumed that her logic is that 'the Virgin Mary was a virgin; I am a virgin; therefore, I am the Virgin Mary.'[47] Here complex thinking is displayed, but the schizophrenic seems incapable of recognizing that the conflation of herself with a religious figure is an illusion. Yet psychosis can often coexist in schizophrenics with phases of apparent rationality. One explanation is that the schizophrenic is still relying on a framework of language established before the onset of their condition, so that even their rationality may hide the fact that word meaning may be quite altered in such an individual, but be partially hidden because of the common language that they share with 'normal' individuals in the population. But of course it may also be the case that the degree of regression from conceptual thinking in a schizophrenic individual will vary greatly depending on changes, for instance stressful incidents, in their environment.

If different biological origins for schizophrenia might manifest themselves in common regressive behaviour at the conceptual level, can Vygotsky's views about this mental disorder help us better understand the social triggers for schizophrenia? A study of schizophrenics and their families by Bjorn Rund of the University of Oslo, published in 1985, found that non-paranoid schizophrenics, but notably not paranoid schizophrenics, generally came from families that displayed abnormalities in their patterns of communication.[48] For instance, the parents of such schizophrenics were often highly egocentric in their behaviour, and they tended to ignore what their children were saying, interrupt them, or send mixed messages.

Rund related his findings to Vygotsky's claim 'that higher mental functions are internalized social relations.'[49] Rund claimed that this implies that 'many of the processes involved in the individual's cognitive activity are direct reflections of processes and patterns which characterize interpsychological activity' and particularly focused on attention, the process whereby an individual's awareness becomes focused, like a spotlight, on a subset of what is going on in their head or their environment.[50] Vygotsky thought that voluntary forms of attention emerge as 'people who surround the child begin to use various stimuli and means to direct the child's attention and subordinate it to their control.'[51] However, in a family situation in which normal communication has broken down, Rund argued that the child has 'no opportunity to establish a firm way of focusing attention on the relevant stimuli in a given situation, because it will never know which are the relevant stimuli. Instead, the attentional styles that are internalized will be characterized by a steady wandering from one stimulus to another, in a search for the most relevant one.'[52] And in line with what I said previously about the importance of emotions, Rund also believed that the 'emotional climate' within a family could affect the development of schizophrenia within it.[53] This is surely even more the case if actual physical and sexual abuse are a part of family life. So anxiety, insecurity, and instability tend to be a feature of the families in which schizophrenics grew up.

Rund left open the possibility that if biological factors play a role in the genesis of schizophrenia, this could explain not only why the condition may run in families, as shown by the fact that unaffected family members can still display abnormal forms of behaviour, but also why such behaviour may help precipitate the condition in a vulnerable individual in the family.[54] However, this also means that other factors outside the family—racism, sexism, homophobia, problems at school or work, or simply the pressure of living in a modern capitalist society—might combine to precipitate the disorder in a specific individual within that family. Finally, it is also worth considering Laing's proposal that the apparent irrationality of schizophrenic behaviour might be an attempt to try and make sense, in however distorted a fashion, of a social environment that may itself be seen as irrational.

What does all this imply for finding new treatments for schizophrenia? Since this disorder seems to be as much an environmental as a biological condition, then treatments should address the social environments that may have triggered the disorder, as well as finding ways to address the dysfunctional aspects of brain function that underlie it. However, the situation is complicated by environmental pressures manifesting themselves through biological mechanisms, which may occur long before the condition becomes apparent.

So Daniel Weinberger of the Lieber Institute for Brain Development in Baltimore, Maryland, recently studied the genomic regions linked to schizophrenia and made the surprising discovery that some appear to play a role not in the brain, but in placental function.[55] Moreover, many had previously been linked to a condition during pregnancy called 'pre-eclampsia' in which foetal growth is restricted due to a malfunctioning placenta. Weinberger believes that 'the placenta is the missing link between maternal risk and foetal brain development.'[56] A less efficient placenta might not allow the foetal brain to receive all the nutrients or oxygen it needs. Ultimately the hope is that such studies could lead to treatments to mitigate effects that a malfunctioning placenta might have on a foetus's brain.

Yet lest we suppose that all cases of schizophrenia are linked to problems that first affect an individual's brain while they are still a foetus, it is important to note other possible environmental links to this disorder. These include exposure to the parasite *Toxoplasma gondii*, transmitted to humans via cats; excessive use of high-strength cannabis; and, particularly important, physical and sexual abuse. Acknowledgement of multiple potential environmental triggers for schizophrenia, possibly each leading to a psychotic state via different biological mechanisms, is combined with increasing recognition that 'schizophrenia' as a discrete condition may itself be an illusion. Indeed, psychiatrist Robin Murray of Kings College London believes we may soon see 'the end of the concept of schizophrenia', because the clinical definition of the disorder 'is already begin-ning to break down, for example, into those cases caused by copy number [genetic] variations, drug abuse, social adversity…presumably this process will accelerate, and the term schizophrenia will be confined to history, like "dropsy" [the former term for complications of heart disease].'[57]

Coupled to this acknowledgement of the potential diversity of causes is increasing evidence that the severity of schizophrenia may differ in particular individuals. This is important because some health profes-sionals have tended to only view this condition as a 'hopeless chronic brain disease' and ignore individuals who make a full recovery, by argu-ing that this 'mustn't have been schizophrenia after all'.[58] Yet a recogni-tion of schizophrenia's heterogeneity suggests that individuals with a milder form of the disorder may not show the same total breakdown of conscious awareness found in someone with a severe form. It may also be the case that the extent of such a breakdown is affected by the varying levels of stress in an individual's life, explaining how some people can 'recover' from such a disorder.

Importantly, not only might different types of schizophrenia show dis-tinct responses to different types of drugs, but also some may be more amenable to psychotherapy than to drug treatment. Indeed, recent studies have suggested that people with a history of childhood trauma diagnosed with schizophrenia are less likely to be helped by antipsychotic drugs.

One psychotherapy approach claimed to have a very good level of suc-cess in the treatment of schizophrenia is called 'Open Dialogue'.[59] This approach, first pioneered in Finland—which in the 1980s had one of the worst incidences of schizophrenia in Europe—has over the past 30 years of application in this country led to 74 per cent of patients with psychosis being back at work within two years, compared with just 9 per cent in Britain.[60]

An important feature of the Open Dialogue approach is that as well as the patient, other members of their family are also involved in the therapy sessions.[61] Such an input could be important given the potential role of dysfunctional family relations in the generation of the condition, but also because I have been arguing that individual thoughts themselves form a kind of dialogue. According to this perspective, not only conscious but also unconscious thoughts may reflect tensions between an individual and other members of their family. It might therefore be beneficial for an affected individual to undergo therapy with other family members. Engaging in a dialogue with them in a forum mediated by a therapist may provide a space to tackle some of the familial tensions that may underlie schizophrenia.

Another key aspect of the Open Dialogue approach is that it is not anti-medication. Treatment, from drugs to different kinds of psychotherapy, is agreed by everyone at the therapy sessions. But whereas the mainstay of standard treatment is usually medication, in the Open Dialogue approach the core is psychotherapy. As Russell Razzaque, a psychiatrist using this approach in East London, has explained, 'in normal treatment you explore what has led to the crisis, but then the response is usually to prescribe medication. Whereas with Open Dialogue the service user takes the driving seat in understanding what are the factors that have led them to be the way they are. That's a very healing thing.'[62]

Perhaps the most effective aspect of the Open Dialogue approach is the time it affords the patient to develop a dialogue about their condition with the same set of therapists—and also with family members—over an extended period of time. The importance of such continuity was expressed by Annie Jeffrey, a British woman whose son committed suicide in 2014 after suffering from psychosis for five years: 'Many service users say they feel like a parcel passed from one team to another: community services, in-patient services, crisis teams, psychiatric liaison...The number of times I went to meetings with my son to see a team of people we'd never seen before and we would never see again. How are you supposed to start talking to someone you don't know? My son just felt that he wasn't listened to.'[63]

One feature of the Open Dialogue approach is that it involves sustained, high-quality interactions with other human beings. And given how much I have stressed the importance of language in the formation of human conscious awareness, and its role in our unique capacities of creativity and imagination, it is maybe not surprising that such enhanced social interactions can have a therapeutic impact. Intriguingly, there is also increasing evidence that psychotherapy can lead to positive changes in the neural chemistry of the brain,[64] suggesting that it is a mistake to only equate such changes with pharmacological interventions.

Here though we currently face a problem. As a consequence of the ongoing economic crisis that has had such a devastating negative impact on healthcare and other welfare services, provision of high-quality psychotherapy for those with mental disorders is getting worse, at the same time as societal pressures such as job insecurity, stagnating salaries, and attacks on pensions are increasing. The result has been a dramatic rise in the number of individuals suffering from such disorders. And this combination of rising social pressures leading to increased mental distress, and a healthcare system increasingly unable to cope with the effects of such distress, is having particularly tragic consequences with regard to the next type of mental disorder I want to discuss—clinical depression and anxiety.

CHAPTER 13

DEPRESSION AND ANXIETY

———◆———

At Barcelona airport, on 24 March 2015, things were not going to plan for two teachers of the Josef-König Academy, a school in Haltern, Germany. The teachers, and sixteen 14- and 15-year-old students who were in their care, were flying back to Düsseldorf, after an exchange trip with a school in the Catalan town of Llinars del Vallès. But having managed to get all the students to the airport for the early morning flight, the teachers found that one schoolgirl had left her passport at the home of her host family.[1] Initially resigned to missing their flight, everyone was delighted when the father of the host family offered to drive the 50 kilometres back to his house and pick up the passport. After an anxious wait, it was returned to its owner, and the school party boarded the plane on time. The students set off at 10 a.m. as scheduled, excited at seeing family and friends again, tinged no doubt with regret at leaving the students they had made friends with, and maybe fallen in love with, in Llinars del Vallès.

They never reached their destination. The lives of the students, their teachers, and the other passengers and crew on the flight, numbering 150 people in all, ended at 10.41 a.m. when the plane crashed into a mountain in the Alps at more than 400 mph. As well as the school party, the passengers on the plane included two opera singers, a pair of newlyweds, and three generations of the same family—grandmother, mother, and daughter.[2]

The first sign that something was amiss began with air traffic control noticing that the plane was descending far too early in its flight. Frantic attempts to contact the pilot and co-pilot met with a total lack of response, suggesting a major flaw in the plane's communication systems. However, after the crash, and the recovery of the airliner's cockpit voice recorder from the wreckage, the shocking actual unfolding of events on the plane emerged.[3]

The recorder showed that at 10.30 a.m. the pilot, Patrick Sonderheimer, left the cockpit to use the toilet, leaving co-pilot Andreas Lubitz in charge. But when Sonderheimer returned he was unable to get back into the cockpit, and Lubitz failed to respond to pleas to open the door. In the minutes before the crash, Sonderheimer became increasingly frantic, even using an emergency axe to try and break down the door, but to no avail. Ironically, after the 9/11 atrocity, cockpit doors were strengthened to prevent terrorists storming a plane's cockpit and using it as a weapon of mass destruction. Mercifully, most of the plane's occupants seem to have only become aware of the gravity of the situation just before the crash, for only then did the cockpit recorder register sounds of passengers' screams.[4]

The most obvious initial possibility was that Lubitz had somehow fallen unconscious and could not therefore open the door. However, while searching his apartment, detectives found a note from Lubitz's doctor declaring him unfit to work. Further investigations revealed that the co-pilot was taking anti-depressant drugs and had been diagnosed with suicidal tendencies. Meanwhile a study of his computer found web searches that included 'ways to commit suicide' and 'cockpit doors and their security provisions'.[5] It was looking like Lubitz had become so depressed that he set out to kill not only himself but also everyone else on the flight. One question was why Lubitz's doctors had not informed his airline of his condition. It transpired that under German patient confidentiality laws, although doctors could advise the co-pilot not to fly, they could not pass on their concerns to his employer.[6]

For campaigners against the negative stigma often associated with mental disorders, the disaster represented a nightmare situation. Masuma Rahim, a British psychologist, noted that coverage of the disaster was associated with 'a barrage of stigmatizing, fear-mongering media reports, both in the UK and internationally.'[7] As Rahim pointed out, 'there is virtually no evidence to suggest that the depressed [generally] pose a danger to others as a result of their illness. [Indeed], this is true of the full range of mental health problems: the scientific literature is clear that people with schizophrenia, long demonized and reviled by the press, are far more likely to be harmed by others or themselves than to enact violence.'[8]

This is an important point given that media reports tend to focus on atrocities like mass shootings, which particularly in the USA are now almost commonplace and are often ascribed to mental disorder. Yet concentrating on the murderous actions of one individual against other human beings obscures the fact that a far greater cause of death in the USA is suicide, with the numbers of those deciding to end their lives this way increasing over recent years. More than 100 US citizens commit suicide each day, this being the tenth leading cause of death overall—third among 15 to 24 year olds and fourth among 25 to 44 year olds.[9] As well as the tragic loss of the individual, there is the pain and heartbreak for relatives and friends of that individual, with potential effects reverberating for years to come.

I know this from personal experience. As I mentioned before, in 2015, my sister Kay committed suicide, after a rapid decline into a severe depression that proved impervious to either drug treatments or cognitive behavioural therapy. Yet she was the last person one might have expected to take her own life. Sociable and outgoing, she was devoted to family and friends, and capable of extraordinary generosity to others. A highlight of my teenage years was when Kay received her first pay cheque from a part-time job, and took me, her younger brother, for dinner at the Box Tree restaurant in Ilkley, Yorkshire, then in possession of its first Michelin star.[10] Later as a health visitor in Bradford, my sister's caseload included heroin addicts who worked as prostitutes to fund their habit. It was typical of Kay that she helped one prostitute to both kick her drug habit and start a new life at university.

In 2013 my sister was diagnosed with skin cancer. As well as the cancerous cells being removed surgically, Kay was given low doses of interferon—a chemical produced naturally by the body, but also used as a drug to treat viral infections and some types of cancer. For a while it seemed that the surgery had been successful, but then my sister found a lump under her arm. In response, her interferon dose was dramatically increased. The first I realized something had noticeably changed in Kay was when we met for a family holiday in Sicily at Easter in 2015. She was completely different from her normal sociable and upbeat self; instead she seemed to take no pleasure in either Sicily or our company.

This was the start of a descent into a severe clinical depression. My idea of depression at this time was of a condition in which a person suffers low mood, feels hopeless and helpless, and has low self-esteem. But I was also aware that depression could be treated by drugs, Prozac being the most well known. However, my sister's depression proved a very different type of condition. Increasingly concerned, we arranged for Kay to be admitted to a psychiatric ward. She remained there for just a few weeks before being discharged. Neither drugs nor CBT-style therapy seemed to have had any impact on her condition, but with a pressure on beds and staff, she was not judged a serious enough case to stay on the ward.

A few weeks later, my brother-in-law, who had taken compassionate leave, had to return to work briefly to talk to his boss. Arriving home, he announced his presence but hearing no reply, he went quickly upstairs. There, in the bathroom, he found Kay hanging from the shower attach-ment, an overturned stool at her feet. He immediately called the emergency number, but attempts by the paramedics to revive my sister proved useless.

It was only later that we learned that interferon can trigger severe depression. In fact, 40 per cent of patients have suffered from this disorder after treatment. When journalist Niall McGee was treated with the drug his doctor told him, 'If you become depressed, please let me know.' McGee initially disregarded such advice, yet following the treatment, he rapidly showed signs of severe depression. 'I was put on an antidepressant and a benzodiazepine, which, despite being highly sedating, did little to ease my out-of-control anxiety and insomnia,' he recalled. 'My depres-sion...culminated in a complete loss of self.'[11]

Most worryingly of all, McGee wanted to end his life: 'I thought about hanging myself from the rafters in my basement, asphyxiation via carbon monoxide poisoning in my car...Every time I rode the subway, I thought about jumping in front of a train.'[12] Such feelings persisted long after McGee stopped taking the interferon treatment. Thankfully McGee's suicidal thoughts eventually subsided, either because of the interferon leaving his system, or because he found a way to personally deal with his depression.[13] Sadly, not only did we get no prior warning that interferon might have such a severe effect on my sister's mental state, but we received little help on how to deal with her situation. At the inquest, the coroner concluded that my sister had been 'discharged too informally by those responsible for her with no paperwork and no advice to the family on how to help her.'[14]

My sister's depression was triggered by a specific adverse drug reaction, but a more general problem in trying to understand why some people, sometimes with no prior warning, decide to end their own lives, and in very rare cases take other people with them, is that as with other mental disorders like schizophrenia and bipolar disorder, we just do not have enough knowledge. We still have only a partial understanding of what makes people become so depressed that they become suicidal, or even of the condition of depression as a whole. This lack of understanding has a particular resonance for me given that my sister's unnatural death is far from the only one in my family. My first such experience was the death of my paternal grandmother, who was found drowned in the nearby canal when I was 15 years old.

My recollections of my grandmother are of a very large, opinionated woman, who loved her food, her drink, but above all the joy of a social gathering of family and friends. Only many years later—a month after the death of my sister—did I learn from my mother that my grandmother had suffered from depression in the last years of her life, often turning up at our house in the early hours of the morning in a distressed state. Unaware of this, as a teenager I was tortured with the uncertainty about whether my grandmother had meant to drown herself or instead had somehow slipped, accidentally, into the canal.

The death of my father, which occurred the year I graduated, left even more unanswered questions. In September of that year I went to Scotland for a walking holiday with my girlfriend. Arriving back at my parents' house, we were surprised to see dozens of cars parked outside. Puzzled, we entered the house only to be told that my father had been found dead at the foot of the entrance shaft of Gaping Ghyll, a pothole whose interior dimensions are larger than St Paul's Cathedral. Having abseiled down the shaft as an amateur potholer, I was completely aware of the depth of the drop. I learned that rather than going to work, my father had travelled by bus and train to the Yorkshire Dales, and then walked to Gaping Ghyll. How he had ended up dead at the foot of the cavern was unclear, as nothing found on my father's body or at our home gave any indication why he had gone to the Dales, or what had happened to result in his death in such a manner.

Finding meaning after the sudden death of a loved one, particularly if due to definite or likely suicide, is surely one of the biggest challenges of life. While my grandmother's death occurred too early in my life to consider how I might have prevented it, the number of times I have mulled over the different ways I might have acted to forestall the deaths of my father and sister are too many to mention. Of course, such thinking leads nowhere, since only with hindsight is it possible to see how such a tragedy might have been prevented, but also because there is no way of reversing what has already happened. But this obvious fact has not stopped me from indulging in this incessant going over of past events.

How might we better understand depression and thereby prevent deaths such as those in my family, in the future? One problem we face is the lack of consensus in even defining this disorder. As a human condition, something similar to depression has been recognized for thousands of years. In ancient Greece, people with 'melancholia' were thought to have too much 'black bile' in their bodies, and the condition was treated by altering their diet.[15] Later, in the nineteenth and early twentieth centuries, those wealthy enough to be able to afford treatment for their depression were subjected to 'rest cures', which mainly consisted of being confined to their room for six weeks in isolation while being grossly overfed.[16]

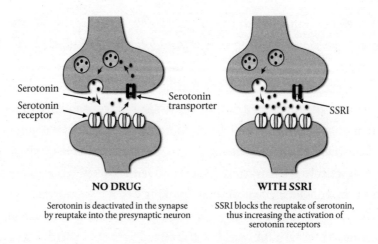

Serotonin

Serotonin receptor

Serotonin transporter

SSRI

NO DRUG

Serotonin is deactivated in the synapse by reuptake into the presynaptic neuron

WITH SSRI

SSRI blocks the reuptake of serotonin, thus increasing the activation of serotonin receptors

Figure 24 Proposed mechanism of action of SSRIs like Prozac.

What seemed like a major step forward in both our understanding of the disorder and how to treat it occurred in the 1950s when evidence emerged that depression might be due to low levels of the neurochemical serotonin. In particular, neuroscientists discovered that certain chemicals that alleviated the symptoms of depression appeared to work by inhibiting the reuptake of serotonin into certain neurons in the brain (Figure 24).[17] Prevention of such reuptake, after serotonin has been released at neuronal synapses, is thought to enhance the ability of this neurotransmitter to continue having its effects upon the brain.

The first set of drugs thought to work by enhancing levels of serotonin in the brain were discovered completely by chance. Scientists at the Munsterlingen Asylum in Switzerland, searching for a treatment to control the mood swings of schizophrenics, identified a class of drug— called tricyclics because of their three-ring chemical structure—that triggered bouts of euphoria in their patients.[18] This only made the schizophrenic individuals worse, but it suggested that tricyclics might be a treatment for depression. And indeed when given to depressive patients in 1955, some patients became newly sociable and energetic, leading to claims that tricyclics represented a 'miracle cure' for depression.

Unfortunately, these drugs, which helped about three quarters of patients, had serious side effects like lethargy, weight gain, and occasionally death from overdose. But a search for chemicals with a similar effect led to the discovery of the selective serotonin reuptake inhibitors (SSRIs), the most well-known being Prozac, released on to the market in 1987.[19] As their name implies, these drugs also increase brain serotonin levels by preventing their reuptake into neurons. SSRIs provided relief for the same percentage of patients as tricyclics, but they were easier to prescribe without risk of overdose and had fewer side effects.

SSRIs have been a huge success for the pharmaceutical industry, with Prozac earning Eli Lilly, the company that developed it, $1 billion a year by 1990. The initial excitement caught the media's attention, with *Time* magazine putting the drug on its front cover and asking, 'Is Freud finished?', and describing SSRIs as 'mental health's greatest success story'.[20] By 1993, Prozac had been taken by some 10 million people worldwide and in 1994 *Newsweek* noted that 'Prozac has attained the familiarity of Kleenex and the social status of spring water.'[21] SSRIs were now used to treat far more conditions than clinical depression, with some doctors prescribing them to everyone from pensioners to pre-teens for everything from premenstrual stress to fear of public speaking. Yet recently, doubts have surfaced about the appropriateness of doctors prescribing SSRIs for such different conditions, the effectiveness of these drugs, and the mechanisms by which they affect brain function.

The American filmmaker Sarah Gross, who produced the documentary *Prozac: Revolution in a Capsule*, has pointed out that while the widespread availability of SSRIs 'has benefited the general depressed population, and more people are receiving treatment, [it] may also have created a backlash, as some people may feel like everything is being pathologized.'[22] Even more controversial is evidence that SSRIs may not be as effective as claimed. For instance, a recent study found that the ability of SSRIs to combat mild or moderate depression appeared no better than a placebo, although this did not generally appear to be the case if the treated individual is suffering from a more severe form of depression.[23]

Another worrying aspect of SSRIs is that while these drugs are only meant to be taken for a limited period, in practice many patients remain on them for years, and for some it can be very difficult to stop taking them, such are the severe effects of withdrawal, which for one patient included 'severe shakes, suicidal thoughts, a feeling of too much caffeine in my brain, electric shocks, hallucinations, insane mood swings.'[24] Another patient noted that 'while there is no doubt I am better on this medication, the adverse effects have been devastating when I have tried to withdraw—with "head zaps", agitation, insomnia and mood changes. This means that I do not have the option of managing the depression any other way.'[25]

One problem in determining the true effectiveness of SSRIs for the treatment of depression is how little is understood about the biological basis of this disorder and how drugs like SSRIs alleviate it. In particular, doubts have been raised about whether the primary defect in depression is really low levels of serotonin in the brain. So Alan Gelenberg, a depression and psychiatric researcher at Pennsylvania State University, believes that 'there's really no evidence that depression is a serotonin-deficiency syndrome. It's like saying that a headache is an aspirin-deficiency syndrome.'[26] The idea that reduced serotonin levels are the primary cause of depression has been undermined in a number of ways.

For instance, a study in 2003 suggested that people with a variant of the gene that controls serotonin uptake were not as well equipped to deal with stressful life events and more likely to develop depression.[27] Since its publication, this study has been cited more than 4,000 times, and around a hundred other studies have been published about links between this gene, stressful life events, and depression risk, with some confirming a link, others finding no such evidence. Yet a 'meta-analysis' of these different studies failed to confirm the findings of the original study. According to Laura Jean Bierut, who led the meta-analysis, this means that 'we still know that stress is related to depression, and we know that genetics is related to depression, but we now know that this particular gene is not.'[28]

If doubts are being cast on a simplistic 'chemical imbalance' model of depression, are we getting any closer to better understanding this disorder? One new theory of depression is that it is due to a lack of new neurons being regenerated. We saw earlier how the idea that such 'neurogenesis' only occurs during development of the brain in the embryo, foetus, and child, but not in the adult brain, has recently been challenged. And a study by David Gurwitz and colleagues at Tel Aviv University has indicated that rather than a shortage of serotonin, a lack of growth of new synapses and reduced generation of new neurons could cause depression.[29] The researchers found that cells in culture exposed to the SSRI drug paroxetine produce a protein called integrin beta-3, which plays a role in cell adhesion and connectivity, new synapse formation, and neurogenesis. It may be that lower serotonin levels are more a consequence than cause of the disorder, since when the brain stops making new neurons, or fewer neuronal connections are formed, this leads to a decrease in serotonin release.

Intriguingly, the idea that depression is associated with problems of neurogenesis may explain the mechanism of action of a new drug used to treat depression—ketamine. This was developed as an anaesthetic, but it is also used as an illegal recreational drug, particularly on the nightclub scene, because of its tendency to make the user feel 'out of themselves' by producing an apparent dissociation between mind and body. Ketamine has recently been shown to have very positive effects in some people with severe depression who had not responded to other treatments. John Krystal, a psychiatrist at Yale University, sees such findings as 'a game changer'.[30] And ketamine appears to act, according to Krystal, by triggering 'reactions in your cortex that enable brain connections to regrow'.[31]

Other recent studies have suggested a link between depression and inflammation.[32] We saw earlier how abnormalities in how glial cells respond to brain inflammation may be one factor in the development of schizophrenia. In depression, inflammation may affect the brain in a different way, sparking feelings of hopelessness, unhappiness, and fatigue. Such feelings may be caused by the immune system failing to switch off after a trauma or illness, and triggering a severe version of the low mood people can experience when fighting a virus like flu. Indeed, recent studies suggest that treating inflammation can alleviate depression, while treatments that boost the immune system to fight illness can be accompanied by a depressive mood—for instance, many people feel down after a vaccination.[33]

Based on such findings, Ed Bullmore, a psychiatrist at the University of Cambridge, believes a new field of 'immuno-neurology' is on the horizon. For him, it is clear that inflammation can cause depression: 'We give people a vaccination and they will become depressed…In experimental medicine studies if you treat a healthy individual with an inflammatory drug, like interferon, a substantial percentage of those people will become depressed. So we think there is good enough evidence for a causal effect.'[34]

If such a viewpoint is correct, it comes tragically too late for my sister, whose ultimately fatal depression may have been triggered by the interferon that was prescribed to protect her from skin cancer. Yet hopefully, such new insights will mean that we should not only be able to devise better treatments for depression, but also identify potential problems with drugs that may have dangerous effects in certain individuals.

Other studies have supported the idea that in a depressive brain, the ability of neurons to connect appropriately becomes abnormal, creating or enhancing low mood, destructive thinking, and other symptoms. Increasingly, neuroscientists believe that the key to understanding depression lies in identifying how the disorder affects neural circuitry and the way that widely separated brain regions communicate through the long-range projection of nerve fibres.[35] According to this view, mental disorders result from the disruption of the larger circuit wiring of the brain. And indeed recent studies have shown that certain people with depression have reduced volume in brain regions used for emotional processing, and fewer neural 'couplings' within these regions.[36]

Yet confusingly, some depressed brains may in contrast use their connections too much, as shown by other studies that indicate that the limbic and cortical regions of the brain in certain depressed patients, which are related to emotional processing, send a 'barrage' of messages at much higher volume than normal.[37] Not only are the connections haywire, but also the messages coming down the lines are abnormal. Intriguingly, given what I have said in previous chapters about evidence showing the importance of brain waves as coordinators of consciousness, there is evidence that such waves may be abnormal in depressed individuals.

What about the idea that certain individuals are biologically prone to depression? This idea has more than scientific interest for me, given my family's history of unnatural deaths, which also includes my maternal grandfather, who died in miserable circumstances in what was then known as a 'lunatic asylum', after he lost first his hairdresser's business and then his sanity in the Great Depression of the 1930s. Yet attempts to find a genetic basis for susceptibility to depression have proved even more confusing than similar investigations for other mental disorders, with, until recently, no clear gene variants being linked to depression.

One problem for such an investigation is that depression is a very common disorder, with more than 350 million people worldwide affected.[38] The disorder's symptoms and severity can vary widely from one person to the next, and also between men and women. This means that there may be a whole spectrum of different conditions that are being grouped together that have quite different biological causes, complicating genetic analysis.

To address this, a recent study led by Jonathan Flint of the University of Oxford focused on depressed women in China.[39] Flint reasoned that because depression tends to be underdiagnosed in that country, women identified as clinically depressed might have a particularly severe form of the disorder. And, indeed, those taking part in the study were at the most severe end of the spectrum. The study identified DNA sequence variations in two genes—LHPP and SIRT1—as being clearly different in many of the depressed women.

In fact, the findings did not initially cast much new light on the biological basis of the disorder, since LHPP's function in the cell remains unclear, while SIRT1, one of a class of genes called sirtuins, is known to play a role in mitochondria, the subcellular structures that produce energy in our cells. Although this led Stanford University psychiatrist Douglas Levinson to say this is 'an appealing bit of biology for a disorder that makes people tired and unmotivated', there is surely more to depression than simply a lack of energy.[40] But more recent studies suggest that SIRT1 may act to protect neurons from ageing and death, so this could be important given that, as we have seen, depression has been linked to inflammation and also to defects in the brain's ability to regenerate itself through neurogenesis.[41]

Another problem in identifying a biological basis for depression in humans is that there is a limit to the type of experiments that can be carried out in our species, for both practical and ethical reasons. Therefore, studies of brain biology and function in animals will surely continue to be central to experimental psychology. And the development of new gene-editing technologies is dramatically expanding the range of species in which the role of genes in brain function can be investigated, as well as increasing the ease with which genetically modified versions of such species can be generated. It is important to restate the importance of such 'models' of human brain function and disorder, because while valuable insights have been gained through the study of human neurons and glial cells in culture, including human brain organoids, these remain pale substitutes of the actual living brain. Yet the view of human consciousness developed in this book also raises questions about the extent to which animal studies provide a route to understanding that consciousness.

One such issue is human uniqueness and how it affects the interpretation of studies of animal brains. For in contrast to views of consciousness that see human consciousness as one end of a spectrum, the view I have been developing suggests that while we can learn much from studies of living animal brains and animal behaviour, the qualitative difference between us and other species means there will be significant limitations to what can be learned from animal studies alone. Unfortunately, some neuroscientists have a tendency to talk about studies of 'depression' or 'schizophrenia' in, for instance, rodents in a rather uncritical way.[42] Now it is quite likely that new molecular and cellular mechanisms that play a role in these disorders in humans can be uncovered in such studies, and that is why throughout this book I have made my own references to animal studies. Nevertheless, it is also important to be cautious about the interpretation of such studies, because whatever interesting insights we might make about abnormal brain function and behaviour in a rodent, it is a great leap to then say we have identified something equivalent to a human mental disorder.

This caution seems justified when we consider what is being measured in a study of say 'depression' in mice. Assessing such a condition in mice is hampered by the fact that, as Michael Kaplitt of Cornell Medical College puts it, 'a mouse can't tell you how it's feeling.'[43] Instead, mice are typically subjected to a 'forced swim' test, in which they are placed in a water-filled cylinder. The longer the mouse swims, the stronger it is judged to have a will to survive. If it quickly gives up—at which point it is rescued—it is deemed 'depressed'.[44] In another test, the mouse is held upside down by its tail. The longer it struggles, the less it is judged to show 'depressive' behaviour.[45] Such studies have been used to identify the role of genes in depression, and also to test drugs, for instance new types of SSRIs—the drug class mentioned earlier of which Prozac is a member.

Now clearly there may be differences in opinion among readers of this book about the rights and wrongs of carrying out scientific experiments on animals. My personal opinion is that if such studies take us closer to understanding mental disorders such as the one that led to my sister's death, and therefore the possibility of identifying new types of drugs that might save the life of someone suffering from such a condition, they are entirely valid. However, given the emerging picture of the complex genetic basis of depression, and the various social determinants of this condition that I will discuss shortly, it seems wise to be critical about overly simplistic assumptions about the possibility of modelling such complexity in a mouse.

A complicating factor for any attempt to identify a biological basis to depression, or for that matter any mental disorder, is the important role that the social environment plays in the genesis of such disorders. In 1978, George Brown and Tirril Harris of Bedford College studied the incidence of depression in London and concluded that the best predictor of this disorder is being a working class woman with an unstable income and a child, living in a tower block.[46] One conclusion is that such a stressful environment is far more likely to precipitate depression than a less stressful one. But another factor that might trigger depression, given what I have said previously about the importance of social interaction in human society, is some disruption to such interaction. And indeed Brown and Harris found that 89 per cent of the depressed women they interviewed had suffered a life-changing event such as family bereavement, a marriage break-up, or loss of a job after a period of full employment.[47] And while all of these events would presumably be highly stressful, they could also lead to a big reduction in an affected individual's social interaction.

Another study that investigated the incidence of depression in New York in the early 1960s found that this condition was more prevalent in people of lower socioeconomic status, but the 'loneliness, the isolation, the lostness…of urban life' was also an important factor.[48] But this does not mean that biological susceptibilities for depression can be ignored. Moreover, it would be a mistake to believe that the influence of biology and environment is easily separable, for as we have seen, it is becoming clear that the environment can directly affect the genome through epigenetic mechanisms. Not only can such environmental influence affect the epigenome of a particular individual during their lifetime, but also more controversially there is evidence that environmental effects can affect the epigenomes of their offspring.

A dramatic illustration of how anxiety—a condition often associated with depression—might be passed down through the generations in an epigenetic manner came from a recent study led by Tracy Bale of the University of Pennsylvania.[49] This found that anxiety induced in male baby mice by exposing them to stressful stimuli—such as fox odour, restraint in a conical tube, and white noise—was passed down to future generations. Bale and her colleagues showed that the anxiety is transmitted to the offspring of the stressed mice by specific micro RNAs—one of the classes of regulatory RNAs. Not only were nine specific micro RNAs found at higher levels in the sperm of the stressed male mice, but also injection of purified forms of these micro RNAs into a fertilized mouse egg generated from unstressed parent mice led to the birth of anxious mice in the next generation.[50]

A further study by Bale's team showed that stress causes an increase of the hormone cortisol in the blood, which triggers the release of stress-inducing micro RNAs from a structure called the epididymis in which sperm are stored after production by the testicles.[51] In an intergenerational feedback loop, the micro RNAs transmitted to the next generation affect genes that regulate blood cortisol in that generation. Bale hopes that having identified 'signals' that carry stress to future generations, it will be possible to study 'how would you reverse those signals, how dynamic are the cells to this environment…and how the interaction of mom's and dad's genes affects [the transmission of stress].'[52]

Of course, people are not mice. And an obvious obstacle in trying to study whether similar molecular mechanisms transmit anxiety across generations in our species is that it would be hardly ethical to induce stress in human baby boys and then see whether this is transmitted to the next generation. So investigations of the role of epigenetic mechanisms, acting across generations, as a possible factor in the genesis of human mental disorders have had to focus on individuals who have been exposed to extremely stressful situations and then assess what might be different about their epigenomes and those of their children compared to unaffected individuals and their offspring.

Undoubtedly, the most horrific event in twentieth-century history was the Holocaust of World War II, in which the Nazis murdered as many as 15 million people—mainly Jews, but also Slavs, gypsies, LGBT+ individuals, the disabled, and political opponents of Hitler and his party—in concentration camps across Germany and occupied Poland.[53] Those who survived the Holocaust were deeply traumatized by their experience, resulting in an increased incidence of anxious and depressed individuals among the survivors. But it has been claimed that Holocaust survivors' children are also more likely to develop post-traumatic stress disorder, depression, and anxiety, despite having grown up in a non-stressful environment.

Moreover, a recent study led by Rachel Yehuda at the Icahn School of Medicine at Mount Sinai, New York, claimed to have found evidence that this apparent higher incidence of mental disorder is associated with epigenetic differences in the DNA of Holocaust survivors' children.[54] In particular, Yehuda claimed that the children of male Holocaust survivors who suffered post-traumatic stress disorder had higher methylation of a gene involved in the stress response. Recall that such methylation changes can affect the degree to which a gene is turned on or off. According to Yehuda, 'the gene changes in the children could only be attributed to Holocaust exposure in the parents.'[55]

Subsequently though, Yehuda's findings and conclusions have come in for criticism from various directions.[56] One criticism is that only a small number of individuals—32 survivors and 22 of their offspring—were studied. Another potential flaw is that the study did not consider the possibility that survivors might—deliberately or inadvertently—have communicated the horror of their experience to their children while the latter were growing up, thereby exposing them to stress. As Josie Glausiusz, a participant in Yehuda's study, and whose father survived the Bergen-Belsen concentration camp put it, 'I was troubled by a question: How does one separate the impact of horrific stories heard in childhood from the influence of epigenetics?'[57] Finally, there has been criticism of the fact that the study only investigated a small number of genes and only found a small amount of change in the methylation states of the genes studied. Reflecting such concerns, John Greally, a geneticist at the Albert Einstein College of Medicine in New York, has remarked that 'the story being told by the Holocaust study is indeed fascinating as a scientific possibility, and will no doubt prompt others to pursue similar studies. Unfortunately, the story is typical of many in the field of epigenetics, with conclusions drawn based on uninterpretable studies.'[58]

Claims about the role of epigenetic influences in the genesis of mental disorders in humans will always need to be scrutinized carefully, and with consideration of potential alternative explanations. Nevertheless, the increasing number of robust findings emerging from animal studies mean that we should pay attention to the possibility that social influences might have impacts reaching beyond the individuals exposed to them. But the way in which such influences might affect the mental state of an individual, or future generations, also needs to take into account the particular structure of current society. This we will do shortly, but before we do, I would first like to look at another important question relating to mental disorders: whether some mental conditions characterized as disorders in the past are purely negative phenomena, or whether we run the risk by such a judgement of discounting what may reflect important diversity in the human population and might indeed be a vital component of what makes us human and enriches our species.

CHAPTER 14

NORMALITY AND DIVERSITY

————◦————

Throughout this book, I have mentioned two main ways of viewing mental disorders. First, there is what is often known as the 'biomedical model', which sees such disorders as due to inherent biological differences in individuals, and ultimately to differences in the genomes of such individuals.[1] Second, there is the view that mental disorders are a product of disturbing events that happened to an individual in the past or are currently a problem, whether trauma, problematic family relations, or the general pressures of living in an advanced capitalist society.[2] This latter stance on mental disorders takes a number of different forms, ranging from Freudian psychoanalysis, which views such disorders as products of the repression of disturbing feelings—primarily sexual ones—to behaviourism, which sees them as dysfunctional responses to past and current events. To complicate matters further, we have seen that recent findings in epigenetics make it likely that such a separation of biology from environment is an illusion, since not only can an individual's environment lead to epigenetic changes in their genomes, but also these may even be passed down to future generations.

Despite their differences, these approaches are united in viewing such mental conditions as abnormal. Yet is it really the case that everything termed a mental 'disorder' is indeed so? Or could it be that in diagnosing conditions that affect the mind, we are ignoring the possibility that some 'disorders' may be part of the normal spectrum of human diversity? To take this argument further, could it even be the case that such diversity is an important component of human society and a source of some of its richness and achievements? In this chapter I will look at these possibilities,

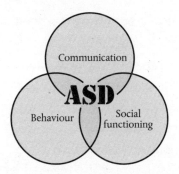

Figure 25 Autism spectrum disorder and three main areas affected.

with particular reference to two quite different mental conditions—autism spectrum disorder and bipolar disorder.

Autism spectrum disorder is estimated to affect 1 in 100 people of all ages in Britain.[3] Autism spectrum disorders have been described by Uta Frith, a developmental psychologist at University College London, as characterized by 'impairments in social interaction and both verbal and non-verbal communication, along with restricted, repetitive or stereotyped behaviour' (Figure 25).[4] Studies in the 1980s by Frith and her colleagues Simon Baron-Cohen and Alan Leslie led to the proposal that autistic individuals differ in their 'theory of mind'. This mental ability allows us to anticipate other people's behaviour in terms of their intentions—what they think and/or want another person to do—and more profound mental states: thinking, believing, knowing, dreaming, cheating, and so on.

Theory of mind allows us to explain and anticipate the behaviour of those around us. Frith and colleagues argued that autistic individuals find it much more difficult to anticipate the behaviour of others.[5] This seems to be due to the fact that most people can infer information about social behaviour that is not explicit, but a person with autism has more difficulty doing so. Autistic individuals need to break the totality of social behaviour into small units in order to understand and learn its often unstated rules and inconsistencies.

Other theories have been put forward to try and explain the restricted, repetitive, or stereotyped behaviour characteristic of many autistic individuals. For instance, it has been proposed that the cognitive skills that allow us to organize ourselves, be flexible, anticipate, plan, set objectives and goals, control our impulses, and so on are altered in autistic individuals. In addition, such individuals seem more attuned to processing local information before global information. Because of this, people on the autistic spectrum tend to first focus on specific details of images or objectives rather than on their entirety. But their tendency towards repetitive behaviour may also reflect their need to look for and create patterns in order to find a safe space within a social world that they may find threatening.

Autism was first described independently in 1943 by two paediatricians of Austrian origin: Leo Kanner, based in Baltimore, USA, and Hans Asperger, in Vienna.[6] For a long time, it was thought that Kanner's autistic children were far more disabled by their condition, since they had delayed or non-existent speech. In contrast, Asperger described his charges as 'little professors'—fluent if pedantic speakers with idiosyncratic interests.[7] In reality, not all of Asperger's autistic patients were so academically gifted and they displayed a wide range of deficits as well as strengths, but for tactical reasons he chose to focus on the four most talented children, stressing their potential for great achievement, given appropriate education.

Asperger may have been trying to protect his autistic charges at a time when Nazi eugenic policy viewed physically and mentally disabled children as 'lives unworthy of life'.[8] Under the T2 'euthanasia' programme, more than 5,000 such children were murdered by means of lethal injection or starvation. However, Asperger's precise historical role is currently a matter of controversy, with recently uncovered evidence suggesting that he also helped identify some children with disabilities and sent dozens to Spiegelgrund, a children's ward in Vienna where adolescents were euthanized or subjected to brutal experimentation.[9]

Notwithstanding these new revelations, Asperger was definitely positive about 'autistic intelligence', seeing traits in both artists and scientists, and observing that 'not everything that steps out of line, and is thus "abnormal", must necessarily be "inferior".[10] However, his findings languished in largely unread German texts until their translation into English in 1981. In contrast, Kanner's views on autism became well known soon after their publication in the USA in 1943, and for a long while dominated discussion about the topic.

Initially, Kanner suggested that the basis for the social difficulties associated with autism lay in a child's inability to form emotional ties with their parents.[11] This view—particularly influenced by Freudianism—reflected a long-held assumption in psychology that children's initial relationship with their parents forms a blueprint for all other relationships. Kanner, at first, suggested that an autistic child was biologically impaired in this capacity. But subsequently he began laying a large degree of blame for autism on 'toxic' parenting. He described the parents—particularly mothers—who brought their children to him as 'emotional refrigerators', whose lack of warmth contributed to their sons' and daughters' retreat into their own worlds. Although Kanner eventually moved away from this position and acknowledged that autism had a strong inborn element, his initial viewpoint unfortunately shaped ideas about autism for decades. Parents of autistic children had to deal with not only the challenges of bringing up an autistic child in an era when few schools would accept such children, but also battling their guilt on the psychoanalyst's couch.[12]

Nowadays it is rarer to hear parents of autistic children blamed for their child's condition. However, recently a spotlight has been directed on what has been called the 'state scandal' of treatment of autistic individuals in some parts of France. Muhamed Sajidi, president of the French organization Conquer Autism, argues that 'today everyone knows that autism is a neuro-developmental problem. It is not a psychosis or mental disorder. But in France it is the psychiatrists—heavily influenced by Freudian psychoanalysis—who remain in charge. And they have shut themselves off from all the changes in our knowledge of autism.'[13] Sajidi set up the organization after his life was 'destroyed', as he puts it, by the medical establishment's failure to diagnose his son's autism. He claims that in France blame for autism is often still laid at the door of the child's parents, especially the mother. In a similar vein, Candy Lepenuizic, a British woman married to a Frenchman, has noted that 'the first time I went to see a doctor when my (autistic) son Gael was three and we thought there was a problem, the psychiatrist asked me...if it had been a wanted pregnancy! Then she asked what sort of dreams I had had while I was pregnant with him...They thought that if the child was failing to communicate with the outside world, it was because of some trauma in the womb or in very early life. There was a family malfunction, and we had to cure it!'[14]

Such instances apart, in general there is now much more acceptance of the idea that autism has a strong inborn element. However, it remains unclear to what extent this is due to genetic differences, or epigenetic or other effects on the embryo or foetus in the womb. Such uncertainty may reflect the fact that autism spectrum disorder is exactly that—a spectrum—which may reflect different underlying causes for different types of autism. Such complexity is indicated by the multiple regions of the genome that have been linked to autism in the same type of genome-wide association studies mentioned previously, and also by studies of neurons derived from autistic individuals. We saw earlier how by taking a skin cell from an individual, and using this to derive first a pluripotent stem cell, then neurons, it is possible to assess how genetic differences in individuals might affect neuronal and brain structure and function. And two recent studies of this type have illustrated just how varied might be the underlying molecular causes of even the most severe types of autism.

The first study, led by Lauren Weiss at the University of California, San Francisco, derived neurons from people with a deletion, and those with a duplication, of the same region of chromosome 16, both types of muta-tions being associated with severe forms of autism.[15] In both cases neurons have fewer synapses than do those from unaffected individuals. But neurons derived from people with the chromosome deletion were found to have unusually large cell bodies and very long dendrites (the parts of the neuron that receive inputs) while those from people with a duplication in this region were atypically small and had shortened den-drites. The findings are interesting given that people with a deletion of the chromosomal region have macrocephaly (abnormally large heads), whereas those with an extra copy of the region have microcephaly (small heads). It is clear then that, although changes in chromosome 16 lead to some common effects, the specific nature of the changes leads to effects characteristic of a particular type of autism.

The second study, led by Christian Schaaf at Baylor College of Medicine in Texas, derived neurons from three people with a deletion of part of chromosome 15, three with a duplication of this region, and three unaffected individuals.[16] One gene in this region, CHRNA7, codes for a protein channel that allows calcium ions to flow into neurons, such flow being essential for neuronal function. When the researchers exposed the cells to compounds that cause the channels to open, less calcium flowed into neurons lacking a copy of CHRNA7 than flowed into normal neurons. This suggests that children with CHRNA7 deletions have too few functional calcium channels on neurons.[17] However, surprisingly, neurons from children with an extra copy of the CHRNA7 gene also show unusually low calcium influx. In this case, too much CHRNA7 protein impairs neurons' ability to assemble the proteins and transport them to the cell surface. The results illustrate that the effects of mutations are not always predictable. Schaaf sees this as 'kind of humbling. We actually need to do that basic science research to understand the consequences of these genetic changes.'[18]

Such studies are of severe forms of autism with a clear genetic link, yet even so they reveal heterogeneity in molecular and cellular causes. Identifying how these genetic differences manifest themselves will be important in devising new ways to treat such forms of severe autism. Indeed, a study led by Riccardo Brambilla of the University of Cardiff has identified a drug treatment that reverses the effects of the type of autism associated with a deletion in chromosome 16 mentioned above—in a mouse model of this disorder.[19]

Brambilla's team discovered that this deletion causes loss of expression of a gene called ERK1, and this in turn leads to abnormal overexpression of a related gene, ERK2.[20] Drugs have been developed to inhibit ERK2 because it plays a role in certain types of cancer, and using such drugs to treat mice engineered to mimic this form of autism, Brambilla's team reversed the symptoms of the disorder. Moreover, female mice pregnant with foetuses with this genetic defect, but treated with such drugs, gave birth to off-spring without autistic symptoms. Brambilla believes this shows that while it would not be advisable to treat pregnant women carrying an affected child, it might be possible 'to permanently reverse the disorder by treating a child as early as possible after birth. In the case of adults with the condition, ongoing medication would probably be required to treat symptoms.'[21]

Despite these findings, the majority of less severe autistic conditions are not clearly defined genetically. There is therefore the possibility that they have a less obvious biological mechanism underlying them, and are more linked to environmental factors. In recent years there have been highly misleading claims about such a link. Notably, a British doctor called Andrew Wakefield published a paper in *The Lancet* in 1998—since retracted by the journal—claiming a link between the vaccine for measles, mumps, and rubella (MMR) and autism.[22] This prompted large drops in vaccination rates. One consequence was that measles cases were shown to have risen by 300 per cent in Europe in 2017, the result of parents shunning vaccines following Wakefield's claims.[23] In fact, such claims have been totally debunked and Wakefield's UK medical licence was revoked. Yet that has not stopped him building a successful career as a leading figure in the 'anti-vaccine' movement in the USA; this movement was given a boost by supportive statements by President Donald Trump.[24]

One reason why Wakefield's claims resonated with anxious parents is that the reported incidence of autism has increased over recent decades. Parents desperate to find an easy explanation for the disorder that might lead to some kind of 'cure' have become an easy target for people claiming to have a simple remedy. For instance, some parents have resorted to giving their autistic children a solution of chlorine dioxide, sold under the name Miracle/Master Mineral Solution (MMS) but basically diluted bleach.[25] The popularity of MMS is attributed to Jim Humble, who heads the 'Genesis II Church of Health and Healing'. Humble claims that autism is caused by viruses and bacteria, and that MMS kills off those pathogens and cures the condition.[26] Not only has this claim no scientific validity, but giving bleach to a child is highly dangerous. Jeff Foster, a British clinician in Leamington Spa, is highly critical of this treatment: 'Autism is a neuro-developmental disease…developed in the womb or early stages of life. You can't just reverse it and anyone claiming that does not understand the condition. When you have very extreme measures like this to "cure" a condition it's just a roulette game. Eventually someone will die.'[27]

In fact, there is a different approach to autism that, far from seeing it as a disorder to be 'cured', instead views at least less severe cases as part of the natural spectrum of human characteristics. Such an approach highlights the positive, rather than negative, aspects of autism. But it also sees the development of the full potential of an autistic individual as a key issue for society as a whole, rather than simply a problem for that individual or their parents.

In many ways the recognition that autism can have positive aspects is not a new one. Ever since Asperger's identification of high-achieving autistic children became more widely known about in the 1980s, there has been increasing recognition that some people with autism can be highly creative individuals.[28] For instance, actor Darryl Hannah has spoken about her struggles with autism since she was a child. Tito Mukhopadhyay is virtually mute, but the eloquent poems he writes by hand or types out have provided a window into how an autistic person can experience life. The artist Stephen Wiltshire has become well known for his ability to draw landscapes in exact detail even after seeing them only once. And many IT experts in California's 'Silicon Valley' have been noted to have autistic traits.[29]

However, there is a potential problem with a focus only on high-achieving autistic individuals, since this can lead to the idea that people who do not succeed in such conventional terms are somehow less worthy of attention. Instead it is important that we consider ways in which the potential of every autistic individual might be boosted. To do so, we need to recognize the positive aspects of autism, but also consider how educational measures might address the more negative features of the condition. To some extent this means widening our definition of what it means to be creative.

In general, autistic individuals have been perceived as being more rigid in their behaviour. This rigidity is linked to the difficulty they often have in judging the actions and intentions of others, and their tendency towards restricted, repetitive, or stereotyped behaviour. Because of these characteristics, it is often assumed that autistic individuals will also be more rigid in their imaginative and creative processes. Yet recent studies have tended to undermine this idea, and instead have suggested that given the right opportunities people with autistic traits can come up with unusually creative ideas.

For instance, a study led by Catherine Best at the University of Stirling assessed creativity in autistic and non-autistic individuals by asking them to generate ideas about static objects such as paper clips and abstract images.[30] Participants were expected to create their own context; the objects were not defined by their use in society. Those with autistic traits offered fewer responses to problems presented, but their solutions were more original and creative than those without such traits. For instance, when asked to list possible uses for a paper clip, creative responses of autistic individuals included using the wire to support flowers, creating a paper airplane weight, or making a token for a board game.[31]

The finding that autistic individuals are better able to display their creativity and imagination in situations that are less dependent on social context makes sense given previous evidence that people with autism often have trouble interpreting their experiences within such a context. Most of us habitually identify objects by their relationship to other elements of a social situation. For example, we might distinguish between a salt and sugar shaker by the presence or absence of a pepper shaker or a coffee pot. Autistic individuals can find it more difficult to use social cues in this way. Intriguingly, this association of 'opposites' that is a characteristic of much 'normal' life—think also of the way we tend to picture a knife and fork together—is reminiscent of what Lev Vygotsky termed grouping into complexes, which he distinguished from true conceptual thought. So autistic individuals may be identifying contradictions in a lot of what passes for conceptual thought in many everyday situations. What the findings do indicate is that when autistic people are shown images on a context-less background, they can visualize contexts not usually associated with those objects, which may be one, but not the only, explanation for their higher degree of creativity.

Jolanta Lasota, chief executive of the charity Ambitious about Autism, is enthusiastic about the findings of Best and her colleagues, remarking that 'there are many misconceptions and myths about autism, the biggest one including being antisocial and having a lack of empathy. However, what people with autism struggle with is fitting their feelings of sympathy and caring into everyday interactions. While it is true that some people with autism can have very specific interests and may struggle with abstract concepts, this research helps to highlight the fact that seeing the world in a different way can be a positive trait too.'[32]

It is one thing to recognize the potential of autistic individuals, but this still leaves the question of how we develop such potential. For although discussion of autism often focuses on highly talented individuals with this condition, the reality is that just 16 per cent of autistic adults in the UK are in full-time work.[33] What is more, this figure has remained the same for the past decade, showing that autistic people are not benefiting from British employment programmes. A similarly large majority of autistic individuals fail to find full-time work in the USA.[34] This suggests that either people with autism are being failed by their schools or there is something else preventing such individuals from entering the world of work.[35]

In terms of education, there is much of potential value in Lev Vygotsky's approach to learning disability.[36] For although his interests in human consciousness were many and varied, one particular area in which Vygotsky played a key practical role was in developing educational programmes for children with special needs in schools in the Soviet Union in the 1920s and early 1930s.[37] He argued that children learn best through shared social activities and when education is geared towards everyday experiences. Such an approach is especially important for children with learning difficulties.[38] An apparent lack of motivation and intellect in some children may conceal an inability to connect the abstract content of school lessons with their own experiences. Vygotsky also believed that a vast potential exists in every child and was fond of mentioning the example of Helen Keller, who, although she became a blind deaf mute at nineteen months old as a result of scarlet fever, nevertheless grew up to become a famous philosopher, scholar, and social activist.[39]

Vygotsky focused on three different areas of special needs education. First were the children labelled 'learning disabled' but whose difficulties stemmed from a deprived social environment.[40] This was particularly important in the nascent Soviet Union in the 1920s as the First World War, then civil war and famine, had left as many as seven million homeless, orphaned, abandoned, and neglected children, many of whom were severely disturbed in their mental development. In extremely forward thinking for this period of history, Vygotsky argued that addressing social problems was central to solving children's learning difficulties.

Second, there were children whose learning difficulties were an indirect consequence of a disorder like deafness or blindness.[41] According to Vygotsky, despite the physical origin of these conditions, it is nevertheless their effect upon the child's integration into society which is most important. He campaigned for entry of disabled children into mainstream education and for their participation in the state youth movement, but also stressed that such integration should not come at the expense of extra resources for special education.

One of Vygotsky's central ideas is that human consciousness has been transformed throughout history by cultural innovations like reading and writing. He believed that society has a duty to develop and propagate the sort of innovations that would allow disabled children to benefit from the wealth of human culture. In Vygotsky's time there were only Braille and various forms of sign language, but computers now surely have revolutionary potential as a powerful interface between severely disabled children and the outside world.

That Vygotsky started from the potential of disabled children rather than from their disability is shown most clearly in his work with children with severe mental disabilities, the third area he dealt with.[42] Such children are often 'written off' as incapable of complex reasoning and thus are given only the simplest of learning tasks. Yet in Vygotsky's opinion, such an approach only reinforced their handicaps. Instead, if children lack well-elaborated forms of abstract thought, the school should make every effort to guide them in that direction.

Vygotsky found that such children best acquired the motivation to take on abstract learning tasks in a cooperative learning strategy involving a division of work in which children tackle different tasks in a group with a shared motive for the entire activity.[43] This provided both the opportunity and the need for cooperation and joint activity by giving the children tasks that were beyond the developmental level of some, if not all, of them.

Such an approach to learning disability has informed more recent approaches to the education of autistic individuals, for instance that developed by Pamela Wolfberg and colleagues at San Francisco State University.[44] Vygotsky viewed the individual and environment as factors that mutually shape each other in an integrated process of growth. So as a child grows older, they interpret the social environment differently according to their development. Since children develop consciousness through the mediation of culture, it is important through education to create conditions in which culture becomes accessible to each child, no matter what their particular needs. Wolfberg and colleagues argue that 'one obstacle in achieving such a goal with autistic children is that their peer interactions, and indeed social interactions in general, are characterized by low rates of both initiation and response.'[45]

The use of non-verbal communication, including gestures and emotional expressions, is also different in autistic children, both expressively and receptively.[46] Such children use fewer non-verbal gestures and a more limited range of facial expressions in their communications than children with other types of developmental disabilities of the same developmental and chronological age. Children with autism appear to pay less attention to other people's emotional displays than do comparison groups and demonstrate fewer acts of empathy or shared emotion. They also engage in less imitation of other people's actions, movements, and vocalizations. Finally, their relatively weak verbal skills, coupled with the social-interactional differences characteristic of autism, can be factors that prevent participation in the type of linguistically mediated social interaction upon which inner speech is dependent, and which Vygotsky saw as so important for development of consciousness.

Wolfberg believes that such factors mean children with autism may benefit if included in cooperative learning groups working together with their peers.[47] Because of their social deficiencies an autistic child can be stimulated with the help of a talented teacher, so that the child can develop his or her skills which might otherwise be unrealized. Vygotsky criticized psychologists and educators who are chiefly concerned with counting and tabulating a child's weaknesses, particularly when they use those measurements as the whole basis for placing the child in one part of an educational programme.[48] In contrast, he argued that the only point of tests should be to develop a child's strengths and talents, which differ for every child.

As an autistic child grows up they have specific differences that affect their behaviour and contact with others in a social environment.[49] Learning in a cooperative group could eliminate the social problems of an autistic child as the child learns how to be an active member of a team. The child in such a group can begin to learn how to use speech in a meaningful environment. The development of speech leads first to the child's social-ization at a cultural level and subsequently to a natural development. Unfortunately, there are still classrooms in which students with disabilities are segregated physically, instructionally, and socially.[50] In such class-rooms, students with disabilities may follow different schedules and have separate physical spaces set aside for them. All of this can inhibit the devel-opment of social interactions between autistic children and their peers.

One particular challenge that autistic children face is during play.[51] While typically developing children need little motivation or guidance to play with peers, autistic children encounter significant obstacles gaining equal access to, and benefits from, inclusion in peer-play experiences. Without explicit support, they are likely to remain isolated and therefore deprived of consistent interactive play experiences that encourage developmental growth and meaningful peer relationships. Developmental delays and differences in underlying capacities for joint attention, imitation, and social interaction are all closely intertwined with an emerging capacity for play. In sharp contrast to the richly diverse social and imaginary pursuits of typical children, the play of children with autism is typified by restricted, repetitive, and stereotyped patterns of behaviour, interests, and activities, which they often pursue in isolation.[52] With few friendships available to them and various social competency deficits, autistic students may have limited ability to cope with the myriad of social interactions present in the school setting, making them susceptible to teasing and bullying. As a result, they can often learn to avoid social contacts, leading to more social rejection. For these students, time with typical peers needs to be organized and planned into the school day.

A completely different mental condition to autism but one in which there is increasing recognition of its positive aspects is bipolar disorder. This condition affects 1 in 100 people in the UK and is characterized by periods of depression followed by episodes of mania, during which affected individuals can feel extraordinarily happy, ambitious, and creative, but also sometimes in this phase they can become psychotic (Figure 26).[53] The high and low phases of bipolar disorder are often so extreme that they interfere with everyday life; consequentially this can be a highly debilitating disorder. However, there is also an association of this mental disorder with great creativity and high levels of intelligence.

For instance, a study by Daniel Smith and colleagues at the University of Glasgow highlighted the links between bipolar disorder and intelligence.[54] The researchers tested the IQ of individuals at age eight. These same individuals were then assessed for manic traits when aged

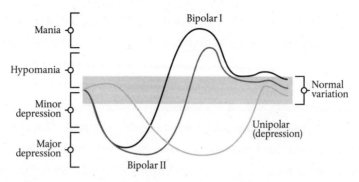

Figure 26 Mood changes in bipolar individuals compared to normal variation.

22 or 23. Each person was given a score related to how many manic traits they had previously experienced. The study showed that individuals in the top 10 per cent of manic characteristics had a childhood IQ 10 points higher than those who scored in the lowest 10 per cent. This correlation was strongest for verbal IQ. Smith claims that the study offers a possible explanation for how bipolar disorder may have been selected for through generations.[55] He hopes the findings will help improve the earlier detection of the disorder, which could then lead to measures 'to help someone at high risk of bipolar, such as making certain lifestyle changes, protecting sleep patterns, [and] avoiding certain stresses'.[56]

Bipolar disorder entails dramatic mood swings between extreme happiness and severe depression. How might this be linked to creativity? James Fallon of the University of California, Irvine, notes that people with bipolar tend to be creative when they are coming out of deep depression.[57] When a bipolar patient's mood improves, their brain activity does too: activity decreases in the lower part of the frontal lobe brain region and increases in a higher part of that lobe. Interestingly, the same shift happens when unaffected individuals have creative bouts.[58] Interpreting how the brain patterns translate into conscious thought, Elyn Saks, a mental health law professor at the University of Southern California, believes that individuals in the manic phase of bipolar disorder do not filter stimuli as well as other people. Instead, they can entertain contradictory ideas simultaneously and become aware of loose associations that most people's unconscious brains would not consider worthy of sending to the surface of our consciousness. Saks believes that while the invasion of nonsense into conscious thought can be overwhelming and disruptive, it can also be quite creative.[59] For example, word association studies, which ask participants to list all the words that come to mind in relation to a stimulus word, have demonstrated that bipolar patients undergoing mild mania can generate three times as many word associations in a given time period as the general population.[60] As for how this leads to strokes of genius, it could be that the sheer bounty of unsuppressed ideas means a greater probability of producing something profound.

Of course, no one is bursting with creative energy during a severe bout of depression. Above all, conditions like bipolar are debilitating and even life-threatening, and although society benefits from the productivity of its tortured geniuses, those individuals do not always consider their moments of brilliance to be worth the extensive suffering. Indeed Saks believes that 'the creativity is just one part of something that is mostly bad.'[61] This is an important point to make lest it be thought that there is anything glamorous about suffering from a mental disorder. As journalist Hannah Jane Parkinson, who suffers from bipolar disorder, has said about the link between the condition and creativity, 'the concept itself isn't a problem; rather, it's the danger inherent in the tortured-genius trope. What about those with mental illness who work in retail, or are civil servants, or pull pints? Do these people not get a look in? How many albums does a person need to make, how many prizes need to be won, before mental health issues become acceptable?'[62] Parkinson also notes how Vincent Van Gogh wrote, 'If I could have worked without this accursed disease, what things I might have done.'

The fact is that the more we learn about mental conditions such as bipolar, but also psychiatric disorders like schizophrenia and clinical depression, as well as autism, the more we are beginning to realize that such conditions seem to be linked to a broad spectrum of biological and environmental triggers, but they are also highly varied in their effects on the individual. What is clearly required, therefore, is more understanding not only of the positive as well as the negative aspects of mental 'illnesses', but also about the diversity of such conditions, and the fact that such diversity overlaps with that of supposedly 'normal individuals'. And, indeed, one positive aspect of the past few years has been an increased willingness for people to talk about their experience of having a mental 'disorder' as well as increased acceptance of people with such conditions. There is, however, one exception to such acceptance, and that is when it comes to the criminally insane, a subject we will now explore.

CHAPTER 15

CRIME AND PUNISHMENT

M y home city of Bradford, Yorkshire, is famous for a number of things. There is its former textile industry, so important in past times that the city was known as the 'wool capital of the world' and T. S. Eliot could write in his 1922 poem *The Waste Land*, about people 'on whom assurance sits, as a silk hat on a Bradford millionaire', satirizing the city's textile barons.[1] Although that industry has now almost disappeared, the people of the Indian subcontinent who came to work in its mills made such an impression that now the city is often referred to as 'Bradistan', and bills itself as 'the curry capital of Britain'.[2] One could also point to the city's artistic heritage, ranging from the novels of the Brontë sisters, who were born in Bradford before moving to nearby Howarth, and the nineteenth-century 'model' workers' village in Saltaire—now a UNESCO site, to David Hockney's artwork.[3] However, one association that Bradford people are less happy to be reminded of is the city's link with serial killers.

In the late 1970s and early 1980s, fear gripped Bradford citizens due to the murderous activities of an individual the media dubbed the 'Yorkshire Ripper', because of the horrific way he mutilated his female victims. In 1981, he was identified as Peter Sutcliffe, a Bradford lorry driver—but only after he had murdered 13 young women and attempted to kill seven others who survived, but were left scarred, both physically and emotionally, for life.[4] There has been subsequent criticism of the time it took to catch Sutcliffe, and especially the fact that he was interviewed by the police nine times as a potential suspect, but let go each time.

One problem was that the police saw the killer as a murderer of prostitutes, despite the fact that many of the victims had no connection to the sex trade. This meant that detectives excluded evidence of attacks on women unconnected with the trade, even if they shared characteristics with the Ripper murders. In the end it was luck—he was stopped for a traffic offence—rather than detective work, that led to Sutcliffe's arrest, prompting Joan Smith, a reporter who covered the case, to recently say that 'the original investigation was a fiasco, and women died as a result.'[5]

Unfortunately, Sutcliffe was not the only serial murderer to plague Bradford in the 1970s. Donald Neilson was a builder who lived in Thornbury, a suburb to the east of the city. I spent the first five years of my life in Thornbury, in a house just around the corner from the street in which Neilson lived, and my sister went to one of his daughter's birthday parties. Yet some years later Neilson was identified as a masked killer named the 'Black Panther' who had become Britain's most wanted man because of his brutal murder of three postmasters during armed robberies of post offices across West Yorkshire.[6]

Neilson acquired his nickname due to his habit of dressing all in black but also because the wife of one of the murdered men described the killer as 'so quick, he was like a panther.'[7] It was also surely not a coincidence that the media adopted this nickname because of its association with the Black Panther Party, the US black activist group that were particularly prominent at the time.[8] Indeed, the white, but masked and gloved Neilson played along with this association by mimicking a Jamaican accent on one raid.

While the murders of the postmasters were horrifying enough, it was the manner of the death of Neilson's final victim that shocked the nation. Lesley Whittle, a wealthy heiress, was only 17 years old when she was abducted by Neilson from her Shropshire home and imprisoned in a drainage shaft while he sought money for her release.[9] Finally, after various attempts to deliver the ransom failed, Whittle was discovered, but in horrific and tragic circumstances. She had been left by Neilson naked, in the dark, on a ladder in the drainage shaft, without food or water, with a wire noose around her neck. What happened next is unclear, but Whittle was found dead, hanging from the noose. On Neilson's eventual arrest—again, mainly due to luck rather than detective skills—he was sentenced to life in jail.[10]

So what was going on in the brain of Peter Sutcliffe when he killed and mutilated women in such a horrific fashion? Or in that of Neilson who— while his main motive was material gain—had such disregard for his victims that he could so brutally dispatch them, while leaving a teenage girl the same age as his own daughter in such torturous conditions, leading ultimately to her death? Was it some biological defect in these individuals, or an event in their past lives, that led them to behave in such a vile and despicable way towards their fellow human beings? These are questions I will explore in this chapter, beginning with the question of whether brain biology is significantly different in a serial killer.

One of the first people to suggest that criminality arose from abnormalities in the brain was Cesare Lombroso, an Italian surgeon who in 1871 conducted a post-mortem of a serial murderer and rapist. Lombroso discovered an unusual hollow at the base of the man's skull, which he claimed was similar to that found in apes. For Lombroso, 'at the sight of that skull, I seemed to see all of a sudden, lighted up as a vast plain under a flaming sky, the problem of the nature of the criminal—an atavistic being who produces in his person the ferocious instincts of primitive humanity and the inferior animals.'[11] He subsequently identified other features that confirmed his prejudice that criminality was a consequence of a reversion to animal and primitive human behaviour. According to Lombroso, thieves had an expressive face, manual dexterity, and small, wandering eyes, murderers had glassy stares, and rapists had 'jug ears'.[12] And to support his belief that criminals display animal-like characteristics, he claimed that they often had irregularly shaped earlobes like those of apes, or a hooked nose, which 'so often imparts to criminals the aspect of birds of prey.'[13]

While Lombroso's theories can be dismissed as a product of a misunderstanding of evolutionary mechanisms and the prejudices of his time, this does not mean that the belief that criminality has a biological basis is no longer held. Indeed, there has been a recent resurgence in this view, focusing on the claim that criminality is a product of biological differences in the brain. The claim is based on genetic analysis and study of the brain itself.

One gene linked to criminal behaviour is monoamine oxidase A (MAOA), which encodes an enzyme that destroys a particular type of neurotransmitter as a way of controlling its activity. The link emerged with a study of a Dutch family which, through five generations going back to 1870, has been full of male criminals, including arsonists, attempted murderers, and a rapist.[14] Han Brunner, a doctor in Nijmegen, examined the family and found a shared defect in the MAOA gene, which is on the X chromosome. Since men have only one X chromosome, such a defect could affect males, but be passed down by unaffected female carriers. When Brunner tested the men's urine, he found low levels of MAOA metabolic breakdown products. A further study suggested that violence associated with the defective variant was significantly worsened if the perpetrator had been sexually abused as a child, whereas abused children with a normal MAOA gene were less likely to be criminals.

The link between violent crime and MAOA was strengthened by another study, by Jari Tiihonen of the Karolinska Institute, Sweden, who investigated almost 900 violent repeat offenders in Finland. Tiihonen thinks one way MAOA deficiency might work is that it could result in 'dopamine hyperactivity' especially if an individual drinks alcohol or takes drugs such as amphetamines (Figure 27).[15] However, Jan Schnupp of City University, Hong Kong, has criticized claims of a strong link between MAOA and criminality, commenting that up to half the population could have one of the gene variants involved. He notes that 'to call these alleles "genes for violence" would therefore be a massive exaggeration. In combination with many other factors these genes may make it a little harder for you to control violent urges, but they most emphatically do not predetermine you for a life of crime.'[16]

Studies like those above highlight the potential pitfalls in trying to relate a complex human behaviour—in this case criminality—to one or a few genes that play roles in the brain. But what about studying the brain itself— can this lead to insights into the mind of a criminal? One scientist

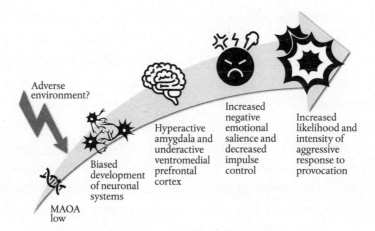

Adverse
environment?

MAOA
low

Biased
development
of neuronal
systems

Hyperactive
amygdala and
underactive
ventromedial
prefrontal
cortex

Increased
negative
emotional
salience and
decreased
impulse
control

Increased
likelihood and
intensity of
aggressive
response to
provocation

Figure 27 Proposal for how low levels of MAOA are supposed to be linked
to crime.

dedicated to finding such a link is Adrian Raine, born in Darlington,
England, but now a professor at the University of Pennsylvania. Raine did
his PhD in Oxford with Richard Dawkins, author of *The Selfish Gene*, who
convinced him of the 'all-embracing influence of evolution on behaviour'.[17] Eager to apply this idea to the study of criminality, but finding little
support in Britain for such a project, in the 1980s Raine moved to the USA.

In 1994, Raine used positron emission tomography (PET) to scan the
brains of convicted killers and those of a control group of people of a
similar age and profile. Intriguingly, the analysis—which reveals changes
in metabolic activity in different parts of the brain—showed what
appeared to be a significant reduction in the development of the prefrontal
cortex in the murderers' brains.[18] This is a region which, as we have seen,
has been proposed to mediate higher mental functions. Raine suggested
that such differences may indicate that murderers have less control over
the limbic system that generates primal emotions such as anger and rage;
a greater addiction to risk; a reduction in self-control; and poor
problem-solving skills—all traits that might lead a person to violence.

Yet the difficulties in demonstrating that such differences determine an
individual's likelihood of becoming a criminal became evident when
Raine scanned his own brain. For, somewhat to his alarm, he found that

he shared more characteristics with the murderers than with the control group![19] Not only that, but Raine has other indicators that he believes predict future criminal behaviour, one being a markedly low heart rate—the idea being that people with a slow resting heartbeat have lower levels of what psychologists call arousal, or the feeling of being awake and alert, and therefore seek out stimulating experiences, which could include risky behaviours and criminal activities, to boost their arousal.

Indeed Raine has admitted that 'from age nine to 11, I was pretty antisocial, in a gang, smoking, letting car tyres down, setting fire to mailboxes … [but then at] 11, I changed schools, got more interested in studying and really became a different sort of kid.'[20] But this surely raises questions about whether biology can 'determine' criminal behaviour, if the same biological and juvenile behaviour profile can occur in criminals, but also in a neuroscientist studying such individuals. And make what you will of the fact that I share with Raine not only our academic profession, but also aspects of his youthful behaviour: as a teenager I too was a juvenile delinquent, being a prolific shoplifter for several years. And I also have at least one of the biological features that Raine has linked to criminal behaviour—a pulse rate that is under 57 beats per minute.[21]

One problem in looking for biological differences as the primary cause of criminality is that this could mean less acknowledgement of the role of environment in the development of a criminal personality. Indeed, this is admitted by Raine, who has said that 'the sociologist would say if we concentrate on these biological things, or even acknowledge them, we are immediately taking our eyes off other causes of criminal behaviour—poverty, bad neighbourhoods, poor nutrition, lack of education and so on. All things that need to change. And that concern is correct. It is why social scientists have fought this science for so long.'[22]

But it also means that both Raine's and my own social circumstances could have had a major impact on how we turned out in life. While I cannot speak for Raine, certainly my family's love and support helped turn me away from petty crime as a teenager. My sister in particular played an important role, for when I was caught shoplifting by a store detective, it was Kay's pleas that persuaded her not to hand me over to the police. If I had been charged with a criminal offence, who knows how different my life might have turned out. And this raises important questions about the role of biology versus environment, but also that of chance and circumstance, as important determinants of criminality and anti-social behaviour.

Of course, trauma during childhood might itself affect brain function. This idea has received support from anecdotal evidence that several notorious US serial killers experienced physically traumatic situations that may have affected their brains.[23] For instance, Ed Gein, a murderer who fashioned clothing and furniture out of women's body parts, was repeatedly beaten around the head by his father. Gary Heidnik, who kidnapped, tortured, raped, and killed women whom he held prisoner in a pit in his basement in Philadelphia (an activity recreated in the film The Silence of the Lambs), fell from a tree at six years old and hit his head hard enough to deform his skull. Jerr Brudos, who killed four women and had sex with their dead bodies, was electrocuted as a 21 year old and complained of recurring headaches and blackouts thereafter. And Edmund Kemper, a murderer and necrophiliac who killed ten people, suffered a head injury after crashing his motorcycle.

To investigate whether brain trauma might make an individual more likely to become a criminal, a recent study led by Ryan Darby of Vanderbilt University in Nashville, Tennessee, imaged the brains of individuals whose criminal activities ranged from fraud and theft to assault, rape, and murder and who had also suffered injuries to their brains because of trauma, stroke, or tumours.[24] Darby claimed to have found evidence that the injuries caused impairments in brain regions involved in 'morality, value-based decision-making, and theory of mind'—these being mental attributes that allow us to understand the mental processes and perspectives of others.[25] Commenting on the findings, Winnifred Louis, a psychologist at the University of Queensland, Australia, believes they suggest that 'the lesions are biasing people towards more utilitarian decisions', with the brain impairments leaving intact the ability of the brain to make reward calculations for the self, but having a negative impact on the ability to understand how it would feel to be in the position of the victims of crime.[26]

Yet before we get too carried away with the idea of a link between brain injuries and criminal behaviour, it is worth re-examining the lives of some of the US serial killers mentioned earlier. For as well as suffering injuries to their heads, causing possible brain damage, many also experienced substantial psychological trauma as children. Take Ed Gein. As well as being regularly beaten by his father, Gein had an unhealthy relationship with his mother, the nature of which partly inspired Alfred Hitchcock's film *Psycho*.[27] Gein's mother verbally abused her son and taught him that all women, apart from her, were prostitutes and instruments of the devil. Yet Gein worshipped his mother, referring to her as a 'saint'. Edmund Kemper's mother belittled, humiliated, and abused her son, often making him sleep in the basement. She refused to cuddle him for fear she would 'turn him gay', yet also told Kemper no woman would ever love him.[28] Gary Heidnek was punished by his father for bed-wetting by being dangled from a window by the ankles.[29]

In my home city of Bradford, Peter Sutcliffe was apparently deeply affected by growing up in a religious Catholic background at the same time as seeing his alcoholic father often abuse his wife, Peter's mother, including calling her a cheat in public.[30] It seems possible that such abusive environments may have contributed to shaping the individuals that carried out such heinous acts, even if brain injury may also have played a role in some individuals. And while the examples above are anecdotal, a systematic study by Heather Mitchell and Michael Aamodt of Radford University, Virginia, has found that serial killers are six times more likely to have suffered abuse as children than the general population.[31]

Besides serial killers, maybe nothing is as likely to generate revulsion in society as paedophilia. Although girls typically begin puberty at age 10 or 11, and boys at age 11 or 12, paedophiles are generally defined as individuals attracted to children under 13. A person must be at least 16 years old, and at least five years older than the prepubescent child, to be diagnosed as a paedophile.[32] This behaviour triggers revulsion in others not only because of the perceived unnaturalness of the attraction, but also because if paedophiles act upon their desires, they may wreck the future lives of the children they target, with all the physical and psychological trauma that can thereby ensue. But most of all, the revulsion is based on the fact that children are seen as innocent and vulnerable compared to adults and the paedophile threatens to exploit that vulnerability and destroy that innocence. Important questions to address then are what causes paedophilia, and whether a better understanding of it could help us diagnose and treat those with the disorder.[33]

Understandably, some believe we should not even debate this issue. As Paul Jones, father of April Jones, a five-year-old girl from Machynlleth, Wales, who was abducted and murdered by a paedophile in 2012, has said, 'people, they think "why should we help the paedophile? We should be prosecuting them, throwing them in jail, having them castrated".'[34] Perhaps surprisingly, Jones does not believe in this approach. As he notes, 'if we offer help to paedophiles we might save children who might have been abused.'[35] If we take this approach, we need to establish, as for other mental disorders, whether paedophiles are a product of biology or environment. And as also for other disorders, increasingly the evidence is that both nature and nurture play a role in the genesis of paedophilia.

Certainly some people think the condition is primarily a product of defective biology. For instance, James Cantor, of the Centre for Addiction and Mental Health in Toronto, who uses magnetic resonance imaging to scan paedophiles' brains, believes that 'paedophilia is a sexual orientation, [it's] something that we are essentially born with, does not appear to change over time and it's as core to our being as any other sexual orientation is.'[36] Cantor claims his studies show that the brains of paedophiles are wired differently to non-paedophiles. According to him, 'it's as if, in these people, when they perceive a child, it's triggering the sexual instincts instead of triggering the nurturing instincts.'[37]

But we need to know whether any differences detected in scans of paedophiles' brains are a cause of their behaviour, or its consequence.

It is worth mentioning in this context the case of a US citizen who suddenly developed a sexual attraction towards children. This man, who was living with his second wife and his stepdaughter in 2000, became overcome by troubling urges.[38] Although harbouring a strong interest in pornography since his teen years, he claimed to have never been attracted to children. Yet now he began collecting magazines and websites focusing on paedophilia and making sexual advances towards his prepubescent stepdaughter. She told her mother, and the man was thrown out of the family home. However, at this point, he began experiencing violent headaches and loss of balance.[39] Rushed to hospital, a scan revealed that he was suffering from a brain tumour.

Amazingly, once this was surgically removed, the man's headaches and balance problems disappeared, but so did his paedophilic urges. Seven months later, he completed a sexual offenders' rehabilitation programme and was allowed to return home to his wife and stepdaughter. But he became concerned when he again started to develop persistent headaches and a renewed interest in child pornography. The tumour that apparently had caused his paedophilia had returned. When it was removed, he was once again cured. Scientists concluded that the tumour interfered with the orbifrontal cortex which helps to regulate social behaviour and likely exacerbated the man's pre-existing interest in pornography, 'manifesting sexual deviancy and paedophilia'.[40] Obviously this is an unusual case, and paedophilia is clearly not generally caused by brain tumours, but it does show that brain structure and biology should not be disregarded—as a 'blank slate' view of the mind might do—when seeking to explain why some individuals show such horrifying sexual urges.

At the same time, it seems that an individual's environment, particularly when they themselves are still a child, can also have an important impact on whether they will later develop paedophilic urges. Sadly, evidence shows that many people who sexually abuse children were themselves abused as a child. Take for example Nicholas, a married father-of-four who admitted to being attracted to boys aged 12 to 14 in a 2016 report about paedophilia, but says he has never acted on his urges. According to him, 'no-one chooses to be sexually attracted to children. And those of us who are unlucky enough to be sexually attracted to children can't [make it go away]. But many of us can and do successfully resist our attraction.'[41] Yet Nicholas knows what it is like to be a recipient of such abuse. At 12 years old, he was molested by a holiday camp counsellor. In fact, about half of child sex offenders appear to have been sexually abused themselves as children.[42]

It is important to note that the vast majority of sexual abuse victims do not grow up to become abusers. However, some scientists believe a biological predisposition to the condition, combined with the experience of being molested as a child, might lead to someone developing the condition as an adult. According to James Cantor, 'it could be that biology causes paedophilia, but that environment makes a person more likely to act on that sexual interest and molest a child.'[43] Similarly, a combination of both nature and nurture might be seen as producing a serial killer. A common form such a viewpoint takes is that biology and environment are separate, albeit able to produce a combined effect in an individual. However, as we saw earlier, the idea that human behaviour is determined by either biology or social environment, or a simple combination of the two, is being challenged by new findings in epigenetics, a subject that a recent article in Nature claimed provides 'a molecular middle ground in the centuries-old debate over nature versus nurture.'[44]

Epigenetic changes, recall, include anything that modifies a gene's activity without changing its DNA sequence. Such changes, which include DNA methylation, chemical modifications in the histone proteins that wrap around the DNA, and alterations in the proteins and RNAs that regulate gene expression, explain why different cell types—for instance, a heart cell or a neuron—differ greatly in their functional properties despite having the same genome. More controversially, there is emerging evidence that epigenetic changes can be passed down through several generations.[45] So what evidence is there that epigenetic effects might be involved in the development of criminal or anti-social behaviour?

Exposure to violent or criminal behaviour as a child might result in a tendency for that individual to be drawn towards such behaviour as an adult. That might explain why criminal behaviour often runs in families, and why individuals abused as children may grow up to be abusers themselves. Of course, such patterns of repetition could simply reflect the fact that crime tends to be concentrated in less affluent layers of society, or that criminal behaviour can be learned. This latter point is after all what a simplistic 'blank slate' view of human behaviour would predict. Yet a recent study in rats suggests that exposure to trauma during adolescence can also lead to epigenetic changes associated with aggressive behaviour in adulthood. This study, led by Carmen Sandi at the Brain Mind Institute in Lausanne, Switzerland, briefly exposed 28- to 42-day-old adolescent rats to a fox odour or a bright light every day for seven days.[46] Rats exposed to such fearful stimuli grew up to become what Sandi termed 'bullies' who violently attacked any new rat introduced to their cage. In contrast, rats not exposed to such stimuli as adolescents were significantly less aggressive.

Imaging studies showed that rats stressed in adolescence had increased activity in the amygdala—the brain region associated with fear responses—and lower activity in the prefrontal cortex, involved in social behaviour and conduct.[47] In addition, stress-exposed rats displayed hormonal irregularities (high testosterone and low corticosterone levels) which are associated with aggression and violence in humans. Studies of the rats' DNA showed epigenetic changes in the cells of the prefrontal cortex, while adult rats exposed to the stressful stimuli did not show such changes. This suggests that trauma incurred early in life—but not in adulthood—can set the stage for aggression in adulthood. Sandi believes her findings highlight 'the importance of developing social programs and scientific research initiatives to offer valid treatments for individuals that have been victimized early in life.'[48]

Of course, this is a study of rat behaviour, and it is important to consider the many differences between aggression in humans and that in rats, and the danger of using terms such as 'bullying' to describe aggression in an animal. Yet there is some evidence that epigenetic mechanisms may also be at work in the development of violent and anti-social behaviour in humans. For instance, Richard Tremblay of University College Dublin has taken blood and saliva samples from 'chronically violent offenders'—that is, people who have committed murder and other serious crimes—and used these to study epigenetic 'marks' across the genome.[49] His finding that several hundred genes, including some linked to aggression, were marked by patterns of epigenetic activity that differed from those he saw in a control group has led him to say that 'it looks like there is an epigenetic basis for the transmission of violence, and not only does it look like it but it makes sense.'[50]

Not everyone accepts such a conclusion. Responding to claims about the influence of epigenetic changes such as DNA methylation on the body, Greg Miller, a psychologist at Northwestern University in Chicago, has argued that 'it's a long way from differential methylation to behaviour, physical health or mental health.'[51] Critics point out that while epigenetics may be one reason why people abuse drugs or turn to crime, it cannot explain the detailed trajectory of a person's life, or their ability—or inability—to change its course. For instance, Jaleel Abdul-Adil, a psychologist at the University of Illinois in Chicago (UIC), who works with at-risk youth, thinks that 'if the funding stream for research and services moves towards trying to classify people, then you're going to evaluate people based on their physical experiences to the detriment of not just how these people look but how they suffer.'[52]

However, others think the findings should force a broad rethinking about the origins of violence. For instance, Gary Slutkin, also at UIC, who used to study infectious diseases in East Africa, but now runs Chicago for Cure Violence—an organization he co-founded in 1995—argues that 'we used to think of people with leprosy as bad people because we didn't understand what was happening. Epigenetic damage is invisible, and neuronal circuits are invisible—so until we start to talk about violence as science we're still in the Middle Ages.'[53]

If these are some of the factors that may affect the development of a serial killer or paedophile, how can modern science inform us what might be going on inside the mind of such a person? Trying to imagine how a human being could not be inhibited from carrying out such despicable deeds is made difficult by the revulsion that the overwhelming majority of us feel when we hear about them. Psychopathic killers characteristically have an apparent absence of remorse or empathy for others, coupled with a lack of guilt or ability to take responsibility for their actions.[54] At the same time, many serial killers can be superficially charming, which is one way they lure potential victims into their web of destruction.[55] This dissonance between the violence of their acts and the ability of such individuals to integrate themselves into wider society as apparently rational, considerate citizens is one reason why serial killers are rarely considered to suffer from a mental disorder to such a debilitating extent that they are considered insane by the criminal justice system.

To be classified as legally insane, an individual should be unable to comprehend that an action is against the law at the moment the action is undertaken.[56] Because of this, many reports about serial killers present such individuals as rational—but totally evil—individuals. Yet surely a truly scientific viewpoint needs to move beyond notions of original sin, or narrow legal definitions of sanity, and consider instead how an individual human can appear rational for much of their existence, yet commit such horrendous crimes against other people.

One explanation for such cognitive dissonance is that psychopaths are individuals in whom two minds coexist—one a rational self, able to successfully navigate the intricacies of acceptable social behaviour and even charm and seduce, the other a far more sinister self, capable of the most unspeakable and violent acts against others. This view of psychopaths has been a powerful stimulus in fictional portrayals ranging from *Dr Jekyll and Mr Hyde*, to Hitchcock's *Psycho*, and a more recent film, *Split*.[57] For instance, in *Psycho* we learn that the serial killer Norman Bates was so dominated by his dead mother's influence that 'he began to think and speak for her, give her half his life, so to speak,' as explained by a 'psychiatrist' at the end of the film.[58] Yet there is little evidence that real-life serial killers suffer from such a form of split personality—now generally referred to as dissociative identity disorder (DID)—in which an individual has two or more personalities cohabiting in their mind, apparently unaware of each other. Instead, DID is a condition more associated with victims, rather than perpetrators, of abuse, who adopt multiple personalities as a way of coming to terms with the horrors they may have encountered, and the effect this has upon their psyche.[59]

Of course a perpetrator of abuse may also be a victim, as we have seen, but in general, serial killers appear not to be split personalities, but rather people conscious of their acts. Despite this, there is surely a dichotomy in the minds of such individuals perhaps best personified by US killer Ted Bundy, who was a 'charming, handsome, successful individual [yet also] a sadist, necrophile, rapist, and murderer with zero remorse who took pride in his ability to successfully kill and evade capture.'[60] In Bundy's case, this allowed him to evade capture long enough to kill at least the 30 young women whose murders he confessed to, but the real figure may be substantially higher. So what was going on inside the head of this 'charming, handsome, successful individual' as he planned his next appalling crime?

One puzzling aspect of serial killers' minds is the fact that they appear to lack—or be able to override—the emotional responses that in other people allow us to identify the pain and suffering of other humans as similar to our own, and empathize with that suffering. A possible explanation of this deficit was identified in a recent brain imaging study by Michael Koenigs and colleagues at the University of Wisconsin.[61] This showed that criminal psychopaths had decreased connectivity between the amygdala—which amongst its other functions as a primary mediator of emotional responses processes negative stimuli and those that give rise to fearful reactions— and the ventromedial part of the prefrontal cortex, which interprets responses from the amygdala. When connectivity between these two regions is low, processing of negative stimuli in the amygdala does not translate into any strongly felt negative emotions. This may explain why criminal psychopaths do not feel nervous or embarrassed when caught doing something bad, or feel sad when other people suffer.

Yet serial killers also seem to possess an enhanced emotional drive that leads to an urge to hurt and kill other human beings. This apparent contradiction in emotional responses needs to be explained at a neuro-logical level. At the same time, we should not ignore social influences as important factors in the development of such contradictory impulses. It seems possible that serial killers have somehow learned to view their vic-tims as purely an object to be abused, or even an assembly of unconnected parts. This might explain why some killers have sex with the dead victim, or even turn their body into objects of utility or decoration,[62] although that still does not explain why they seem so driven to hurt and kill their victims. One explanation for the latter phenomenon is that many serial killers are insecure individuals who feel compelled to kill due to a morbid fear of rejection.[63] In many cases, the fear of rejection seems to result from having been abandoned or abused by a parent, often their mother, which as already mentioned has been a feature of many infamous male serial killers' lives. Such fear of rejection may compel a fledgling serial killer to want to eliminate any objects of his affections. He may come to believe that by destroying the person he desires prior to entering into a relation-ship with them, he can eliminate the frightening possibility of being abandoned, humiliated, or otherwise hurt by someone he loves, as he was in childhood.

Another factor to consider about the minds of serial killers is that such individuals appear to lack a sense of social conscience. We saw earlier how an individual human consciousness can be viewed as a kind of dia-logue. Part of this dialogue involves the representation of acceptable social behaviour in the mind of the individual. Through our parents, sib-lings, teachers, peers, and other individuals who influence us as we grow up, we learn to distinguish right from wrong. It is this that inhibits us from engaging in anti-social behaviour. Yet serial killers seem to feel they are exempt from the most important social sanction of all—not taking another person's life. Not only that, but some such killers have even gloated about being exempt from normal social sanctions. For instance, Richard Ramirez, named the 'Night Stalker' by the media, who killed at least 13 people and possibly six others, claimed at his trial that 'you don't understand me. You are not expected to. You are not capable of it. I am beyond your experience. I am beyond good and evil…I don't believe in the hypocritical, moralistic dogma of this so-called civilized society.'[64]

Meanwhile, British TV personality Jimmy Saville, who has not been shown to have murdered anyone, but who sexually abused and raped over two-hundred vulnerable individuals aged between 5 and 75, described his behaviour as pursuing the 'ultimate freedom.'[65] Therefore any discussion of the consciousness of the criminally insane needs to consider why they might feel exempt from social norms. Additionally, to better understand how such individuals arise, we need to move from just considering the individual mind of the serial killer or paedophile, and also look at the specific type of society in which such behaviour occurs. Indeed, overly focusing on the individual criminal, whether in terms of their biology or life experience, runs the risk of neglecting the fact that individuals live within a specific society. This matters because the values of that society may influence not only which types of behaviour are judged acceptable, but also the different types of distorted behaviour that may develop in certain individuals within it. We need to consider again the link between mind and society, but in the context of the advanced capitalist societies that dominate the world in the early twenty-first century. In particular, that means recognizing that we do not live in a homogeneous society, but one divided in very specific ways and particularly by class, that is a class society. This is important because class may affect the behaviour of those at the bottom of society, but also those at the very top of the current economic system.

In this respect, my decision to mainly focus on the criminally insane in this chapter, and therefore to look at serial killers, paedophiles, and the like, may have skewed the discussion about the links between crime and human consciousness too much away from a consideration of more mundane forms of crime. Returning to the topic of my home city of Bradford, I was saddened to read that a 2014 YouGov survey found that it was judged by those questioned to be Britain's 'most dangerous city'.[66] This view may have partly been influenced by the riots that took place in Bradford in 2012.[67] But there is also an ongoing problem of crime in the city, often linked to widespread use of drugs such as heroin, and the street gangs that make money selling such drugs to addicts. Yet lest anyone tries to claim such a state of affairs reflects a natural disposition towards crime in my fellow Bradfordians, it is worth pointing out the social backdrop for this situation, namely that a third of adults in Bradford are out of work, 40 per cent of the city's wards are among the poorest 20 per cent in Britain, and the city has the highest level of child poverty in the country.[68]

In this context it is worth mentioning the words of the US writer and social reformer Studs Terkel, who has said 'you know [the phrase] "power corrupts, and absolute power corrupts absolutely?" It's the same with powerlessness. Absolute powerlessness corrupts absolutely.'[69] Indeed, the link between crime and poverty was recently highlighted by one of Britain's most senior police chiefs—London Metropolitan Police Assistant Commissioner Patricia Gallan. In an interview in 2018, she argued that 'children are not born bad' and called for more effort to deal with inequalities that leave people feeling like 'they do not have a stake in society'.[70] Asked about the link between poverty and alienation and people committing crime, Gallan replied that 'I think if you are a young person and you haven't got opportunity necessarily—and this isn't an excuse for it, it is explanation—what's your risk? You've got a sense of belonging if you are in a group or a gang…and you get the material aspects that you would like, so that's part of the challenge.'[71]

Moving from drug-linked crime in places like Bradford to a different set of activities, consider a recent article by Charles Ferguson, a former adviser to the US White House, about individuals who have 'tolerated, and in some cases aggressively courted, money laundering by rogue states, terrorist organizations, corrupt dictators, and major drug cartels' over recent decades.[72] Furthermore, these people also 'created special handbooks on how to evade surveillance, created special business units to handle money laundering, and actively suppressed whistle-blowers who warned of drug cartel activities.'[73] Finally, Ferguson states that 'we now possess overwhelming evidence of massive securities fraud, accounting fraud, perjury, and criminal…violations by [the individuals in question] during the housing bubble that caused the financial crisis [of 2008].'[74] At this point you may be wondering who these individuals might be. Ferguson's answer is senior executives of some of the world's main banking organizations. As he puts it, 'over the past two decades, the financial services industry has become a pervasively unethical and highly criminal industry, with massive fraud tolerated or even encouraged by senior management.'[75] Meanwhile, recent studies conducted by Canadian psychologist Robert Hare have suggested that while 1 per cent of the general population can be categorized as psychopathic, the prevalence in the finance industry is 10 per cent.[76] Yet the individuals mentioned above are among the richest and most powerful people in the world. So what defines one set of activities as criminal, yet another as a major route to riches and the height of high society? And is it really true that so many people who work in high finance can be classified as having a mental disorder? To look at these questions further, it is necessary to bring a new factor into our discussion of human consciousness, namely the question of class, and the divisions that this gives rise to in current society.

PART V

THE SOCIAL MIND

CHAPTER 16

CLASS AND DIVISION

⸻◈⸻

W e currently live in remarkable, but troubling, times. At the start of this book, I mentioned some recent, amazing technological achievements of *Homo sapiens*, ranging from our ability to send robotic probes to the far reaches of the Solar System, to new techniques in biology that allow us to both read and precisely edit the genomes of living cells. A defining feature of modern society is its continual transformation by new technologies. Think for instance how advances in computing have changed the world over the past few decades.[1] A decade ago, smartphones did not exist. Three decades before that, hardly anyone owned a computer. Yet today, nearly everyone in the developed world, and large numbers of people beyond it, seem to be gazing at some form of glowing electronic device. Meanwhile, the Internet, with its rapidly diversifying new types of social media, connects billions of people across the planet in a way that would have been unimaginable to past generations.

Yet despite such technological progress, there are increasing concerns about where modern civilization is heading.[2] Undoubtedly the biggest concern is climate change, which raises questions about the very future of civilization because of its likely impact on sea levels—threatening many major cities—and agriculture, which we rely on to feed the almost eight billion people on the planet.[3] Others include rapidly escalating extinctions of natural species, resistance of bacteria to antibiotics, increasing military tensions between the major powers, and large numbers of refugees on the move, fleeing war and famine. Even the coronavirus disease COVID-19 that swept through human populations across the globe in 2020 is claimed to have first spread from bats to humans because of the erosion of natural bat habitats and poor safety standards in the wild animal markets that are now a billion dollar industry in some parts of the world.[4] Meanwhile, in many countries, political parties representing the neoliberal consensus that has dominated politics over recent decades are losing ground to organizations that offer more extreme solutions to the current crisis.[5]

One factor fuelling unrest has been the continuing stagnation of the world economy, with wage freezes and cuts in public services proposed as the only solution to the crisis.[6] Yet alongside this, the gap between the super-rich and the majority of people is widening, with a recent report predicting that by 2030, the richest 1 per cent will own two-thirds of global wealth.[7] All of this is likely to affect the minds of individual human beings. At the same time, the current scenario raises questions about the collective consciousness of humanity—as compared to that of individuals—given the existence of such major divisions in society. In this chapter I will look at such issues, but to do so, we need first to consider how the current organization of society arose, and how it compares to past human societies.

Homo sapiens is around 200,000 years old, but proto-human species first became distinguished from apes around 6 million years ago.[8] For the vast majority of that time, our ancestors lived in so-called 'hunter-gatherer' societies. In such societies, most or all food is obtained by foraging for plants and berries and hunting wild animals, in contrast to agricultural societies, which rely mainly on cultivated crops and domesticated animals.[9] Although only small numbers of people still live in hunter-gatherer societies, studies of such existing societies, as well as a few past ones, have identified some key features.

A common assumption about such societies is that they probably live on the brink of starvation. Yet Richard Lee of the University of Toronto studied the Ju/'hoansi 'bush people' of the Kalahari desert in the 1960s, and found that they made a good living from only 15 hours' work per week.[10] One source of this success is skill in exploiting natural resources in a sustainable manner. The Ju/'hoansi consume over 150 plant species, and can hunt and trap effectively any animal they choose, but they only work to meet immediate needs, do not store surpluses, and also never harvest more than they can eat in the short term.

Another feature of hunter-gatherers is their egalitarianism. Cambridge-based anthropologist James Suzman believes that 'the evidence of our hunting and gathering ancestors suggests we are hard-wired to respond viscerally to inequality.'[11] A common characteristic of hunter-gatherers is that they hate individuals showing off; for instance, bush people typically 'insult' a hunter's kill. A Ju/'hoan man explained to Richard Lee why they do this: 'When a young man kills much meat, he comes to think of him-self as a chief or a big man—and thinks of the rest of us as his servants or inferiors. We can't accept this…so we always speak of his meat as worth-less. This way, we cool his heart and make him gentle.'[12]

Such societies also have no formalized leadership institutions.[13] Men and women enjoy equal decision-making powers, children play largely non-competitive games, and the elderly, while treated with affection, are not afforded any special privileges. This means that in these societies no one sees any point in accumulating wealth or influence, or seeks to over-exploit their marginal environment. Of course, it is important not to idealize hunter-gatherer societies. Without the benefits of modern medi-cine such societies are far more vulnerable to disease, whether this be viral or bacterial infections, or disorders with a more genetic basis.

Yet while many people today live in an infinitely richer society than hunter-gatherers in material terms—and this includes access to cutting-edge advances in medical diagnosis and treatment—modern civilization is deeply divided. Not only wealth divides the world; racism, sexism, and homophobia can also incite people to hatred, and even murder, of other human beings because of their skin colour, gender, or sexual choices.[14] Yet although such divisions are often assumed to be fundamental features of humanity, both the immense power of modern society and its divided status originated only around 12,000 years ago when certain human groups first began tending cattle and growing crops—the so-called 'agri-cultural revolution'.[15] In the 'fertile crescent' extending through what is now Lebanon, Syria, and Iraq, the domestication of plants led to a more sedentary life and fixed-field agriculture that, in turn, led to the development of cities and the rise of the state.

The development of civilization was once thought to be very rapid; however, recent studies have indicated a gap of 4,000 years between the first domestications and the rise of the city as a dominant way of ordering human affairs.[16] Rather than embracing farming with enthusiasm, communities initially adopted subsistence strategies that combined hunting and gathering with a low level of domestication and cultivation. This seems to be because, at least in its early days, agriculture was a rather fraught process.[17] Not only did crops sometimes fail, but with less time to hunt and gather, diets became more restricted—which adversely affected health—and with more people living in close proximity to their animals, there were more possibilities for disease to spread. For this reason Jared Diamond of the University of California, Los Angeles, has declared agriculture to be the 'worst mistake in the history of the human race.'[18] Yet although an important corrective to the idea that the agricultural revolution was a smooth ladder of progress, this statement neglects the fact that civilization, and all its later technological marvels, could only have come about through this revolution.

In particular, agriculture allowed the growth of city states, which led to the development for the first time of the civilizations that we take for granted today. However, we should also be aware of the brutality involved in the formation of such city states. James Scott of Yale University has argued that for a state to exist it needed a crop that was easily taxed, and wheat proved ideal in this respect.[19] Because the fields were fixed and this crop ripened over a short period of time it was impossible for the farmer to avoid the tax collector. Communities elsewhere in the world reliant on tubers or root vegetables such as yams and manioc as their crop were more able to avoid tax since these plants can be left in the ground and harvested over a long period, and indeed such societies seldom developed into city states.

Another advantage of wheat was that it could be stored within the city, from where it was doled out to slaves and soldiers or used to feed the whole population during a siege.[20] Through taxation the state became the quartermaster and the producers subjects. Early states functioned as 'population machines' designed to control labour, with slavery a key element. Maintaining the number of slaves was vital and if this fell a new batch could be gathered through warfare. Raiding to acquire goods and manpower became a normal part of life. Far from an aberration, slavery was built into the first civilizations.[21] Above all else, the rise of civilization led for the first time in history to human beings being divided into 'classes'.

Other classes besides slaves existed in ancient societies, such as 'free' workers—who tended the land or did other jobs in the cities—merchants, priests, and rulers: either a king or pharaoh, or as in the case of the Roman Republic, a governing body like a senate. However, the surplus labour that provided the income and wealth of the ruling class was primarily extorted from slaves, and there was a continual need to define them as inferior beings. This was despite ancient societies like those in Greece or Rome first establishing the principles of democracy: the democracy was only for a small portion of society. According to Oxford historian Geoffrey de Sainte Croix, 'slavery increased the surplus in the hands of the propertied class to an extent which could not otherwise have been achieved and was therefore an essential precondition of the magnificent achievements of classical civilisation.'[22]

Slaves, but also women of all classes, were excluded from democracy in the first city states.[23] This had an impact on general perceptions of what it means to be human. For while in hunter-gatherer groups individuals might have different roles in the group, but no one was seen as superior to anyone less, in the new city states the idea grew that some human beings were inferior to others, and that slaves were in many senses actually sub-human. For instance, the Greek philosopher Aristotle argued that 'the use made of slaves and of tame animals is not very different; for both with their bodies minister to the needs of life.'[24]

Class society also led to the idea—for the first time in history—that women were inferior to men. Through their dominance in the warlike city states, some men began to accumulate wealth and power. And as an individual's possession of property became a defining feature of their status, so it became important to track the inheritance of property rights. As a consequence, among the rulers, women became redefined primarily as producers of children in a monogamous relationship, and as maintainers of a stable household and family environment, rather than an equal partner in the process of producing food and other resources, as they had been in hunter-gatherer societies.[25] While lower class 'free' women remained an important part of the production process, the relegation of a woman's role in the upper layers of society also influenced perceptions of women's roles in the lower layers.

The next form of class society in history, feudalism, was different from ancient societies in that absolute slavery was abolished in such a society. While the specific features of this type of society varied in different global regions ranging from Europe to India and China, a common characteristic of feudal society was that while the peasant in theory owned their own patch of land, in practice they were 'tied' to a particular lord, who also took a proportion of everything they produced.[26] Moreover, any peasant who tried to rebel against the system could be quickly punished or even killed for daring to challenge their status in society. Another crucial justification for this system, which we will consider in more detail later, was the idea that everyone had their place in society, ordained by religion.

In contrast to such past situations, the ideology of our current, capitalist society is that all human beings are born free and equal, as expressed in the US Declaration of Independence of 1776, which states 'that all men are created equal, that they are endowed by their Creator with certain unalienable rights, that among these are life, liberty and the pursuit of happiness.'[27] The French revolutionaries a decade later expressed this idea even more succinctly with their slogan 'Liberté, égalité, fraternité'.[28] Yet a report in January 2018 by the Oxfam charity estimated that the richest 42 people in the world hold as much wealth as the 3.7 billion who make up the poorest half of the Earth's population. Mark Goldring, Oxfam UK's chief executive, commented that 'the concentration of extreme wealth at the top is not a sign of a thriving economy, but a symptom of a system that is failing the millions of hardworking people on poverty wages who make our clothes and grow our food.'[29]

So what is it that prevents the bottom half of the world's population challenging the privilege of those few individuals at the top? One reason is the imbalance of power in society. The example of Chile in 1973, when an army coup headed by General Augusto Pinochet overthrew an elected left-wing government led by President Salvador Allende and murdered Allende and an estimated 40,000 other individuals, shows that parliamentary democracy is not as secure as one might assume, if it threatens to directly challenge the interests of a capitalist ruling class.[30] But it would also be equally mistaken to assume that the maintenance of the power structures of capitalism is primarily by force.

Karl Marx and Friedrich Engels—probably capitalism's most well-known critics—argued in 1848 that 'the ideas of the ruling class are in every epoch the ruling ideas... The class which has the means of material production at its disposal has control at the same time over the means of mental production, so that thereby, generally speaking, the ideas of those who lack the means of mental production are subject to it.'[31] By this, they had in mind ideas put forward in schools and universities and by the media, but also more generally the assumptions that govern the workplace, the justice system, and other state institutions.

Yet such a state of affairs is far from being just driven by simple propaganda. Those propagating ruling class ideas may sometimes be cynical about their role, but in general they are as likely to believe in the ideas they are propagating as the people they propagate them to. In this respect, an important feature of capitalism is that unlike previous class societies, the source of wealth in current society, and therefore its driving mechanisms, are far less transparent than was the case in ancient slave civilizations, or even in feudal society.

Marx believed that a key feature of capitalist society is that people within it exist in a state of alienation. In its general sense, alienation refers to a feeling of separateness, of being alone and apart from others.[32] However, Marx's more specific definition refers to the contradictions that stem from workers in capitalist society selling their labour in return for wages. The worker creates something—a plate of food, a piece of clothing, or for that matter an office report or spreadsheet—and receives a wage for doing so. The object is then sold for more than it cost to make, or used to boost a firm's overall income, which is the source of the employer's profit. At the end of the process the object produced no longer belongs to the worker but to the employer; it now appears to the former as akin to an alien object. This contradiction was expressed by Marx's statement that, 'the worker feels himself only when he is not working; when he is working, he does not feel himself…His labour is, therefore, not voluntary but forced, it is forced labour…Its alien character is clearly demonstrated by the fact that as soon as no physical or other compulsion exists, it is shunned like the plague.'[33] Adam Smith, one of the first theorists of capitalist economics, believed that work required the worker to give up 'his tranquillity, his freedom, and his happiness.'[34] Yet in terms of the propagation of capitalist society, alienation has the virtue that workers may often underestimate—or even be totally unaware of—their centrality to the production process.

One reason for the alienation of workers from the labour process is a principle Smith identified as a key characteristic of capitalism—a tendency to break down work into simpler sub-tasks, which he famously illustrated through a detailed study of the manufacture of pins.[35] The industrialist Henry Ford later perfected such a division of labour in his car plants, which is why this practice is sometimes referred to as 'Fordist'.[36] An important aspect of this approach is a continual drive to increase the speed of production by putting pressure on each worker to perform their part of the manufacturing process at the maximum speed.

The consequences of such pressures in a modern context were revealed in a recent report in *Business Insider*, which interviewed workers at delivery warehouses of the Amazon tech-giant.[37] Amazon 'pickers' move around such warehouses on a predetermined route to collect items for delivery, scanning each with a handheld scanner, which monitors the length between scans. Because of the pressure this puts on workers, one US

Amazon employee described the 'awful smell' that often emanated from warehouse trash cans, due to workers urinating in these instead of using the bathroom, because of fears about missing targets.[38] Other workers confirmed a general picture of constant surveillance and a crippling fear of missing targets. Given what I have said previously in this book about the link between stress and mental disorders, the likely risks of such work practices should be obvious.

It might be argued that the conditions in an Amazon warehouse represent an extreme example, given that this is a non-unionized workplace whose employees are often relatively unskilled, and on low wages and temporary contracts. However, a similar increasing pressure to perform, plus low pay and job insecurity, have been an increasing feature in recent years even of such skilled occupations as my own area of work—university teaching. Increasingly UK universities employ junior lecturers on insecure, non-permanent contracts, which range from ones that typically elapse within nine months to those in which lecturers are paid by the hour to give seminars or mark essays and exams, with over 50 per cent of university teaching now done in this way.[39] How it feels to teach students in such a context was recently summed up by one lecturer who said, 'we are seasonal labourers like fruit pickers. You have to e-mail every September, cap in hand, saying: "Is there any work for me this year?"'[40]

Alienation affects more than just workers under capitalism. Previously, I mentioned claims that people with a 'psychopathic' personality are ten times more likely to be found at the top of the financial industry than in the general population.[41] Yet even if true—and it is important to distinguish such a personality from that of a violent psychopath—there is a danger of seeing such individual behaviour as the cause of the cut-throat world of high finance, rather than as a consequence of the values of capitalism at its most intense. It may well be that individuals with 'superficial charm, conning, and manipulative behaviour, lack of empathy and remorse, and a willingness to take risks'—which is how psychopaths were defined in this case—might thrive in such an environment.[42] But it could also be the case that such behaviour tends to emerge because this industry tends to encourage competitive, short-term gain over long-term benefit, aspects of capitalism at its most extreme. And the fact that venture capitalists may take decisions that might seem deplorable when assessed in human terms—for instance, the takeover and asset-stripping of a company, with its workforce 'relieved' of their jobs in a 'down-sizing' operation—might seem more permissible if judged only as figures on a balance sheet, rather than something affecting real, living human beings.

Meanwhile, since under capitalism everything can be viewed as a commodity to be bought and sold, including human labour, this can lead workers themselves to view their colleagues primarily as competitors for a job, rather than fellow exploited workers. This tendency can become particularly strong if linked to the idea that workers who are seen as 'different' pose a particular challenge to native workers. This difference might be skin colour, but even a white worker from Ireland or Poland can be viewed in this way.[43] Such feelings are often stoked by sections of the media, with headlines about 'immigrants' taking jobs, housing, and social services, often laced with crude stereotypes about such immigrants.[44]

Alienation can also shape perceptions of the position of women in society. I said earlier that because possession of wealth and property was key to a person's standing in ancient societies, the inheritance of such wealth and property had to be along clearly defined lines, leading to the rise of monogamous marriage. Because of this, upper-class women became increasingly viewed as possessions, rather than subjects in their own right. This view extended to feudal society, as shown by the way that ruling class women often had little choice in their choice of husband and could be betrothed as young as nine years old and be married by the age of twelve.[45] Although boys could also be betrothed at a young age, in general girls of the ruling order tended to be married to substantially older men.[46]

In peasant society, women were expected to take part in the same backbreaking work as their menfolk. But the view that women were the lesser sex extended to peasant life. In Russia, a country in which many ancient feudal traditions continued up to the 1917 revolution, a popular 'joke' went along the lines of 'I saw what I thought was two people but then I realized it was a man and his wife.' A common object above the marital bed in many Russian peasant households was a whip, placed there as a warning to—and sometimes used on—women who started voicing opinions deemed inappropriate to the status of their sex.[47]

In modern society, the status of women has changed in fundamental ways, although such a change was far from automatic, nearly always involving bitter struggles. A key step forward was the right to vote, granted to women in Britain in 1918.[48] Women are also increasingly seen as key members of the workplace in modern capitalist society, at both the highest and lower levels. However, women are still frequently judged as inferior for two main reasons. First, they are expected to be the primary individuals who care for children and look after the home.[49] Second, alienation under capitalism means that everything can be treated as a commodity.[50] Women's bodies are used in advertising to sell everything from cars to soft drinks. In the sex industries, a woman's body is more directly offered for sale. This affects the perception of women in society in general, not only of prostitutes and porn stars. The allegations against the movie producer Harvey Weinstein of the assault and rape of a number of women in the film industry, for which he was later convicted, unleashed a flood of others, not just in the entertainment industry, but in a whole number of other sectors.[51]

So how might alienation manifest itself in the individual mind under capitalism? One feature of the human brain that might explain why many working people can accept and even acquiesce in their state of exploitation, and even sometimes see other workers as the primary threat, rather than as allies in campaigns for better pay and conditions at work, is the way our brains have evolved to respond to living in a potentially hostile natural environment. In such an environment, from an evolutionary point of view, any species, whether this be fish, fowl, bacterium, or even a virus, has two main imperatives: one is to survive, the other is to reproduce. But to do both effectively, two, sometimes opposing, tendencies need to be at work. One is the tendency to keep doing all the things that have kept a species alive and reproductive across many generations; the other is the ability to respond rapidly to any perceived threat to that status quo. The first tendency, which encourages sticking with a certain set pattern of behaviour and is rooted in the more primeval parts of our brain,[52] may explain why it can take quite a considerable change in socioeconomic forces for the majority of people to think that there can be any point in challenging the status quo.

As for the second tendency, two brain regions that we have already mentioned, the amygdala and the prefrontal cortex, are relevant. As a primary mediator of emotional responses, the amygdala is one of the main ways that our brains respond to potentially threatening situations.[53] Imagine walking down a dark alley and suddenly there is a noise. In such a situation, the amygdala will immediately become activated. Yet as soon as we realize that the noise is just a rustle of leaves, or a dog barking in a neighbouring garden, our rational side, personified by our prefrontal cortex, suppresses such irrational fears and the activity of our amygdala returns to normal. Yet despite this, there is increasing evidence that many unconscious fears about a perceived threat of people of a different colour skin, sexual persuasion, or even male fears about female equality, involve the amygdala.[54] Because of this, although inflammatory speeches by politicians or media articles may enhance such fears, thereby dividing people who have much to gain by working together for better pay and conditions, they are helped by being able to trigger deep-seated impulses in the human brain.

Another way that the structure of current society is relevant to discussions about human brain function is through its impact on mental health. Earlier in this book, I looked at how differences in individual biology, but also tensions and stresses in family relationships, can influence whether a particular person succumbs to a mental disorder. However, it would be a major omission to ignore the role that modern society plays. People are taught from an early age that the route to fulfilment in life is to make themselves acceptable for the job market. Such concerns play upon the idea that success in future employment is primarily down to individual merit. Yet a person's gender, ethnicity, and sexual persuasion still have far too much impact on their success in the labour market. And another factor affecting people's status is the rapid and dramatic changes that occur in the economy following economic crisis.

Demonstrating the impact of economic recession on mental health, suicide rates in the USA jumped from 12.1 to 18.9 per 100,000 people after the 1929 Wall Street crash. More recently, a study has found that after the recession of 2008, suicide rates surged in the European Union, Canada, and the USA. Aaron Reeves, of the University of Oxford, who led the study, notes that 'there has been a substantial rise in suicides during the recession, greater than we would have anticipated based on previous trends.'[55] David Stuckler, a researcher on the study, drew particular attention to Greece, a country in which austerity policies imposed by EU institutions and the International Monetary Fund have meant that the country 'has gone from one extreme to the other. It used to have one of Europe's lowest suicide rates; it has seen a more than 60 percent rise.'[56] Each suicide corresponds to around 10 suicide attempts and between 100 to 1,000 new cases of depression. In Greece, according to Stuckler, 'that's reflected in surveys that show a doubling in cases of depression; in psychiatry services saying they're overwhelmed; in charity helplines reporting huge increases in calls.'[57]

The COVID-19 pandemic and subsequent social distancing measures as part of a general 'lockdown' in countries across the world in 2020 also had major impacts on mental health.[58] This was both due to the lockdown's effect on people's ability to make a living and because lack of social interaction proved very difficult for some individuals to deal with and led to mental instability, which is maybe not that surprising given that, as we have seen, social interactions are key to the development and maintenance of human consciousness.

The divisions in current society, by ethnicity and gender as well as class, also exert a heavy toll on mental health. The possible impact of racism on mental health was suggested by a recent study that found that Afro-Caribbeans living in Britain are nine times more likely to be diagnosed as schizophrenic than white people.[59] Arguing against this difference having a biological origin, the incidence of schizophrenia among black people in the Caribbean is identical to the white population in the UK. The researchers also found no evidence that the increased incidence of schizophrenia amongst Afro-Caribbeans in Britain was due to brain injury or drug abuse. One possibility is that racism could explain the difference; however, the study found no evidence that other ethnic groups were more susceptible to schizophrenia.[60]

There are a number of possible explanations for the study's findings. The experience of racism by Afro-Caribbeans might be more intense than for other ethnic minorities, although given the way that many Muslims have been subjected to racism in the guise of 'Islamophobia' in recent years, it is not clear that this would be the case. There might also be differences in those other ethnic minority communities, for instance in terms of family structure, cultural or religious differences, or other factors that might somehow insulate individuals of such other ethnic groups from succumbing to this particular mental disorder.[61]

Another possibility is that the greater tendency for Afro-Caribbeans to be diagnosed as schizophrenic itself reflects racism in British society. London-born psychiatrist Kwame Mackenzie believes one problem is that the diagnosis of schizophrenia is based mainly on the experience of white people. According to him, 'the concept of schizophrenia was a concept that really was generated out of the white, European tradition. It is believed that you can use the same bunch of symptoms to diagnose schizophrenia in African-Caribbean people in the UK. But nobody has ever bothered to find out whether that is true, whether the diagnosis of schizophrenia is as valid in the African-Caribbean community as it is in the white community.'[62] This shows the dangers of assuming that increased diagnosis of a mental disorder in one ethnic group reflects increased incidence of the disorder, rather than differing perceptions of what constitutes 'normal' and 'abnormal' behaviour in a particular group.

The position of women in capitalist society is also likely to affect the minds of both sexes in that society. Nowadays in countries across the world, women have an equal right to vote and are increasingly represented across various occupations. The election of heads of state such as Sirimavo Bandaranaike in Sri Lanka, Golda Meir in Israel, Indira Gandhi in India, and Margaret Thatcher in Britain first showed that a woman could reach this pinnacle of political power, and women are now found at the highest levels of professions such as finance and law. Yet a recent report showed that British women still earn an average 18.4 per cent less than men.[63] Globally, women receive only 63 per cent of the pay that men get.[64]

Unequal pay is only one of the challenges that women face. The #MeToo movement spawned by the Weinstein case illustrated the scale of sexual harassment women face at work and in society in general.[65] A recent study has found that even non-physical sexual harassment can cause psychological harm.[66] The study, by researchers at the Norwegian University of Science and Technology, collected data from almost 3,000 high school students and showed that derogatory comments, unwanted sexual attention, and unsolicited explicit images can lead to anxiety, depression, negative body image, and low self-esteem. Leif Edward Ottesen Kennair, a researcher on the study, notes that 'people usually don't consider the effects of non-physical harassment without a certain amount of physical harassment. [But] even when you control for [physical coercion and force], you still find an effect on mental symptoms due to non-physical harassment.'[67]

Participation in social media is an increasingly important aspect of people's lives, and this is particularly true of young people. A recent study of the effect of social media on mental health in 14 year olds in Britain found that girls are twice as likely to show signs of social media-linked depression compared to boys. One possible reason is that girls are using social media at higher rates, with two in five of them spending three or more hours a day on social media as opposed to one in five boys. But girls were also more likely to have suffered online harassment or online bullying than boys. Teenage girls are also more likely than boys to be dissatisfied with their body image and being continually bombarded with images of unattainable physiques on social media may play an important role in this.[68]

The treatment of women as sex objects also affects male minds. Since 1999, feminist activist and writer Julie Bindel has interviewed men who pay prostitutes for sex, as part of a re-education programme for such men in my home county of West Yorkshire. Many men explicitly compared their actions to buying goods at a shop, with one stating that 'I made a list in my mind. I told myself that I'll be with different races, e.g. Japanese, Indian, Chinese...Once I have been with them I tick them off the list. It's like a shopping list.'[69] Another claimed that 'selecting and purchasing has something to do with domination and control.'[70]

Tragically, the vulnerable positions of many prostitutes means that they are also far more likely to be beaten or even killed in the course of their work, by men whose idea of 'domination and control' can extend to murder.[71] Previously I mentioned Peter Sutcliffe, the 'Yorkshire Ripper', who murdered 13 young women and attempted to kill seven others who survived, but were scarred, both physically and emotionally, for life. After his arrest, Sutcliffe was diagnosed as a 'concealed' paranoid schizo-phrenic, mainly based on his claim that a 'divine voice' had told him in 1967 that it was his mission to kill prostitutes.[72]

In fact, some believe that Sutcliffe was merely tapping into the domin-ant image propagated particularly by the media that his primary target was this group of women. Yet a look at Sutcliffe's proven and suspected victims—for this may be a larger group than he has admitted to—under-mines the idea that he only attacked prostitutes. For instance, one victim, who escaped although with severe head injuries, was a 14-year-old school-girl, Tracey Browne.[73] What interviews with Sutcliffe have illustrated—and this echoes a theme that has almost become a cliché in many reports about male serial killers who prey on the opposite sex—is that in many such people there seems to be a mental battle going on between seeing women either as sacred objects, often demonstrated by an adoring but uneasy relationship with their mothers, or as agents of vice or contamin-ation who must be eradicated like vermin.[74] Tellingly, when Sutcliffe was asked by his brother Carl why he had committed such horrendous crimes, his response was that 'I were just cleaning up the streets, our kid.'[75]

It is important though that while we may seek partial explanations for the murderous activities of individuals like Sutcliffe in the objectification of women that is a general feature of capitalist society, we also see this as only one aspect of an explanation. For most men do not pay for sex with prostitutes, and only a tiny number end up as murderous serial killers. One good reason for this is that people are not machines. While we may all be influenced by the values of a particular society, how we interpret those values is very much affected by other factors in our lives, plus the influence on our behaviour of our own specific personality.

That this is a complex question is shown by the many varied ways in which different individuals have reacted to major crises in class society. For at certain times in history, ordinary men and women have radically challenged the society they live under. Not that such challenges are always progressive. The twentieth century is a good example of how different reactions to crisis can mean seeking a more progressive society, but may also lead to the most horrendous abuse of other human beings in ways that are the stuff of nightmares. Whatever the case, what is clear is that at such times, far from individual consciousness being a static affair, instead we find evidence that ideas and beliefs can rapidly, and radically, change.

CHAPTER 17

RESISTANCE AND REBELLION

Since the birth of civilization, people have at times come together to resist exploitation and oppression. In the first recorded instance, in ancient Egypt in 1155 BC, during the rule of Ramesses III, the skilled artisans who worked on the Egyptian rulers' tombs—the pyramids and other monuments that to this day draw visitors from all over the world—downed tools because of delays in receiving the wheat rations that were the payment for their work. The scribe Amennakhte described how they complained that 'the prospect of hunger and thirst has driven us to this.'[1] Such resistance had apparently never been heard of previously in ancient Egypt, but after a lengthy dispute, the artisans were eventually granted their delayed rations; indeed, these were now increased. Clearly, these workers had risked a great deal in taking action in what was essentially a military dictatorship. Yet having done so, they established a principle seen as important to this day—the right to withdraw one's labour.

Fast forward over three thousand years to Britain in April 2018, and another group of skilled workers—lecturers and other university staff—celebrated an important victory after their successful strike prevented attempts to substantially downgrade their pension scheme.[2] As a participant in that action, but also someone interested in human conscious awareness, in this chapter I will explore how ideas can change in a mass movement such as a strike, and why in some periods of history such activities can seem few and far between, while in others widespread industrial militancy is a central feature. I will also look at moments in history when resistance to the status quo reached such a scale that a fundamental shift occurred in the structure of society—so-called revolutions. In so doing, I will aim to relate the actions of a mass movement to what might be going on in the heads of its individual participants, and ask what mental processes might underlie the apparently rapid shifts in social consciousness that seem to be a key feature of any genuine mass progressive movement of the people.

A survey of past protest movements reveals a number of typical characteristics that may provide clues as to how human consciousness—here particularly meaning conscious awareness of oneself and one's place in society—develops in such circumstances. One is that a sustained protest may begin about a specific matter but can end up addressing much wider issues in society. The artisans in Egypt in 1155 BC initially stopped work due to concerns about delayed wheat rations, but as the dispute progressed, they began to raise much wider questions about the nature of ancient Egyptian society.[3] A key guiding principle at this time was *ma'at*—a concept of universal, communal, and personal balance in society, so that the latter could function in accordance with the will of the gods. While artisans refusing to work might be seen to violate the principle of *ma'at*, the strikers turned this argument on its head by arguing that they had been forced into such drastic action by a serious breach of this principle, namely a failure of state officials to deliver the workers' rations in a timely manner.

It may not be a coincidence that the strike occurred following a series of major battles with neighbouring powers that severely taxed Egypt's resources.[4] The strikers might therefore have recognized that all was not right generally in the Egyptian state; however, it took the concrete fact of hunger to lead them to more general questioning of ruling ideology. Meanwhile, the state officials' realization that the dispute had tapped into a more general discontent may have dissuaded them from trying to brutally crush the strike movement.

Similarly, a feature of the recent British university strikes was how quickly these generalized from a specific dispute about pensions to discontent about other issues. Previously I mentioned the increasing use of temporary, low-paid, junior academics for teaching in British universities.

This, and other recent negative trends—lack of funds for research, an investment in new buildings but not in staff salaries, student fees, and the general marketization of higher education—all became topics of discussion during the strike.[5]

One way that strikes can lead to generalization is when strikers begin to realize who supports them, and who does not. This can lead to changes in perceptions that go beyond the demands of a specific dispute, as in the year-long strike by British miners in 1984–5, precipitated by Margaret Thatcher's decision to close down a substantial section of the UK mining industry.[6] As the strike progressed, the miners began to realize that the might of the establishment—from the police to much of the media—was being marshalled against them.[7] This led to changing perceptions in many individual miners' minds about their relationship with their local policeman, or the newspaper they were accustomed to buying each day.[8]

At the same time, the miners received support from groups who—while their support was welcome in material terms—presented a challenge to the many prejudices typical of an overwhelmingly male workforce. Before the strike, the mining community was largely defined by traditional values, with a miner's wife expected to have her husband's dinner on the table when he returned from work, despite herself often having a full-time job. But during the strike, through organizations like Women Against Pit Closures, miners' wives and other female supporters played a highly active role, and this affected their mindset.[9] As one miner's wife, Marie Collins, later recalled, 'for the first time the women had some control over their own lives. They weren't just appendages of men—their views mattered. You could see women's confidence growing. They began to challenge the men, to go picketing themselves…I went on speaking tours around Kent and Scotland and even Germany.'[10]

Perhaps the most unexpected pairing during the strike was between the miners and the LGBT+ community.[11] The organization Lesbians and Gays Support the Miners fundraised in gay pubs and clubs in cities and took the proceeds to mining districts. The experience challenged homophobic prejudice in the coalfields and increased understanding of the strike in the LGBT+ community. Miners returned the solidarity by marching with their union banners on the 1985 Lesbian and Gay Pride demonstration in London.[12]

Instances like these raise questions about what is happening in the brains of individuals taking part in social movements. Recall here Valentin Voloshinov's view of human consciousness as a boundary phenomenon, having its origins in social discourse but situated in the individual mind, as well as being a kind of dialogue.[13] This view has important implications for how the meaning of words develops and changes in different circumstances. As Voloshinov put it, 'meaning does not reside in the word or the soul of the speaker or the soul of the listener. Meaning is the effect of interaction between speaker and listener produced via the material of a particular sound complex. It is like an electric spark... Only the current of verbal intercourse endows a word with the light of meaning.'[14]

One possibility this raises is that social interactions that occur outside normal experience may have transformative effects on individual consciousness. For instance, during the British university strikes, one striker noted how on the picket lines 'conversations arose and we had space to discuss issues and talk together. It was the kind of time and space we don't get in our busy lives in the university.'[15] Another described how 'the local community centre became our alternative university', with free seminars and workshops offered to all about a range of social and political issues. Not only did such initiatives result in 'interdisciplinary' interactions, but also they involved individuals of different ages and status within the university.[16] Perhaps most importantly, they created the sense of a 'public university open to everyone', which contrasted with the current system in Britain in which students now pay fees, and are increasingly defined as 'customers'.[17]

The creation of a more open and fluid space for debate may be one reason why discussions among participants in this dispute often moved from the specific issue of pensions to debates about other matters, particularly those affecting women and black, Asian, and minority ethnic (BAME) individuals. So at the University of Oxford where I work, there were debates on the picket line and in 'teach-ins' about the gender gap in pay and lack of childcare facilities, and also the fact that BAME individuals are underrepresented as both students and senior staff in top British universities like Oxford. Another issue discussed was the sense of an increasing lack of democracy and representation in governance structures. Notably, the very different social environment of the strike seems to have helped create a space in which the nature of such institutional structures could be questioned and challenged.

While clearly a very different dispute, the prominent role that women played in the miners' strike also led to transformative experiences.[18] Before the strike, miners' wives were mainly situated in the home. But by becoming involved in the support networks that played a crucial role in sustaining the strike, theirs became a collective experience, and also brought them into contact with much wider groups of people than would normally have been the case. This in turn created a space in which some women began to contest assumptions about their place in society and to raise questions in their minds about issues like gender and sexuality.

While the miners' strike was ultimately unsuccessful, in other historical eras individual struggles have acted as a catalyst for mass movements that successfully challenged the very idea of what is possible in society. One example was the 'new union' movement of the late nineteenth century in Britain, partly because the workers who triggered it—the 'match girls' as they were known—were initially seen as unlikely to act as a catalyst for anything at all.

If anyone had been asked at the start of 1888 which workers were most likely to go on strike, the employees of Bryant and May's match factory in east London would probably have been very low on their list. The age of the workers, who could be as young as 13 years old, their lower status as females, and the 'unskilled' nature of their work—at a time when trade unionism was the preserve of skilled, male workers—made the match girls an unlikely spearhead of a new movement.[19]

In June 1888, a journalist called Annie Besant wrote an article exposing the conditions at the factory.[20] She drew attention to the fact that employees worked 14 hours a day for less than five shillings a week, while exposure to the white phosphorus used to make the match heads could cause 'phossy jaw', a degenerative disorder in which the whole side of the face turned green and then black, discharging foul-smelling pus, and which could lead to death. Initially, it seemed as if the article's publication—meant to highlight the match girls' plight—had backfired, for certain girls who had spoken to Besant were promptly sacked.

However, the company had not reckoned on the response of its young workforce. Immediately, hundreds of girls stopped work and soon all 1,400 workers were on strike. Particularly interesting was the way the girls themselves described their action. As one striker noted, 'one girl began, and the rest said yes, so we all went. It just went like tinder.'[21] Looking to express what she and her fellow workers had done, the striker had found a metaphor for their actions in the product—the fire-starting match—produced in the factory.

This metaphor acquired an even greater resonance the following year. For the success of the strike—after three weeks of intense activity involving regular mass meetings, a strike committee, and collections from other workers—led to other unskilled workers also striking for better pay and conditions. A report in the *East London News and Advertiser* in September 1889 gave a flavour of the movement: 'The present week might not inaptly be called the week of strikes...parcels postmen; car men; rag, bone and paper porters and pickers. The employees in jam, biscuit, rope, iron, screw, clothing and railway works have found some grievance, real and imaginary, and have followed the infectious example of coming out on strike.'[22] But it was the initial action of the match girls, the report concluded, that had precipitated this huge strike wave, and picking up on the girls' own metaphor, their action was described as the 'proverbial small spark' that had 'kindled a great fire'.[23] Many British trade unions still in existence today came into being during this period of action.

The speed with which people's generalized mass actions can spread means that some of the biggest strike movements in modern history have often surprised even those sympathetic to the idea of such action.

In January 1968, the left-wing philosopher Andre Gorz wrote that 'the working class will neither unite politically, nor man the barricades, for a 10 per cent rise in wages, or 50,000 more council flats. In the foreseeable future, there will be no crisis of European capitalism so dramatic as to drive the mass of workers to revolutionary general strikes.'[24] Yet only months later in May, unrest began in France with a series of student protests which then spread to factories with strikes involving 11 million workers—more than 22 per cent of the population at the time. Eventually, order was restored, but only following major concessions to the strike movement, and to this day, the events of May 1968 are considered as a cultural, social, and moral turning point in the history of France.[25]

Successful revolutions—events by which a society's whole structure is transformed—happen rarely, but when they do their consequences are profound, so a common term used to describe the English revolution of 1642–9 was 'the world turned upside down'.[26] A sense of exhilaration is also common in revolutions' initial stages. The English poet William Wordsworth expressed the hopes of a generation about the French revolution of 1789 when he wrote, 'Bliss was it in that dawn to be alive. But to be young was very heaven!'[27] After the 1917 Russian revolution, Alexander Luria, a psychologist who worked closely with Vygotsky, wrote, 'I began my career in the first years of the great Russian revolution. This single, momentous event decisively influenced my life and that of everyone I knew.'[28]

Such views reflect the fast-moving pace of revolutions, the way they can challenge long-held views about society and human relationships, but also the unique space they can often offer to younger people to make their mark on the world. Note that the frontal lobes—the areas of the brain particularly involved in planning, reasoning, and judgement—are still maturing in young people, and this may be one reason that individuals at this age can make decisions that may seem to be guided more by emotion than cold rationality—behaviour generally considered a sign of immaturity.[29] However, in a revolution, it might mean younger people are far less likely to be influenced by what Karl Marx called the 'tradition of all dead generations [which] weighs like a nightmare on the brains of the living.'[30]

Of course, revolutions can also disappoint, and this reflects the fact that people who drive a revolution forward, especially in its early stages, are not always those who ultimately benefit from it. These are complex political issues that I can barely touch upon here. What I would like to do though is try to connect some general features of revolutions to the sorts of changes that may be going on in the consciousness of individual participants.

To an even greater degree than other types of mass movement, a common feature of revolutions is that they rapidly progress from specific and limited demands to ones that can challenge the whole structure of society.[31] So how is such a change, from the specific to the general, reflected in individual human minds? To address this question, we need to return to the notion of thought as a dialogue. This can mistakenly give the sense that thought is as transparent as a conversation between two individuals. But the inner dialogue that drives an individual consciousness is far less defined in terms of both its structure and the meaning of the words and phrases. And we also have to consider the role that the unconscious may play in the human thought process, and how this might be enhanced during periods of upheaval.

All this means that a particular individual's consciousness may contain contradictory elements, and these are likely to be heightened during mass movements, but particularly when such movements coalesce into a full-blown revolution. The reason is that the ideas a person may have grown up with through conversations in the family home, or their education at school or university, can all be challenged by what they experience in a major social movement. But this does not prevent a person having contradictory ideas even at the height of such a movement—a point particularly developed by the early-twentieth-century Italian political activist and thinker Antonio Gramsci—and these can continue to coexist for some time.[32]

One consequence of such a contradictory consciousness is that the majority of people in a mass movement may initially be reluctant to 'go too far' in their demands, and thereby look to leaders who espouse a more limited change in society.[33] However, as a revolution progresses and throws up practical dilemmas and contradictions, this can lead to a more questioning attitude amongst some people, and an increased willingness to follow leaders of a more radical persuasion. This can explain how individuals who are initially minor players in society can come to play a much more prominent role as a revolution unfolds.

For instance, in the revolution of 1642–9 that ended feudalism in England, Oliver Cromwell was initially quite a marginal figure. However, his determination to defeat the Royalist forces, as opposed to the half-hearted efforts of many other Parliamentary leaders, brought him to prominence. One exchange between Cromwell and the Earl of Manchester—the first Parliamentary commander—that showed such a difference in attitude occurred after the battle of Edgehill in 1642. Manchester complained that 'if we beat the king 99 times, yet he is still the king...but if he beat us but once we shall all be hanged,' to which Cromwell replied, 'if this be so why did we take up arms at first? This is against fighting hereafter.'[34]

This determination to succeed led Cromwell to select officers based on ability, and willingness to prosecute the war as ruthlessly as required, rather than on their perceived social status. As Cromwell himself expressed it: 'I had rather have a plain, russet-coated Captain, that knows what he fights for, and loves what he knows, than that which you call a Gentleman and is nothing else.'[35] Such a new approach radicalized the constitution of the army, but itself reflected a radicalization taking place in English society in general. And for a short while, this led to a very radical group indeed, the Levellers, coming to prominence.[36]

Similar progressions in public feeling have typically occurred in other revolutionary situations. For instance, in Russia in 1917, the initial beneficiaries of the revolution which began in February were moderate political parties. However, as these failed to deliver on key demands of many peasants and workers—for bread, land rights, and an end to participation in

World War I—a more radical political group—the Bolsheviks—began to gain support.[37]

Two common mistaken viewpoints exist regarding revolutions. Some see them as purely spontaneous affairs; others consider them to be the work of a few conspirators with the masses as mere spectators. Yet neither viewpoint adequately reflects the reality of how genuine revolutions progress. Regarding spontaneity, it is true that revolutionary situations can progress very rapidly, reflecting a radicalization among the mass of people. This is one reason why revolutions often seem to come out of the blue, surprising even many would-be revolutionaries. The Russian revolution of February 1917 began with a demonstration led by women campaigning for more bread for their families.[38] Initially the women were encouraged to return home by local Bolshevik male leaders, who feared the demonstration might be used as an excuse for a crackdown. However, the women ignored this advice and kick-started a movement that ended the centuries-old Tsarist monarchy.

Important as spontaneity is, successful revolutions have also always involved political leadership. A key aspect in the radicalization of the English revolution of 1642–9 was the agitation carried out by the Levellers. Through the use of petitions, pamphlets, and public demonstrations, this group built a movement that influenced both members of the army and the public as a whole.[39] The Bolsheviks only gained a majority in the 'soviets'—the elected bodies that became an organizing forum for the revolutionary workers and peasants—after months of agitating in the factories and streets.[40] Having then achieved such a majority, the Bolsheviks were able to use the soviets to mount a challenge for control of the state.

If mass social movements involve major changes in the ways people perceive what is acceptable, and possible, in society, how might this relate to brain function? In particular, two parts of the brain that we have already mentioned seem to be important in this context—the amygdala and the prefrontal cortex. As befits our status as rational beings, the prefrontal cortex plays a particularly important role in humans in terms of our interactions with others. Yet as we have seen there is increasing recognition that a lot of what we take for granted about the state of the world—and this can include prejudices about others based on their skin colour, nationality, sexual persuasion, or gender—also involves unconscious feelings that stem mainly from the amydala.[41] But the interaction between these two brain regions is dynamic, and in a situation of rapid social change as in a mass movement, I believe that there is the potential for the prefrontal cortex to play a much more dominant role.

The reason for this is that mass movements force individuals into unfamiliar situations and lead to new types of social interactions with other people in which language plays a key role. And this restructuring of neural impulses in the prefrontal cortex could then start to challenge the type of 'unconscious biases' that are rooted in the amygdala. In fact recent studies have shown that the prefrontal cortex itself is far from homogenous in this respect. Instead, such studies have shown that while the ventromedial prefrontal cortex is involved in guiding our future actions based on things that have happened to us in the past, the dorsomedial part of this brain region in contrast allows us to adjust our behaviour to a rapidly changing situation.[42] This makes evolutionary sense for while past experiences may be a good guide to action in a relatively stable environment, they may not remain so in times of great uncertainty. And therefore the dorsomedial prefrontal cortex is likely to play a more dominant role during the rapidly unfolding events characteristic of major social upheavals.

Previously in this book we have looked at different types of mental disorders, and whether they are a product of a defective biology or social environment. So how is mental disorder regarded in mass social movements? One interesting possibility, given the way revolutions are said to 'turn the world upside down', is that this radical disruption of the established order may provide a space for individuals with views of the world deemed unconventional, and perhaps even highly abnormal, to come to the fore in a way that would be unlikely in a calmer social climate. Théroigne de Méricourt was an activist in the 1789 French revolution. Initially Méricourt was welcomed into the ranks of the radical Jacobin party, so much so that she 'was not relegated to the gallery with her sisters, but [allowed to sit] in the body of the hall with the men…wearing a semi-military costume.'[43] However, this relationship soured when Méricourt tried to organize a radical organization for working women. In doing so, she asked, 'Fellow women citizens, why should we not enter into rivalry with the men? Do they alone lay claim to have rights to glory…let us open a list of French Amazons; and let all who truly love their Fatherland write their names there.'[44]

Such feminist views were too radical even for the Jacobins. However, this souring of relations did not prevent Méricourt being involved in key further events such as the storming of the Tuileries Palace in 1792. Ironically, her fall from grace in the revolution was precipitated by other women, for in 1793, as she crossed the Feuillants Terrace to deliver a speech, she was attacked by female supporters of the Jacobins, who stripped her naked and beat her.[45] After this, Méricourt never made another public appearance. Tragically, in 1794, she was certified insane and sent to the Pitié-Salpêtrière Hospital in Paris, where she was said to crawl on the floor like an animal, eat straw, and strip off her clothes.[46] According to novelist and commentator Hilary Mantel, Méricourt's 'madness was not without eloquence…She denounced her keepers as royalists, and spoke of decrees and government measures, addressing her words to the Committee of Public Safety. But the great Committee was long since disbanded, its members guillotined or in exile. For Théroigne time had stopped, some time in 1793.'[47] Méricourt eventually died in the Pitié-Salpêtrière in 1817, at the age of 54.

Méricourt's life raises questions about madness and the passions aroused during a revolution. Was her psychotic state triggered by the trauma of her public beating? Or did elements of insanity exist in her before this time, but in the context of the revolution made Méricourt a bold and inspiring figure to others, at least for a while? In fact, the very passions that made Méricourt an influential leader during the revolution may in a different context have led to her later descent into madness. Or alternatively, Méricourt's madness may have been a way of dealing mentally with the retreat, as Napoleon Bonaparte took power, from the original egalitarian ideals of the revolution. For someone said to have lavished her love, not 'upon a man or upon men, but upon an idea, the passion for justice, for liberty', perhaps psychosis was Méricourt's way of not having to come to terms with reality.[48]

Certainly, a feature of the French revolution in its first days was its enlightened attitude to mental disorder, illustrated in Robert-Fleury's 1876 painting *Pinel Freeing the Insane*, which shows the doctor Philippe Pinel unchaining female patients after the revolution.[49] As medical historians Elizabeth Fee and Theodore Brown have observed, 'the new "moral therapy" developed by Pinel and his contemporaries in the reformed asylums was fundamentally based on the idea of freeing the mental patients' trapped humanity. In theory this allowed for a therapeutic doctor–patient alliance that was sensitive to the life situations and social circumstances of the "madmen" and "madwomen" who were formerly treated as subhuman.'[50] Unfortunately, the reality turned out somewhat different. According to sociologists David Pilgrim and Anne Rogers, 'like the workhouses, asylums quickly became large regimented institutions of last resort, which if anything were more stigmatizing. Although they were run by medical men, they failed to deliver the cures that a medical approach to insanity had promised.'[51] The new approach had run up against the limits of medical knowledge, but also reflected the determination of those now in power to separate out and segregate those individuals who were able to work from those who were not.

That revolutions can end up disappointing and even murdering some of their key protagonists is one reason why these events are often presented in a negative light in history lessons in schools and universities. Yet those who see failure and disappointment as an inevitable aspect of revolutions tend to ignore the reasons why revolutions often fall short of expectations. One factor to consider is that those who play the most active role in a revolution may not be its ultimate beneficiaries. This was the case in the English and French revolutions in which groups like the Levellers and the Jacobins played key roles in driving the revolution forward, but were later sidelined by other interests. This reflects the fact that at certain stages in history a particular class in society may be able to play a highly important role in getting rid of the old order, but is then not in a material position to reap the benefits of the revolution.

A key force in the English revolution was the so-called 'middling sort', which encompassed shopkeepers, artisans, and the lower ranks of the gentry.[52] But although such people could put forward demands that would reverberate for centuries to come—for instance, for universal suffrage and women's rights—ultimately it was the new capitalist class, represented by individuals like Oliver Cromwell, who were best placed to take control. And an important aspect of taking control in this way was to limit the demands of the revolution.

A particularly radical demand of the English revolution was expressed by the Leveller Colonel Thomas Rainsborough when he said, 'I think that the poorest he that is in England hath a life to live, as the greatest he; and therefore every man that is to live under a government ought first by his own consent to put himself under that government; and I do think that the poorest man in England is not bound in a strict sense to that government that he hath not had a voice to put himself under.'[53] Yet since the working class as we know it did not yet exist, and instead there were only the 'middling sort' and peasantry, such expressions of universal rights could ultimately not compete with more limited, but realizable goals.

In the Russian revolution of 1917 workers and peasants took power in a country for the first time in history. Of course, the hopes for the revolution later ended in tragedy with the rise to power of Stalin. It is not clear exactly how many people died in the Stalinist gulags or because of the Soviet Union's botched agricultural policies, but the total figure may run into millions.[54] This human tragedy is often used to back up the idea that revolutions always end in tyranny, but it could also be used to illustrate the terrible nature of reaction and counter-revolution. The radical break between the original ideals of the 1917 revolution and the regime that subsequently developed under Stalin is shown by the fact that of the 15 members of the first Soviet government, ten were murdered on Stalin's orders, four died naturally, and only one, Stalin, survived.[55] But the new regime also marked a break with 1917 in other ways. For instance, the radical view of human consciousness developed by Lev Vygotsky that has inspired this book was also suppressed by Stalin after Vygotsky's death in 1934, as were the exciting new approaches to education and disability that accompanied this view.[56]

In addition, far from developing new ways to help those in mental distress, the repressive regime that developed under Stalin and his successors began to use the term 'mental illness' to label those who opposed the regime. As Nikita Khrushchev remarked in 1959, since it should be impossible for people in a 'communist' society to have an anti-communist consciousness, 'of those who might start calling for opposition to Communism on this basis, we can say that clearly their mental state is not normal.'[57] In 1969, Yuri Andropov issued a decree on 'measures for preventing dangerous behaviour (acts) on the part of mentally ill persons.'[58] Consequently, psychiatrists began to diagnose and confine anyone who fitted the description of a political agitator. According to the archives of the International Association on the Political Use of Psychiatry, at least 20,000 USSR citizens were sent to mental institutions for political reasons, but the real figure may be much higher.[59]

A parallel tragedy had previously taken place in Germany in the 1930s with the rise to power of Adolf Hitler. In what became a prelude to the Holocaust that would eventually result in the murder of 17 million people, a pseudo-scientific view that labelled people with mental health problems and learning disabilities as 'degenerate' led to the extermination by the Nazis of 200,000 such individuals by the end of World War II.[60] In what British psychiatrist Tom Burns has described as German psychiatry's 'most shameful chapter', the 'extermination of the mentally ill is compounded by several prominent psychiatrists leading it and none vigorously opposing it...The broad mass of the profession probably did not share the extreme views articulated, but they voiced no effective opposition.'[61]

Such historical developments raise the question of how all those people who backed the initial revolution in Russia in 1917 with such high hopes could allow it to end in such a state of terror. Or for that matter, why did so many individuals in Nazi Germany either participate in genocidal activities or remain passive while others carried them out? For some commentators, the answer is simple—all mass movements are inherently likely to end in violence and terror, because of the primeval passions unleashed within such a 'mob'.

This view is particularly associated with nineteenth-century French psychologist Gustave Le Bon who argued that in a crowd the individual yields 'to instincts which had he been alone, he would perforce have kept under restraint'.[62] Like a hypnotized person, 'he is no longer conscious of his act...He is no longer himself, but has become an automaton who has ceased to be guided by his will...In the crowd he is barbarian. He possesses the spontaneity, the violence, the ferocity and also the enthusiasm and heroism of primitive beings.'[63]

In fact Le Bon had in mind a very specific type of individual when he wrote about crowds: the 'lower orders', whom he considered simple minded, as indicated by his claim that 'ideas being only accessible to crowds after having assumed a very simple shape must often undergo the most thoroughgoing transformations to become popular. It is especially when we are dealing with somewhat lofty philosophical or scientific ideas that we see how far-reaching are the modifications they require in order to lower them to the level of...crowds.' Le Bon believed that such crowds are particularly susceptible to the promises of demagogues, since 'to know the art of impressing the imagination of crowds is to know at the same time the art of governing them.'[64] Notably, both Hitler and Benito Mussolini showed interest in Le Bon's book *The Crowd: A Study of the Popular Mind*, but then so did Sigmund Freud, who agreed with much of what it said about the primitive nature of crowds.

Yet recent studies of the psychology of crowds have undermined Le Bon's claim that individuals in a crowd tend towards barbaric forms of behaviour, or that they are dupes who will unquestioningly follow the instructions of the first demagogue that comes along. Investigations over recent decades of crowds at events including soccer matches, public demonstrations, festivals, and riots, conducted by British psychologists

Stephen Reicher, Clifford Stott, and John Drury, have found that individuals in crowds do not abandon their rationality or surrender their identity to a mob mentality.[65] Such individuals do, however, become highly sensitive to what those around them are doing, and become strongly cooperative as a result. In line with this, eyewitness accounts of the revolutions mentioned above confirm the high degree of creativity of individuals in such movements as well as their tendency to show tolerance towards, and promote the rights of, more vulnerable members.[66]

Indeed, an important feature of any genuine revolution—in this case meaning a progressive, fundamental change in society—is the extent to which it tends to challenge divisions between people. Notably the English revolution of 1642–9 was intertwined with a cry for religious freedom. At a time when social relations tended to be defined in religious terms, this was a radical development. The Levellers were particularly forward-thinking in this respect, for they campaigned for the rights of Protestants to worship as they saw fit, but also opposed Cromwell's subjection of the Catholic citizens of Ireland.[67]

The French revolution that began in 1789 expanded rights of citizenship to minorities such as Protestants and Jews.[68] Similarly, in 1776, the revolutionaries who wrote the American Declaration of Independence stated that 'all men are created equal, that they are endowed by their Creator with certain unalienable Rights, that among these are Life, Liberty and the pursuit of Happiness,' although in practice it would take 87 years and a bloody civil war for most African Americans to gain their 'unalienable Rights', and a further 100 years of courageous protests before those rights could be fully exercised.

In pre-revolutionary Russia, persecution of Jews was state policy, with pogroms actively encouraged by the authorities.[69] In contrast, on gaining power, the Bolshevik government abolished all legal restrictions on Jews. The role that Jews played in the Bolshevik party was shown by the fact that the central committee of the party in October 1917 contained six Jews out of 21 members in total.[70] Lev Vygotsky and Alexander Luria were typical of the many Jewish intellectuals who initially thrived after the revolution.

The position of women in society has been a central issue in revolutions. Women made decisive interventions in the English revolution of 1642–9, particularly in radical groups like the Levellers.[71] In the French revolution, women organized to move the revolution forward at some of its key moments.[72] Women also played major roles in the 1917 Russian revolution as mentioned. This was reflected in the measures enacted after the revolution: women were given the vote, abortion and LGBT+ rights were legalized, contraception was made freely available, and state-run nurseries, laundries, and restaurants were set up.[73]

If such examples appear to be from the distant past, it is worth mentioning a more recent example—the Egyptian revolution of 2011. A retired businessman who travelled from Alexandria to join the protesters in Cairo gathered to demand the fall of dictator Hosni Mubarak noted: 'I found [there] something lovely. There were all kinds of people. From universities, secondary schools, preparatory schools. Homeless people. People from every religion. All divisions disappeared. Everyone had one purpose. I was really crying, for this was the first time I saw the Egyptian people unafraid of anything.'[74] Sadly, this lack of fear turned out to be brief, for although Mubarak was indeed deposed by the mass movement, and a democratically elected leader—Mohamed Morsi—installed, a new military dictatorship soon took power. Subsequently, there has been a major crackdown by the new regime on human rights, particularly against women, LGBT+ individuals, and religious minorities.[75]

A similar reaction took place in Russia in the 1930s under Stalin, with women's rights replaced by a celebration of motherhood and the family; indeed, women were even given medals for having large numbers of children. In addition, Russian Jews and other oppressed minorities once more became a scapegoat for society's troubles, and having championed the right of nations to self-determination, the Soviet Union under Stalin reverted to the same 'prison house of nations' that it had been under the Russian Empire of the Tsarist regime.

Ironically, despite their claims to represent opposite poles of the political spectrum, both the Soviet Union under Stalin and the German Nazi regime of Adolf Hitler were remarkably similar in their reactionary attitudes to women, LGBT+ individuals, and ethnic and religious minorities. But perhaps this is not so surprising, given that both regimes were based on a totalitarian system that moved quickly and ruthlessly to quash any dissent. Obviously one factor to consider here is the potential risk in voicing opposition to a dictatorship such as those led by Hitler or Stalin. But it is also worth asking what psychological explanations there might be for such apparent mass acquiescence with evil.

One possibility would be to return to the claims made by Gustave Le Bon about the malleability of crowds, which by definition he considered to be made up of lower class people of an inferior intellect. Yet a problem with this viewpoint is that the Nazis in Germany, as well as Mussolini's Fascists in Italy, derived their main support from those respectable middle class individuals—doctors, lawyers, civil servants, and the like—that Le Bon saw as the primary audience for his speculations about the defects of the lower classes. And lest this be seen as purely a historical matter, studies have shown that the alarming rise in support for far-right and fascist parties in countries across Europe in recent years has been particularly driven by the popularity of such parties among the middle classes.[76] This then poses the question—why might fascism be particularly attractive to this layer of society?

One person who speculated about this issue was the German psychoanalyst Wilhelm Reich. According to him, the middle classes in Germany prior to Hitler's rise to power both craved authority and rebelled against it, this ambiguity being a reflection of a feeling of being caught between wealthy capitalists on the one hand and the powerful workers' movement that existed in Germany in the 1920s.[77] Reich believed that the intense economic crisis of the post-war years brought these anxieties to a head. Reflecting his psychoanalytic background, he also sought to link the tendency of this social layer to submit to an authoritarian regime to the sexual repression that Reich believed was a strong feature of middle-class society at this time.[78] Today, the economic crisis is not yet as severe as in the Great Depression, yet the sociologist Oliver Nachtwey believes that the German middle classes, after decades of upward social mobility, now feel threatened by a 'downward escalator'.[79] Nachtwey believes that 'for the lower middle class...harsh social competition, the struggle for prosperous life, and the disappointed expectations of ascent and security [has led] to a "brutalisation" of social conflict...[where] fear of downward mobility produces a very specific authoritarianism.'[80]

Of course it is one thing to hold racist or authoritarian ideas; it is another to take part in the murderous activities that characterized the Holocaust. So what might have motivated people to take part in such activities? One famous scientific study that has strongly influenced our thinking in this area was carried out by US psychologist Stanley Milgram from Yale University in the 1960s, who wanted to test the claim by the German Nazi Adolf Eichmann, during his war crimes trial, that he and his accomplices in the Holocaust were 'just following orders'. Milgram showed that it was possible to persuade volunteers to 'electrocute' other human 'learners' by telling them that this was a necessary part of the study.[81] Although the learners appeared to scream in pain, in reality they were only acting and there was no electric shock. More recently, Tomasz Grzyb, from the University of Social Sciences and Humanities in Warsaw, carried out a similar study and confirmed Milgram's findings. In this study, 80 volunteers were recruited, including women as well as men, and 90 per cent were shown to be willing to inflict the highest shock level of 450 volts to a complicit 'learner'.[82] Just as in Milgram's experiment, volunteers were spurred on by prompts from the supervising scientist such as 'the experiment requires that you continue', and 'you have no other choice, you must go on'. Interestingly, women were far less likely to obey the orders than men. For Grzyb, the study 'illustrates the tremendous power of the situation the subjects are confronted with and how easily they can agree to things which they find unpleasant.'[83]

On the face of it, studies such as these seem pretty damning in reinforcing the idea that most people will acquiesce with authoritarianism and evil. Yet although troubling, there is surely a danger in extrapolating from a highly artificial laboratory study, without also surveying the actual ways that individuals have reacted to real-life authoritarian regimes like that of Adolf Hitler. What such a survey shows is acquiescence by many, but also resistance and the greatest bravery by some individuals, even in the most difficult of circumstances.

For instance on a recent visit to Munich, I was moved by the monument to Sophie and Hans Scholl, who were part of the White Rose student movement that secretly distributed anti-war leaflets in cities across Germany from 1942 onwards. Captured by the Nazis in February 1943, the Scholls were tried and swiftly executed. Just before her death, Sophie said: 'Such a fine, sunny day, and I have to go, but what does my death matter, if through us thousands of people are awakened and stirred to action?'[84] I could also cite other examples, such as Oskar Schindler, the industrialist and Nazi party member who nevertheless saved more than a thousand Jewish individuals from the death camps, at great risk to himself.[85] Or there is my own personal favourite, the biochemist Jacques Monod, who by day uncovered the mechanisms by which genes are switched on or off, and by night clandestinely fought the Nazis by blowing up railway lines and bridges as part of the resistance in occupied France.[86]

Such examples confirm that not only during progressive social movements, but also at the heights of deepest reaction, individual actions can make a tremendous difference to events. But we also need to explain in terms of brain function how it is that some individuals react to such social events in a progressive fashion, yet others acquiesce with, or even support, reaction. And here, given what I said earlier about the likely changing interaction between the amygdala and the prefrontal cortex in times of social upheaval, there may be some fruitful future research to be done in terms of studying how this interaction differs between individuals who acquiesce and those who rebel during such upheavals.

The main argument of this book is that the human brain is unique compared to those of other species because it has been transformed by culture. What is evident from studying the influence of genuine revolutions in society, compared to less profound transformations, or indeed movements that represent outright reaction, is their effects upon human culture. For a feature of all genuine revolutions is the huge progressive impact they have upon areas of human endeavour ranging from art and literature to science and technology, whereas periods of reaction generally impede such progress. However, it is also important to recognize there is a highly complex relationship between these areas of culture and social change. Human creativity can thrive when society is moving forward in a progressive fashion, but it can also be stimulated by the tensions and problems thrown up by society. In the next few chapters, I will explore how particular human societies, and the tensions within those societies, have stimulated the various arts and sciences, and how these in turn have affected the development of human conscious awareness. And to do so, what better place to start than with what may be one of the first human cultural pursuits—music.

PART VI

MIND AND CULTURE

CHAPTER 18

MUSIC AND RHYTHM

W herever humankind exists on planet Earth, there music will also be found. As musician Dave Randall, guitarist with the bands Faithless and Slovo, has observed, 'Look around any crowded street, bus or subway carriage almost anywhere in the world, and you see people experiencing music. It spills from a sea of headphones and reverberates from shops, cars, buskers, bars, places of worship and homes...It's part of children's play and expresses our identity as we navigate our way into adulthood. It walks us down the aisle and marches us off to war.'[1] Music has been an important part of human culture for a very long time; the earliest examples of musical instruments—pipes made of bone and horns—date to around 40,000 years ago.[2] And it is likely that other forms of music making that leave no physical evidence, such as singing, or musical instruments made of wood or other materials that have long since decayed, date back far further than this, making music perhaps the oldest art form.

At first glance, making music seems only remotely related to human survival. Why, then, is it so important to us? I would argue that to ask this question in such a form betrays a narrowness in thinking about human existence and its motivating factors. For just as a central theme of this book is that human consciousness differs qualitatively from that of other species, so we should not be afraid to look for explanations for human behaviour that go beyond basic needs. Instead, we should consider again the specific features of humanity that distinguish us from other species, and how these might affect different forms of culture.

We have already noted that two key distinguishing features of *Homo sapiens* are our ability to design and use tools to transform the world around us, and our capacity for conscious awareness, which itself, I have argued, is based on the framework of abstract symbols that constitutes human language. At the same time, we should be open to the possibility that our ability to produce and appreciate music might also be based on some evolved characteristics of both mind and body that we share with other non-human species.

Following Charles Darwin's revelation that humans are as much a product of natural selection as other species, there have been plenty of attempts to try and explain our musical skills and appreciation of music with reference to our link with the animal kingdom. Indeed, Darwin himself suggested such a link in 1871. Noting that many male animal features evolved not to aid survival directly, but for successful courtship of the female—a famous example being the peacock's tail—Darwin proposed that some human features might also be selected in this way, including music. As he put it, 'primeval man probably first used his voice in producing true musical cadences as do some of the gibbon-apes at the present day; and...this power would have been especially exerted during the courtship of the sexes.'[3]

As an idea, this has some plausibility. In animals ranging from birds to whales and gibbons, the male's song plays an important role in courtship. And females in these situations can be demanding listeners. For example, in some bird species females prefer more complex songs, putting males under pressure to evolve mechanisms to create and sing them.[4]

The perceived link between love-making and music is not new; indeed, it inspired Shakespeare's famous lines: 'If music be the food of love, play on, give me excess of it.'[5] And a common perception about rock stars is that they are particularly good at attracting sexual partners. Jimi Hendrix reportedly had sex with hundreds of female fans during his brief life, and Robert Plant, the lead singer of Led Zeppelin, has claimed that 'whatever road I took, the car was heading for one of the greatest sexual encounters I've ever had.'[6]

Supporters of a link between music and sex have also pointed to the fact that human musical productivity seems to match the course of an individual's reproductive life. For instance, Geoffrey Miller, of the University of New Mexico in Albuquerque, has studied jazz musicians and found that their output rises rapidly after puberty, reaches its peak during young-adulthood, and then declines with age and the demands of parenthood.[7]

Yet such claims fall apart once submitted to proper scrutiny. In the animal kingdom, only males use song to court females, but in human society women are also excellent singers. Tying music too closely to sex also seems at variance with the complex ways by which this art form influences us.[8] A man does not have to be gay to appreciate an all-male choir, and a heterosexual woman can enjoy listening to a soprano. But undoubtedly the biggest problem with the idea that our musical capacity is analogous to animal courtship sounds is the fixed nature of the latter. In contrast, a unique feature of music is its immense variety.

Another explanation for the centrality of music in human society is that it plays a key role in coordinating activity. For instance, psychologist Annett Schirmer of the University of Singapore has argued that rhythmic sound 'not only coordinates the behaviour of people in a group, it also coordinates their thinking—the mental processes of individuals in the group become synchronized.'[9] This may explain how drums unite tribes in ceremony, why armies used to march into battle accompanied by bugle and drum, religious ceremonies are infused by song, and speech is punctuated by rhythmic emphasis on particular syllables and words.

Schirmer's claims are based on a study her team carried out in which people were asked to study a series of images and then identify when an image was shown upside down, while a drum tapped out a simple rhythm in which the fourth beat was always skipped. Participants identified the inverted image much more rapidly when it was shown at the same time as the missed beat, rather than being out of sync with the beat.[10] This suggests that the brain's ability to make decisions can be enhanced by an external rhythm and heightened at precise points in synchrony with the beat. And an electroencephalogram (EEG) recording of participants' brain waves showed that the waves became synchronized around the external rhythm, while also revealing a more surprising effect of rhythmic sound on brain function.[11] Normally, seeing a picture or hearing a sound generates a brain wave in the cerebral cortex region where such information is processed, and indeed a wave in the visual cortex occurred each time the image was presented. But if the image was presented simultaneously with the missing drumbeat, the wave was bigger than when the image was presented out of rhythm.

Schirmer believes the findings indicate that our brains evolved in such a way that music can coordinate group activities, noting that 'rhythm facilitates people interacting by synchronizing brain waves and boosting performance of perception of what the other person is saying and doing at a particular point in time. When people move in synchrony they are more likely to perceive the world in synchrony, so that would facilitate their ability to interact.'[12] Such findings are particularly interesting given the recent evidence mentioned previously that brain waves can coordinate the actions of different regions of the brain.

How might an ear for rhythm have helped our prehistoric ancestors? One possibility is that it helped coordinate the social production of tools. Another is that it could have aided primitive hunters; when humans walk, we make noise, and the sounds of the footsteps of a group of hunters could potentially mask the sound of a predatory animal or other auditory indications of danger. Our ancestors may have evolved to synchronize their steps to create predictable sounds as a group, improving their ability to recognize external rhythms.[13]

If these are some of the theories about how human ability to produce and appreciate music evolved, what can modern neuroscience tell us about how music affects the brain? Valorie Salimpoor and colleagues at the Rotman Research Institute in Toronto studied this question by using functional magnetic resonance imaging (fMRI) to analyse the brains of participants while they listened to excerpts of songs they had never heard before. The participants were then asked to say how much money they would spend on a given song. The study found a correlation between how much people said they would spend and activity in a brain region called the nucleus accumbens, which is involved in forming expectations.[14] Salimpoor's team also identified a link with another brain region, the superior temporal gyrus. Studies have shown that the genres of music that a person listens to over a lifetime have an impact on neural activity in this brain region. This region alone does not predict whether a person likes a given piece of music, but it is involved in storing templates from what they have heard before. For instance, a person who has heard a lot of jazz is more likely to appreciate a given piece of jazz music than someone with less experience. Salimpoor believes this shows that 'the brain kind of works like a music recommendation system.'[15]

In another study, led by Daniel Abrams at Stanford University, participants listened to music by the late Baroque composer William Boyce, chosen because he was unlikely to be familiar to the participants, while their brains were imaged by fMRI.[16] The study identified similar brain activity patterns in the different participants. For Abrams, this shows that 'despite our idiosyncrasies in listening, the brain experiences music in a very consistent fashion across subjects.'[17] Brain regions involved in movement, attention, planning, and memory consistently showed activation when participants listened to the music. As these regions are not directly involved in auditory processing, Abrams believes this means that when we experience music, many other things take place in the brain beyond merely processing sound.[18] These brain regions may help to hold particular parts of a piece of music—such as the melody—in the mind while the piece plays on. But it might also be the case that the study has identified a neural basis for the capacity of music to unite different listeners.

As well as exploring evolutionary reasons for the centrality of music in human society and the neural mechanisms that mediate this, it would be a major omission in a book about the unique aspects of the human mind to neglect the role that music plays in the creation of meaning. Previously, I pointed to language as important not only for communication, but also as something that allows us to make sense of the world, and even develop radical new ideas about that world. In other words, it allows us to create a unique sense of the meaning of life. Yet while language is clearly the most important of the 'cultural tools' that Lev Vygotsky proposed as mediators of human consciousness and which account for the latter's uniquely 'extended' character, music can be viewed as another such tool. Crucially, music possesses the ability to speak to us in a non-verbal fashion, primarily by manipulating our emotions.

Recent research has identified links between music, emotions, and social behaviour. Autistic individuals, for instance, tend to perceive less emotion in music, while those with a condition called Williams Syndrome show a heightened interest in music.[19] People with this syndrome have learning difficulties, but are very cheerful and associate easily with strangers. There is evidence that they have enhanced activity in the amygdala, which, recall, is a brain region particularly associated with the emotions, and increased brain levels of oxytocin, which as we have seen mediates social interactions.[20] It is interesting then they are also more likely to look for and play music, and have heightened emotional responses to music, which tends to reinforce the idea of a link between music, emotional responses, and sociability.

In addition to expressing emotion, there are other similarities to be found when we compare the characteristics of music across cultures. For example, music of any genre nearly always involves some form of complex sound event, such as structured rhythms, or pitch organization, overlying a regular beat.[21] Through allowing people to create something together via such a regular beat, musical sounds may have provided a means by which people could share experiences, thereby fostering social bonds. Similarly, music's capacity to generate emotions that are felt by everyone, yet remain specific to an individual because of their own particular life history and experiences, suggests that it may have arisen as a way of bringing together individual and group feelings, particularly in times of social uncertainty.[22]

Indeed, this particularly seems to be the case in hunter-gatherer societies, the form of society that human beings have existed in for the vast majority of our time on Earth. Studies of such societies, including the bush people of the Kalahari, whose way of life I have already discussed, the Pygmies of Equatorial Africa, and the indigenous people of America, coupled with archaeological evidence from prehistory such as cave paintings, have allowed us to develop a picture of the role of music in this original type of human society.[23] In fact not only music but also dance should be considered as these two activities are intertwined in hunter-gather societies, unlike the situation in modern society where they can be quite separate.[24]

Music and dance are central to hunter-gatherer societies. Previously I have mentioned how labour—here defined as shared work using tools to transform the environment—played a key role in human evolution, as well as the importance of play in the development of conscious awareness during childhood. In modern society, we tend to view play and work as radically different, and even opposed. But in hunter-gatherer societies there is much less distinction between these two activities, and indeed there is a certain playfulness during work that may involve a musical component such as singing or drumming.[25] The music and songs that accompany work in hunter-gatherer societies help to bind the group members in shared activities and can even play a role in hunting, as mimicry of animal sounds as part of a song may be used to lure hunted animals into a trap.

Song and dance have a particularly important role to play in such societies after nightfall. For instance, studies of the bush people of the Kalahari have shown the importance of firelit activities that 'steer away from tensions of the day to singing, dancing, religious ceremonies, and enthralling stories.'[26] Societal concerns, as well as the personal gripes of individuals, are put aside as everyone gathers to make music, dance, or tell stories. These activities often close social rifts and facilitate bonding. Unlike modern society, where music and dance are generally the job of professional performers, in hunter-gatherer societies all group members are expected to take part in what is very much a communal activity.[27] And in these often mobile societies in which individuals have few possessions, while there are dedicated musical instruments, music can also be made with whatever is available, for instance a hunting bow. In addition music and dance are central to gatherings of separate hunter-gatherer groups, where they help dissipate inter-group tensions and cement alliances.

In terms of the effect of music and dance in enhancing social bonds in hunter-gatherer societies, of central importance is the effect of such activities in stimulating the secretion of neurochemicals such as endorphins and oxytocin—which promote well-being and social interaction, respectively—in the brain.[28] In addition, echoing Vygotsky's statement mentioned earlier that play allows a child to act as if 'a head taller than himself', the role-playing element in song and dance can be an important way for individuals in hunter-gatherer societies to explore deeper themes than their concrete, everyday experience.[29] It is not a coincidence that music and dance in such societies play a major role in spiritual matters. Spirituality is a topic that I will explore in greater detail later, but music's alliance with spiritual experiences seems likely to have added a profound value to human life in prehistory.

If such is the role of music and dance in the original form of human society, how have things changed in the class societies that have characterized humanity since the birth of civilization about five to six thousand years ago? Here two important factors need to be considered: first, the divisions that characterize class societies and, second, the cumulative development of technology to an ever accelerating degree in such societies. I would argue that these factors have affected the position of music in society, and its effects on human conscious awareness, in two notable ways. First, the development of new technologies has increased the possible means by which human beings can produce sounds in ways that have at certain times in history led to profound transformations in both the type of music available to humanity and the effects of these on the human mind. Second, changes in the social relations of human society have often had a major—although often far from direct—influence on music styles and their social impact, and thereby can be argued to have increased the diversity and depth of the ways that music is able to speak about what it means to be human.

That music may act as a barometer of social aspirations, but also discontent, in society has long been recognized. In 380 BC, Plato remarked that 'a change to a new type of music is something to beware of…[f]or the modes of music are never disturbed without unsettling of the most fundamental political and social conventions', while the rulers of ancient China set up an office to supervise court music and monitor the musical tastes of the masses, in case these showed indications of social unrest.[30] In twelfth-century Europe, church authorities placed strict limitations on music, with artists only allowed to play chords and patterns judged to be pleasing to the ear. Intervals between notes that went against this idea of harmony were banned. One banned interval, the flat fifth, or tritone, was even branded *diabolus in musica*—the devil in music.[31] Yet despite attempts to stifle it, the dissonance that the authorities tried to prohibit in music became a growing feature of society in the fourteenth century, a period characterized by a wave of peasants' revolts. To try to address the issues that caused the revolts, concessions were made by the authorities—some material, others cultural. One such concession was that the church removed the ban on certain musical intervals.[32]

The progressiveness of such a move is shown by the diversity of human emotions and situations that even a single musical interval like the flat fifth can convey. For on the one hand, as its nickname suggests, this 'devil's tritone' can create a sense of foreboding, and so has been used to create a tense and sinister atmosphere by artists ranging from Richard Wagner, who employed it in his opera *Götterdämmerung*, to the heavy rock group Black Sabbath, who used it in their song of the same name.[33] However, to only associate the tritone with the sinister underestimates its potential range, since it also features in Jimi Hendrix's song 'Purple Haze', and in two love songs—'The Girl from Ipanema' by João Gilberto and Stan Getz, and 'Maria' from the musical *West Side Story*.[34] For Anthony Pryer of Goldsmith's College in London, these varied uses of the same interval reflect the fact that it 'is something that yearns to be resolved. A very good example would be the opening of … Maria. It wants to resolve into the next note. It is a special kind of tension.'[35] Given the ultimate tragedy of *West Side Story*, its use in this case may even have been a deliberate— or unconscious—anticipation of future sadness to come. Such diversity in the range of what a single musical interval can convey—often in potentially unconscious ways in a listener—illustrates how music can sometimes transcend what it is possible to convey through words alone.

Previously I mentioned how the French revolution that began in 1789 had a huge impact on human consciousness, with its slogan *liberté, égalité, fraternité*. Yet a study of how this event influenced music demonstrates that it is not necessarily revolutions per se, but rather the contradictions within such moments of dramatic historical change that can prove most productive in stimulating great artistic endeavours. A sign of this is that far from being generated at the centre of revolutionary change in Paris, the most profound music at this time was being composed in Vienna, a relative backwater, by one Wolfgang Amadeus Mozart.

In Austria there was no decisive break with the old order as happened in France in 1789.[36] Although social progress was achieved during Emperor Joseph II's rule in the 1780s it came from above in an often half-hearted manner. Yet the tensions produced by this contradictory situation were the backdrop for one of the most fertile eras in the history of music, for Mozart is widely considered by many as the greatest composer of all time.[37]

Unlike Johann Sebastian Bach, who worked for the church, or Joseph Haydn, the servant of a local aristocrat, Mozart decided to attempt a career as an independent musician, something almost unheard of at the time. What helped Mozart in this pursuit of artistic independence was that Joseph II initiated reforms aimed at forestalling a revolution like the one in France.[38] These included an attack on church privileges, legal reforms, and less censorship in the arts. It was this more liberal environment that allowed Mozart to live independently and create his masterpieces.[39] Yet the precarious nature of his existence probably also helped hasten his death, and burial in a pauper's grave, at the age of just 35.

Both the revolutionary spirit of the age and its contradictions influenced Mozart's work. His opera *The Marriage of Figaro* was based on Beaumarchais' play of a servant outwitting his master, which was banned in Austria because of its perceived subversive character.[40] Interestingly, for Figaro's aria of revenge on his master, the Count, Mozart set his words to a minuet—a dance form associated with the elegance of the aristocratic ballroom.[41] In this case form and content are clearly opposed, of note given that Vygotsky believed a central feature of great art is that it always contains an 'intimate conflict' between content and form.[42] Here this conflict adds a powerful element of irony to the situation.

Yet perhaps the greatest achievement of Mozart's operas is the way that music is used to portray the many psychological contradictions within human character, behaviour, and intentions. Ever since *Don Giovanni* was first performed there has been a debate about whether the central character is a villain, a hero, or both of these things at once.[43] And while *The Magic Flute* is a thinly veiled celebration of the ideals of freemasonry—in Mozart's era a progressive movement of which the composer himself was a prominent member—we are continually wrong-footed in our estimations of the character, motives, and trustworthiness of the two ruling protagonists, the Queen of the Night and Sarastro. As critic Luke Howard has noted, *The Magic Flute* 'invites the viewer to look past first appearances, and examine the premises and assumptions on which those appearances are based.'[44] Perhaps, writing in 1791 with the two-year old French revolution already throwing up complex practical and moral dilemmas, Mozart was well aware that not everything is reducible to polar opposites.

Moving to a different century, previously I described how scientists are using brain imaging to study the neural mechanisms underlying the creativity of jazz musicians during an act of improvisation.[45] But we should also consider the social and psychological influences that affected the development of jazz. The political activist Malcolm X linked together jazz improvisation and the civil rights movement at a rally in Harlem in 1964 with his claim that 'when a black musician picks up his horn and starts blowing, he improvises, he creates, it comes from within. It's his soul.

Jazz is the only area in America where the black man is free to create.'[46] In fact, the first truly improvisational jazz, bebop, was heavily influenced by the radicalizing influence that fighting in World War II had upon a generation of young black men.[47] The war had supposedly been fought against Nazi racism, yet returning black soldiers found themselves confronted by the same institutionalized racism and public prejudice as before. This had an important impact on the form and content of jazz in the post-war years.

In contrast to the big bands that characterized jazz in the pre-war years, bebop was played by small groups in which the drummer and bassist were on a musical par with the horn and piano soloists, producing a more democratic structure in the band setup.[48] This democratization was also aided by the improvisational style, in which the music was guided by an equal interaction of different musicians, rather than by a sole band leader. Bebop was also played at a much faster tempo than swing. One reason for this was a desire to be taken seriously. Many jazz musicians felt that during the swing era, they were often reduced to providing ambient background music and forced to take a back seat to their audiences—which were often predominantly white. In contrast, Billy Eckstine, who led a bebop band in the 1940s and 1950s, noted that he deliberately rearranged many popular songs to have fast tempos so that the audience could not dance to them, forcing them to pay more attention to the musicians. Here bebop tried to create an equality that wider society still denied to blacks.

Another important development in music since the birth of civilization and class society, compared to that in prehistoric hunter-gatherer societies, is that technological innovations have transformed music in an increasingly rapid fashion over the centuries. And I would argue that this has had an important impact on the range of ideas, feelings, and emotions that music is able to convey, and therefore the influence that it can have on human conscious awareness. One example of this is the way that changes in musical technology in the late eighteenth and early nineteenth centuries—particularly the transition from the harpsichord to the piano—affected the range and power of first Mozart's music, and then Beethoven's.

These two types of instruments differ in that despite being both keyboard instruments, a harpsichord's sounds are produced when a series of quills pluck the instrument's strings, a mechanical process that allows for only limited dynamic variations, while on the piano the strings are struck by hammers, with the varying force exerted on the keyboard by the pianist providing superior dynamic contrast and responsiveness, allowing for unbroken melodic lines and greater emotional expression.[49] Because of this, unlike its predecessor, the piano was able to transmit a far wider spectrum of dynamics, musical colours, and emotions.

A sign of the rapidly growing importance of pianos in Mozart's lifetime is shown by the fact that in 1756, when he was born, pianos were still rare in Germany, had barely arrived on Britain's shores, and were basically unknown in France, but by 1791, when he died, they were being widely produced in all three countries.[50] And this transition was important in Mozart's musical development, shown for instance in the piano sonatas he composed that reflected the new instrument's potential, leaping exuberantly from loud to soft, high to low.

Ludwig van Beethoven, coming on to the world stage just after Mozart, was even more able to realize the true potential of this rapidly developing instrument. The innovations in piano design that continued after Mozart's death provided Beethoven, along of course with his own musical genius, with the means to express his extraordinary creative impulses and the capacity to further break from the confining musical conventions of the eighteenth century.[51]

Of course while noting the influence of technological developments on the development of musical styles may be seen as important from a sociological point of view, how does this relate to conscious awareness and specifically the central thesis of this book, that human consciousness has undergone a qualitative shift compared to that of an animal? We noted previously that music is another cultural tool besides language that has had an important impact on the development of human conscious awareness. This may include new ways of making music. In the case of the piano played by a virtuoso like Beethoven, it allowed him to radically transform the variety and scope of emotional responses that it was possible to elicit in a listener using a musical instrument.[52]

Another demonstration of how an individual can transform our perceptions of the potential of a musical instrument, with regard to not only the listening experience but also the potential impact on the human psyche, can be found in the music composed and played by Jimi Hendrix in the 1960s. Hendrix pioneered the use of the guitar as an electrical sound source. Players before him had experimented with feedback and distortion, but Hendrix turned those and other effects into a controlled, fluid vocabulary.[53] One of his most famous live performances was at the Woodstock festival, held at a rural location north of New York in August 1969—at the height of the Vietnam War—in which he played an instrumental version of the patriotic US anthem 'The Star Spangled Banner'.[54] However, during his performance, Hendrix used electrical feedback to mimic explosions, machine gunfire and a wailing siren—musical images of the horror of war. Hendrix believed his use of the electrical guitar in this way allowed him to tap into emotional responses in ways that traditional playing had not, and indeed he claimed that 'technically, I'm not a guitar player, all I play is truth and emotion.'[55]

It is one thing to select examples of music linked to social and political unrest. But is this really how music in our advanced capitalist society generally connects with people's consciousness? For isn't music's main function nowadays to make money for the billion-dollar industry that manufactures it on a vast scale and act as a distraction from most people's humdrum lives? This was certainly the view of German philosopher Theodor Adorno, who argued that 'the frame of mind to which popular music [appeals] and which it perpetually reinforces is simultaneously one of distraction and inattention... Distraction is bound to the present mode of production, to the rationalized and mechanized process of labour to which, directly or indirectly, masses are subject. This mode of production, which engenders fears and anxiety about unemployment, loss of income, war, has its "non-productive" correlate in entertainment; that is, relaxation which does not involve the effort of concentration.'[56]

Yet it is surely an oversimplification to claim that music—even generic pop music—acts only as a distraction from social issues and does not involve concentration. For pop songs can engage with their audience in quite profound ways through both music and lyrics. Indeed the message coming from the music in such songs may often be at variance with the lyrical content. Consider John Lennon's song 'Imagine'.[57] Lyrically, the song is a plea for equality, and an end to private property, nations, and religion, yet because this is all wrapped in a sweet harmony, it became one of the most popular songs of the twentieth century. As Lennon himself put it, the song 'is anti-religious, anti-nationalistic, anti-conventional, anti-capitalistic...but because it is sugar-coated, it is accepted.'[58] In fact, 'Imagine' is far more complex harmonically than its air of simplicity suggests, and this helps to create a sense of serenity, yet at the same time of restless musing,[59] but the basic point applies.

While Lennon was making a very direct political point, pop songs can still speak of unease and discontent with existing society even without an obvious political message. In the soul song 'Drift Away', Dobie Gray begins with a plea about the pressures of the modern world and the pain of living in such a world—alienation in other words—but also notes music's potential to heal in this respect, by allowing our minds to 'drift away'.[60] Dave Randall believes that one reason lyrical expression of 'feelings of alienation' can have an impact is that 'by describing them, they reassure us that we are not the only one with those feelings. At the very least, this consoles us making our days more bearable. But at its best, music can prise open the cracks that let in the light...By reminding us how good it feels to emotionally connect with others, it invites us to imagine a less alienated future...It enables us to locate our very personal and subjective feelings within a kind of collective consciousness.'[61]

In fact, this sense of collective consciousness contains many contradictions. One is the power imbalance between musicians and their audience.

After a concert or festival, fans go back to their humdrum existence, while the artist's life is perceived as one of excitement and glamour. In contrast to the general alienated state of society, 'performers seem to express themselves through their work. They appear to combine their physical, intellectual and creative attributes—even their soul—in a tangible, immediate act of production. They present us with the tantalizing mirage of unalienated labour—work that is actually deeply satisfying—and leave us enviously speculating about their glamorous lives.'[62]

In fact Randall sees the reality for a performer like himself as more complicated: 'those of us who tour are actually more removed from the collective—in some senses, more alienated—than most. We may bring people together, but we do so as itinerants removed from society and placed in a bubble of tour-buses, hotels, dressing rooms, and VIP areas.'[63] Even someone as famous as John Lennon remembered the height of the Beatles' success as 'complete oppression. I mean we had to go through humiliation upon humiliation with the middle classes and showbiz and Lord Mayors and all that. They were so condescending and stupid. Everybody trying to use us. It was a special humiliation for me, cause I could never keep my mouth shut and I always had to be drunk and pilled to counteract the pressure.'[64]

One feature of modern popular music is its bewildering number of different genres. To the more traditional categories of rock, country, blues, soul, reggae, and folk, we can add punk, disco, hip-hop, rap, house, rave, drum'n'bass, grime…the list continues. Yet an ideal common to all of these genres is the importance of 'keeping it real'. In other words, to truly connect to the fears, hopes, and desires of their audience, an artist needs to remain rooted in the society that first inspired their music and lyrics. But this poses a problem, since the more fame a musician achieves, the less connected they are to the life that helped create their art.

The fear of losing touch with one's roots is a common one in many genres of music. In his 1975 hit 'Country Boy', Glen Campbell sang about the material benefits of success as a famous musician but contrasted this with the feeling that as a 'country boy', his heart was still in Tennessee, where he once sang his songs for free, but felt freer and happier for it.[65]

In fact, such fears are surely not so different from those voiced by Eminem—the pseudonym of the rapper Marshall Mathers—in his 2017 song 'Walk on Water'. In the late 1990s, Eminem became famous for expressing the hopes, fears, and frustrations of life in the poor white 'trailer trash' society in which he grew up in Detroit.[66] And in 2017 the fears and feelings of inadequacy clearly remained, as shown by Eminem's mention of his insecurity about whether his latest lyrics might tarnish the 'legacy, love or respect of his fans.'[67] The question is whether such lyrics can have the same resonance now that Eminem is a multimillionaire.

Richey Edwards, guitarist and chief lyricist with Welsh rock band the Manic Street Preachers, showed literally how passionately he felt about the need to 'keep it real' when he carved '4 REAL' into his arm with a knife in response to a journalist who questioned the band's integrity.[68] In fact, this was just one of several incidents of self-harm by Edwards, who also suffered from anorexia. Demonstrating the potential of music to provide insight into mental disorder, the Manics' song '4st 7lb', which appeared on their album The Holy Bible, movingly and disturbingly took us inside the head of an anorexic, although its lyrics were also considered problematic because of the risk of glamorizing the condition.[69]

In fact, this double-edged portrayal of mental disorder would become an epitaph for Richey Edwards himself. Soon after the completion of The Holy Bible, his car was found abandoned near the Severn Bridge, which separates Wales and England.[70] With his history of neurosis and self-abuse, an obvious conclusion was that he had committed suicide by jumping into the river Severn. However, because Richey's body was never found and he had withdrawn a significant sum of money from his bank account before his disappearance, for many years his family, the other band members, and many fans have clung to the idea that he simply decided to quit the limelight and start a new identity somewhere else.[71] While an obvious way for those grieving to cling to the hope that Richey is still alive, this belief may also have been influenced by the idea that such an outcome was more fitting for this unique and talented individual than a lonely death in the cold, murky waters of the Severn.

CHAPTER 19

ART AND DESIGN

W hen I was a child, in the school holidays my friends and I would often disappear for the day, intent on finding out more about the slivers of countryside that encroached on the urban landscape of north Bradford where we lived. We explored fields, woods, rivers, and even disused railway tunnels. From my current viewpoint as a parent in the twenty-first century, it seems amazing that my own parents let me roam all day unsupervised, but a positive feature of such an attitude was that it fuelled my sense of independence and, most of all, adventure. These forays were partly driven by a search for freedom from the restrictions of home and school, but also by a sense that we might make a great discovery, or stumble upon a mystery of the sort we had read about in books like Enid Blyton's Famous Five series. Yet while my friends and I had a great deal of fun on our travels, we never did make that great discovery we had hoped for. However, this was not the case for another group of children who, in September 1940, made a discovery that would revolutionize our view of the origins of human culture.

The four boys—Marcel Ravidat, Jacques Marsal, Georges Agnel, and Simon Coencas—lived in Montignac, a village in the Dordogne region of France. Inspired by a local legend about a tunnel containing hidden treasure, they set out to explore the woods near Montignac to try to find the tunnel, and when their dog found an opening at the foot of an uprooted tree, the boys decided to enter it.[1] What followed was a steep and hair-raising descent down 15 metres of tunnel, but at its foot the boys found themselves in a large underground chamber. Holding up the oil lanterns they had brought, they were astounded to see revealed in the flickering light, not gold or silver, but a cultural treasure that would prove far more valuable.

As Marsal described it, the boys had discovered a 'cavalcade of animals larger than life painted on the walls and ceiling of the cave; each animal seemed to be moving.'[2] The chamber, now known as 'The Hall of the Bulls', was just the start of a series of passages covered with amazing cave paintings, which featured not only scenes of hunted animals, but also a figure of a man with a bird's head—perhaps a shaman who carried out rituals in the cave. Later analysis would show that the paintings were created over 15,000 years ago.

Today the paintings of Lascaux are just one of many examples of pre-historic art discovered not only in caves, but also above ground, as in the Côa valley in Portugal where exploration in the 1990s as part of a plan to create a dam revealed hundreds of engravings in the overhanging rock walls of the valley.[3] Thankfully, the construction of the dam was stopped, and the area made a UNESCO protected site.[4] Most recently, cave paintings of animals found on the island of Sulawesi, in Indonesia, were estimated to be at least 35,400 years old, making this the oldest-known example of figurative art anywhere in the world.[5]

Quite what motivated our ancestors to descend into the bowels of the earth or venture into isolated valleys to adorn rock walls with such intri-cate drawings remains open to interpretation. Some believe the art was important for its content, while others assert that its primary significance was in the ritual act of producing it, for since hunting was critical to early humans' survival, such art may have been part of a magic ritual whose aim was to influence the success of the hunt, or increase the fertility of herds in the wild. Images that seem to have been clawed or gouged with spears favour the former idea, while a painting of a pregnant-looking horse in the Lascaux cave supports the latter.[6] What is clear is that the birth of figurative art marked a significant shift in our species' ability to represent the world around us, and, presumably, a step forward in our capacity for conscious awareness.

The ability to draw a familiar object such as a cat or dog develops so naturally in children that it is easy to forget what a remarkable human capacity this is. For unlike tool use, spoken language, or perhaps even the capacity to make music, which may all have evolutionary precedents in the actions or sounds made by other species, the ability to draw appears to have no relationship to anything found in the rest of the animal kingdom. And certainly for the view presented in this book of the uniqueness of human conscious awareness, the ability of humans to both produce and appreciate visual art is something that deserves a wider discussion. In particular, abstract symbols, mathematical systems, planning diagrams, and other forms that constitute visual art can all be viewed as further examples of the 'cultural tools' that shape our minds besides language.

Indeed, recent studies have shown that the production, and appreciation, of visual art can play an important role in the development of conscious awareness in children. For instance, Margaret Brooks of the University of New England in Armidale, Australia, has shown that encouraging children to draw what they see around them can be a major stimulus to the development of conceptual thought in such children. Brooks has noted that when such children were 'encouraged to revisit, revise and dialogue through and with their drawing, they were able to explore and represent increasingly complex ideas.'[7]

We may wonder, given our present understanding of the role of different brain regions in human consciousness, how such different regions might be involved in both the production and appreciation of visual works of art. To investigate this question, Steven Brown and colleagues at McMaster University in Hamilton, Canada, carried out a meta-analysis of 93 different studies that had used neuroimaging to identify regions of the human brain involved in artistic appreciation.[8] They found no evidence that artworks activate brain areas distinct from those involved in appraising everyday objects important for survival. Brown believes this shows that the aesthetic system of the brain evolved first for the appraisal of objects of biological importance, including food sources and suitable mates, and was later co-opted for artworks such as paintings and music.

As he puts it, 'much as philosophers like to believe that our brains contain a specialized system for the appreciation of artworks, research suggests that our brain's responses to a piece of cake and a piece of music are in fact quite similar.'[9] This finding fits with other studies mentioned in this book that failed to find specific brain regions dedicated to language, musical appreciation, or other 'higher' human abilities. However, this does not mean that neuroscience cannot provide important insights about the ways in which the human brain perceives artworks and their impact in our minds.

One recent study, led by David Freedberg of Columbia University in New York, tried to understand more about the neural mechanisms that might underlie an aesthetic response to a work of art. Freedberg's team used an imaging method called transcranial magnetic stimulation (TMS) to monitor the brains of participants as they viewed a detail from Michelangelo's *Expulsion from Paradise*, a fresco panel on the ceiling of the Sistine Chapel at the Vatican. The detail was of a fallen-from-grace Adam warding off a sword-wielding angel with his eyes averted from the blade and his wrist bent back defensively. The study found that viewing this detail stimulated regions in the brain's primary motor cortex that controlled the observer's own wrist movements.[10] Such a feeling could form an important part of the aesthetic response that makes viewers of the fresco feel they are right there with Adam fending off the angel's blows. Freedberg believes this may also explain why some viewers of Edgar Degas's paintings of ballerinas have reported that they experience the sensation of dancing. Given what we learned previously about mirror neurons, which are stimulated when we observe another person carrying out an action and have been proposed to play a role in allowing us to understand the intentions and emotions behind such an action, further research should explore whether these neurons play a role in such aesthetic responses.

Rather than investigating the viewers of artworks, another imaging study looked for differences in the brains of artists compared to non-artistic individuals. The study, by Rebecca Chamberlain and colleagues at the Catholic University in Leuven, Belgium, used an approach called voxel-based morphometry to scan the brains of art students compared to non-artists.[11] This revealed that artists had increased neuronal matter in areas relating to fine motor movements and visual imagery. Notably, the artists had significantly more neurons in a part of the parietal lobe region of the brain that Chamberlain says 'is involved in a range of functions but potentially in things that could be linked to creativity, like visual imagery—being able to manipulate visual images in your brain, combine them and deconstruct them.'[12]

Participants in the study also completed drawing tasks and those best at drawing were found to have greater numbers of neurons in the cerebellum—a brain region that as we have already seen is involved in fine motor control and performance of routine actions, but is also increasingly recognized as playing a key role in creativity and imagination.[13]

Such findings leave unanswered the question of whether the differences reflect innate differences between artists and non-artists, or changes that have occurred as individuals learned to produce their first artistic creations. This is important, given the complexity of the interaction of nature and nurture that, as I have stressed throughout this book, shapes an individual human consciousness. As a follow-up to this particular study, the next step might be to examine changes in the brains of young people as they begin to produce artistic works. Indeed, Chris McManus from University College London, who was also involved in the study, has said that the researchers have plans 'to do further studies where we look at teenagers and see how they develop in their drawing as they grow older.'[14]

Studying how different brain regions are involved in the production and appreciation of art may reveal some of the mechanisms underlying these processes but there is clearly more to art production and appreciation than simple mechanics. In particular, in line with what I have been arguing elsewhere in this book, there is good reason to believe that a central aspect of producing and appreciating art is our human ability to abstract the properties of the world in a symbolic fashion—and thereby create new meaning—that is unique to our species. This symbolism is clear—some might say personified—in the Lascaux cave paintings. With just a stick and some pigment, whoever created what has been referred to as the 'Sistine Chapel of prehistoric art'[15] nevertheless managed to convey, in a few strokes, not only the animals' shape, but also their movements, in such a way that this made a deeply memorable impression on some teenage boys 15,000 years after the paintings' creation.

While this capacity of art to speak to human beings across the gulf of time is one of its key features, it would be mistaken not to recognize that art has also undergone change and development over the centuries, and that what an artwork says to future generations may be radically different from what its original creator intended. Yet whether there can be any sense of 'progression' in the visual arts remains controversial. At the same time, we have other issues to consider, such as whether artistic creativity is primarily a product of the rational mind, or whether it flows more from unconscious impulses, and even from thoughts that in other circumstances might be considered odd or even insane. Finally, I will consider how visual art is being affected by newer technological developments like cinema and the Internet.

On the subject of the development of artforms over time, which now includes moving images, clearly on one level great works of art, whether from prehistory or any subsequent period following the birth of civilization, can be considered equally worthy of our respect. In particular, we can compare an image such as one of the bovine species depicted in the Lascaux cave paintings mentioned earlier with the cows rendered in beautiful detail by sixteenth-century Dutch painter Aelbert Cuyp, or Pablo Picasso's much more abstract images of bulls.[16] Such a comparison shows that all of these artistic creations have been drawn with tremendous skill, and as such it would be hard to justify the claim that any of the images are in any way superior to the others. And indeed a recent study has shown that the Lascaux paintings of animals are more accurate in their depiction of how such animals walk, based on photographic evidence of this process, than similar animal images by later artists.[17]

Yet on another level I would argue that at certain periods of history, there have been important steps forward in the capacity of visual art to say novel things about the state of the world and the human condition. This has particularly been the case during periods of major social upheaval, but may also reflect widening global horizons as humanity has developed, as shown by the work of the fifteenth-century Dutch artist Hieronymus Bosch. When I first saw *The Garden of Earthly Delights*, by Bosch, in the Museo del Prado in Madrid, my initial feeling was how modern the painting seemed, despite being half a millennium old, and that Bosch might have been high on some psychedelic drug when he painted it. With its lurid colours and hallucinatory images—which include a bird-headed creature eating a naked man, a pig dressed as a nun, and a hollowed-out giant with trees for limbs and an inn inside his pale egg-like torso—this tryptich painting seemed closer to the twentieth-century surrealist painter Salvador Dali than to other late medieval art.[18] Yet Bosch was very much a product of his society, while at the same time being highly attuned to radical changes in that society.

Bosch lived in the Netherlands during the transition between the late medieval period and the Renaissance. Trade and urbanization were increasing, weakening the feudal system, and increasing the demand for art.[19] At the same time the gap between rich and poor was expanding in the cities. In the late fifteenth century, perceived misbehaviour by religious leaders led some ordinary people and clergy to set up a religious reform movement known as Modern Devotion. Their aim was to change the relationship between church and society, and they created small communities in which participants shared property. There is evidence that Bosch or his sponsors were influenced by their ideas, which, according to art historian Jos Koldeweij, emphasized that 'all human beings have to think, themselves, about good and evil...[and] to read the Bible in our own language, [and] take the message out of it for our own lives.'[20]

The influence of this way of thinking is shown particularly clearly in another of Bosch's tryptichs, *The Haywain*. In this, he shows people of all classes following a hay wagon.[21] Some are fighting each other in order to seize some hay. Tellingly, members of the clergy are also stuffing their bags. Meanwhile, demons are pulling the wagon towards hell. The message is clear—people's greed is leading them towards a world full of agony. And in a historical period of rapidly changing social relations, and consequential changes in different individuals' material status and wealth, such images may have had a special resonance.

But Bosch was at his most striking when he conjured up fantasy creatures as in *The Garden of Earthly Delights*. In this he merged the characteristics of nature, animals, people, and machines into a new absurd reality.[22] In the first two panels, naked people are frolicking in a heavenly landscape with huge flowers, fruits, and animals. But in the third panel, the dark landscape is of a hell inhabited by demons, and sinners desperately trying to flee their tortures. This vision reflected the real difference that ending up in heaven or hell represented for people in a society defined by religion, and which Bosch portrayed so viscerally.

However, the specific content of this vision also hints at another possible influence, namely the important advances in science and technology, and the fruits of European voyages to the Far East and the New World that were just beginning to occur at this time.[23] For example, the presence in Bosch's paintings of architecture that seems to be modelled on human organs suggests an interest in anatomy; the giant fruits suggest opulent places far away. Meanwhile, the ecstatic nudes may be inspired by stories of 'naked peoples' brought back by travellers to the Americas. In this sense, far from being a vision totally at variance with the spirit of the times, Bosch may have been engaging with a new and exciting reality. And as such, this vision may be seen as the beginning of a truly global human consciousness, although it should be said, one which at this stage was still only available to a small section of the world's population, the privileged few in a colonizing nation, and not to the colonized peoples themselves, or even for that matter, the vast majority of ordinary Dutch citizens.

Previously in this book I have criticized two polarized views about human consciousness: the first sees differences in individual human personality and behaviour as ultimately a product of differences in biology, while the second views human beings as blank slates and the subsequent development of an individual's character as purely a product of social influence. In contrast, I have sought to develop a view of individual human consciousness that sees nature and nurture as equally important, with the two contributing to personality and behaviour in a highly interactive fashion. This is surely highly relevant for our discussion of Bosch's work, for while this reflects a highly personal and distinctive way of looking at the world, it was also clearly influenced not only by ideas within society at that time, but also by new influences on that society. In addition, I have discussed evidence that there seems to be a narrow dividing line between creativity and mental disorder, with many neural processes that enhance the former also often seeming to play a role in the latter. And we will now further look at how such considerations can affect our understanding of art.

Michelangelo Buonarotti's reputation as an artistic genius is based not only on the quality of his creations but also on his versatility, since he excelled not only in painting and sculpture, but also in architecture. And recently, it has been claimed that a key ingredient in Michelangelo's genius is what has been interpreted, retrospectively, as his autistic personality. So Muhammad Arshad, a psychiatrist at Whiston Hospital in Merseyside, England, and Michael Fitzgerald, from Trinity College Dublin, have claimed that 'Michelangelo's single-minded work routine, unusual lifestyle, limited interests, poor social and communication skills, and various issues of life control appear to be features of high-functioning autism or Asperger's syndrome.'[24] Such a claim would fit with the fact that people with Asperger's syndrome have difficulties with communication and social interaction but can show an unusual, often obsessional, talent or skill in a particular area.

In Michelangelo's case, Arshad and Fitzgerald believe that this explains how the artist was able to generate, in such a short time, many hundreds of sketches for the Sistine ceiling. But not everyone agrees with this claim, with the art historian James Hall countering that 'the character traits they are referring to—obsessional behaviour, fiery temper, propensity to be a loner—could be linked to many artists.'[25] In fact there are other reasons for rejecting the idea that a great artist's genius can be seen as primarily a

ART AND DESIGN

product of their biological character—'abnormal' or more conventional in nature—and that is the ample evidence that, as with Bosch, social environment can also play a tremendous role in nurturing such an artist.

In understanding Michelangelo's art, the rise of Florence as an economic power from the twelfth century onwards is of major importance. This rise was connected to the financial and transport facilities of the Crusades being organized from the city; free competition first developed there in opposition to the guild ideal of the Middle Ages; and the first European banking system arose in Florence. The economic growth of the city helped fund the beautiful Florentine art created at this time, but was also accompanied by a new 'humanist' approach to life, which helped to shape this art. This influence is shown in Michelangelo's statue *David*.

The artist himself said that he carved the statue of David from a block of marble by hewing 'away the rough walls that imprison the lovely apparition to reveal it to the other eyes as mine see it', implying that the form was already waiting there in the stone.[26] And the pioneering sixteenth-century art historian Giorgio Vasari argued that Michelangelo's particular genius arose from his peculiar ability to imitate 'the most beautiful things in Nature in all forms, both in sculpture and in painting'.[27] This idea is in line with a view of imagination that I mentioned previously, namely that this human capacity is a process whereby the mind copies objects in the real world and combines the elements in various different ways. However, while it would surely be difficult to stand before the artist's *David* and not be awestruck at the level of sheer technical skill involved in its creation, there are good reasons to question the view that Michelangelo's prodigious skill in shaping stone—or paint—according to his will was merely a case of skill in 'imitating' nature or representing 'reality'.

The social backdrop to the creation of the statue of David is worth mentioning, for while based on the biblical story, the motivation behind the statue was highly secular and political, being commissioned in 1501 by Florence's new republican government to celebrate the recent ousting of its Medici rulers and the city's independence from Milan and the Roman Papacy.[28] The symbolism is obvious, and the social aspirations of Florence at this time also influenced the way Michelangelo carved *David*. For unlike his predecessors Donatello and Verocchio, who sculpted David after the defeat of Goliath with the giant's head at his feet, Michelangelo portrayed him before the battle, thereby making this a work that looks to the future, not the past. What is more, Michelangelo's *David* is not in fact in 'perfect' proportion; the head and hands, especially the right hand holding the stone, are slightly too large, which has the effect of making David a brain as well as a body, and a doer and maker as well as an object of beauty.[29] Therefore, the statue may be seen to celebrate not one specific 'man', but a vision of the ideal that humanity has the potential to become in a new, republican society.

Today we live in the era of contemporary art, which began with modernism. In fact the modernist movement—which is often defined as occurring between the 1890s and the mid-twentieth century—included not only visual art, but also architecture, poetry, novels, theatre, and cinema.[30] While general definitions are difficult, modernist work tends to be formally experimental and highly self-conscious—think of Pablo Picasso's cubist paintings or the 'stream of consciousness' of James Joyce's novels. A key characteristic of modernism is an emphasis on dislocation and fragmentation. This characteristic can be seen to reflect the social events that took place between 1900 and 1930—a time of great upheaval in which dreams of peace and prosperity were shattered by war, slump, revolution, and reaction.[31] The accelerating development of technology and penetration of mass production methods into every sphere of life also added to a deep sense of uncertainty. In the visual arts, a crucial distinction between modern art and 'traditional' art is that the latter tends to be more representational and naturalistic, while modern art either wilfully distorts physical appearances or, in abstract art, abandons them altogether. And just as with music, this has been accompanied by the fact that a successful visual artist in modern times is no longer viewed as a paid servant of relatively lowly stature, but rather an 'artist' in the modern free sense, namely someone that is most interesting when provocative and pushing at boundaries.

A contender for the first, truly 'modern' work of art is *Les Demoiselles d'Avignon*, painted by Picasso between 1906 and 1907.[32] In terms of its impact on future art, *Les Demoiselles* can be viewed as having opened the doorway, first to cubism and subsequently to futurism, synthetic cubism, expressionism, vorticism, abstraction, suprematism, Dadaism, and more besides.[33] Within just ten years artists were producing works, such as Malevich's *Black Square on White* and Duchamp's 'found objects', which would not previously have been regarded as art at all, but which have subsequently achieved classic and iconic status. So what was it about *Les Demoiselles* that made it such a catalyst? And what might have been going on inside Picasso's brain during the production of such a radical vision?

We mentioned earlier that Picasso claimed that his genius as an artist was based upon an ability to view the world as would a child. And certainly some of his works have an astonishing childlike simplicity. Take for instance the 'bull's head' that Picasso created much later in his career in 1942, from a discarded bicycle seat and handlebars.[34] And certainly to someone used to the art of the great masters of the Renaissance, *Les Demoiselles* might also appear at first sight childishly crude in its composition and harsh in its treatment of the subject. Yet nothing could be further from the truth, and it is worth considering some of the complex neurological and psychological factors that might have been involved in the creation of this iconic work.

One factor is what might be called Picasso's 'intensity of vision'. As neuropsychologist Christine Temple has pointed out, Picasso had a fantastic visual memory, but he also spent many hours staring at an object that he wanted to paint, whether this was a famous artwork of the past that he was reinterpreting or a living model.[35] And according to Temple, consequentially Picasso tended to 'overlearn' such objects. As she explained this term, 'When you overlearn something, you don't just look at it, think about it and try to learn from or remember it. You look at it (or listen to it) again and again until it becomes completely familiar. Overlearning enables memory to work much better and in much greater detail.'[36] Now all of this has potential relevance to what we said earlier about the mechanisms of human memory, but also creativity and imagination. For as we noted, the ability to come up with imaginative solutions to problems seems to involve not merely a memory of past images, but rather an ability to combine these in radical and unexpected new ways. And we also saw how the cerebellum—the part of the brain that plays a central role in the unconscious learning of repetitive movements—is increasingly being recognized as also playing an important role in creativity and imagination. So it seems possible that the many hours Picasso spent gazing at an object were all part of his radical reappraisal of that object, by allowing his cerebellum to run through different 'takes' on the object that, in unconscious communion with other parts of the brain, thereby produced a revolutionary new perspective.

So how does any of this relate to *Les Demoiselles*? At its most basic level, the painting depicts five nude women, prostitutes in a brothel, as indicated by the painting's orginal title, *Le Bordel d'Avignon*. However, the women appear rather menacing and are rendered with angular and disjointed body shapes. Notably, given what we have said about the imaginative reinterpretation of seen objects, the depiction of three of the women seems to have been influenced by ancient Iberian sculptures that Picasso had seen the previous year in the Louvre, while the two other women have heads that resemble the African sculptures or masks that the artist had observed in the Musée d'Ethnographie du Trocadéro.[37]

We can speculate about the individual factors that led Picasso to compose *Les Demoiselles* but how did this relate to what was happening in society as a whole? Here we need to consider how that society was changing technologically. An important characteristic of traditional art, closely bound up with its naturalism, was the high level of craft skills that it involved and demanded. These skills, particularly developed in the sixteenth and seventeenth centuries, lay especially in the precise rendering of surfaces: lace, satin, velvet, sable, glass, silver, feathers, flesh tones, the folds in drapery or robes and so on. However, the introduction of photography made such skills largely redundant. In retrospect, the premium on reproducing life in paintings was clearly waning from Claude Monet and impressionism onwards, but *Les Demoiselles* marked the decisive break.[38] In 1907 it would have looked like not just a move away from the traditional skills, but a full-scale assault on them.

However, there was more to *Les Demoiselles* than simply its non-naturalistic approach. In particular one could argue that it represented a radical new approach to the portrayal of women. Previously, I mentioned how the rise of class society led to the subjugation of women, and also the fact that under capitalism, even sex can be treated as a commodity. And because of this, in the tradition of nude painting that stretches through Botticelli, Giorgione, Titian, Rubens, Velázquez, Goya, Ingres, and Renoir the artists' aim was always to present their female subjects as passive, beautiful objects of male desire. This reflected the fact that these paintings were paid for by men and often hung in their bedrooms.

Women's greater entry into the workforce and their growing political demands in the nineteenth century had an impact on the portrayal of women in art. So already there were signs of change in Manet's *Olympia*, Paul Cézanne's *Bathers* series, and Toulouse Lautrec's brothel scenes, but it is with *Les Demoiselles* that the sharpest confrontation with tradition occurs. The art historian John Molyneux has argued that Picasso broke with traditional forms of naturalistic representation and launched an assault on conventional standards of beauty, to smash and eliminate any traces of sentimentality and glamorization. So Picasso not only makes no attempt to make the women 'beautiful' but, by the use of African masks, he broke away from conceptions of conventional European views of beauty.[39] And Molyneux has argued that Picasso created a flattening—an extreme foreshortening of space—in the painting to stage a confrontation between the painting and its audience, representing the client and the prostitutes, to cut through the ingrained habit of evasion of the reality of prostitution.[40]

However, this is only one reading of *Les Demoiselles*. Some feminist critics have responded far less positively, seeing it as highly misogynistic. For instance, Carol Duncan has argued that in this work Picasso 'dredged up from his psyche the terrifying and fascinating beast' that women represented for him: 'whore and deity, decadent and savage, tempting and repelling, awesome and obscene, looming and crouching, masked and naked, threatening and powerless…no other work reveals more of the rock foundation of sexist anti-humanism or goes further and deeper to justify and celebrate the domination of woman by man.'[41]

The possibility of multiple interpretations of modern art is enhanced by its very abstractness. This reached a peak in the work of Kazimir Malevich. Born in Kiev in 1879, Malevich developed a style—'suprematism'—which took abstraction to an ultimate extreme.[42] The idea behind this was that by sticking to simple geometric shapes and a limited range of colours the artist could focus on the painting itself and not be distracted by representing a scene, or landscape or a person. Malevich's *Black Square*, exhibited in Kiev in 1915, is exactly that: a painted black square on a white background.[43] The painting might be viewed as the height of formalistic, non-representational objectivity, yet Malevich himself saw it as liberated self-expression. In his 1919 essay 'Non-Objective Art and Suprematism' he concluded that 'I have overcome the lining of the coloured sky, torn it down and into the bag thus formed, put colour, tying it up with a knot. Swim in the white free abyss, infinity is before you.'[44] So what might be the basis for such an ambitious claim for this abstract work?

In seeking to answer this question, it is worth looking again at what we have learned about how visual perception works, and also the importance of meaning for our species. Recall that Hubel and Wiesel's experiments first showed that a key way by which the visual cortex in our brains allows us to make sense of the world is by breaking it down into lines and edges. Because of this, the neurobiologist Semir Zeki has argued that Malevich's *Black Square*—and all the other coloured squares and rectangles that came after it in abstract art—have a particular resonance because they 'stimulate cells in the visual cortex, and the properties of these cells are, to an extent, the pre-existing "idea" within us.'[45] So there is a potential neurological basis to the power of such artforms to attract our attention. Yet it is impossible to fully understand the power of a work of art like *Black Square* without also recognizing its symbolic content. In particular, Malevich's decision to hang the painting high on the wall against the corner of the room where it was first exhibited might mean nothing to a non-Russian viewer today, but this was the same sacred spot that a Russian Orthodox icon of a saint would sit in a traditional Russian home.[46] And not only would this have symbolic value in its own right, but the fact that it was happening in 1915, against the backdrop of the slaughter of the First World War, and with continuing unrest following the failed Russian Revolution of 1905, all contributed to the painting's power. This shows that *shared* symbolic meaning is vitally important both for ordinary life and for art appreciation.

More generally, modernist art may be seen as a celebration of the new technologies made possible by capitalism and at the same time a critique of the alienation characteristic of this form of society. The challenges of representing new experiences of speed well beyond customary horse power, of rendering multiple other novel ways of viewing the world on an immobile flat canvas or page—absorbing the insights from science—and the advent of photography and cinema focused young artists' minds and provided their motifs.[47] Yet despite their appreciation for new technologies, many modernist artists remained critical of the capitalist system itself. For cultural critic Perry Anderson, this double-edged character of modernism reflected the fact that it 'flowered in the space between a still usable classical past, a still indeterminate technical present, and a still unpredictable political future.'[48]

The unpredictability of such a future, and the ambiguity at the heart of modernism, can help to explain the very different political currents with which modernism became associated in the early twentieth century. In Italy under Mussolini, the modernist art movement celebrated the inhuman power of technology and the destructiveness of war.[49] In contrast in Russia, most modernist painters, composers, and poets supported the 1917 revolution, which seemed to offer untold freedoms, a chance to use bold new forms—cubism, abstraction, street art, film, jazz, satire, fantasy—and share in the making of a new type of society.[50] Not only Malevich, but also the mystical Wassily Kandinsky and Marc Chagall and the constructivists Vladimir Tatlin, Alexander Rodchenko, and Lyubov Popova, all responded with equal euphoric intensity. Kandinsky's sizzling *Blue Crest* and Chagall's *The Promenade* showing himself whirling his bride Bella through the air capture this heady exuberance, while the Cyrillic alphabet itself takes colourful flight in Ivan Puni's *Spectrum: Flight of Forms.*[51]

The poet Vladimir Mayakovsky argued that 'we do not need a dead mausoleum of art where dead works are worshipped, but a living factory of the human spirit—in the streets, in the tramways, in the factories, workshops and workers' homes...Let us make the squares our pallets, the streets our brushes.'[52] He and his fellow writers and artists produced posters and decorations for anniversaries, pageants, street theatre, and agit-prop trains during the civil war that engulfed the country. A glimpse of how all this looked in practice was provided by filmmaker Sergei Eisenstein, who on a visit to Vitebsk was amazed to see the 'strange provincial city' transformed by its artists: 'Here the red brick streets are covered with white paint and green circles are scattered across this white background. There are orange squares. Blue rectangles. This is Vitebsk in 1920. Its brick walls have met the brush of Kazimir Malevich. And from these walls you can hear: "The streets are our palette." '[53]

Eisenstein is of particular interest for the view of human consciousness being developed in this book, because of his close interaction with the psychologist Lev Vygotsky, and then after the latter's death, with Vygotsky's colleague Alexander Luria. This interaction helped shape the theory and practice of Eisenstein's cinematography, and as the filmmaker was one of the most influential early pioneers of this visual artform, it also has relevance for the future of the medium itself, with practical implications that reverberate to the present day.

Today Eisenstein is most remembered for early-twentieth-century films such as *Battleship Potemkin, October, Alexander Nevsky*, and *Ivan the Terrible.*[54] But he also pioneered many radical approaches and techniques in film-making for the first time, one being montage—a film-editing technique in which a series of short shots are sequenced to condense space, time, and information. For Eisenstein, the power of the edit was that film shots could be cut together to create brand new meaning that was not present in the original shots.[55] The practical use of this approach was first demonstrated in the film *Battleship Potemkin*. This tells the true story of the mutiny of that vessel's crew during the 1905 Russian Revolution and the subsequent brutal response of the Tsarist regime. A highlight of that film is the montage sequence showing the massacre of civilians on the steps of Odessa, in which editing created a dynamic, tense, violent, and deeply allegorical set piece.[56] The impact of Eisenstein's ideas can be seen today in the rapid cutting of modern action sequences or the rhythmic edits of music videos. A recent example of the power of this approach is a montage sequence in Bong Joon-ho's award-winning 2019 film *Parasite*, which both rapidly and skilfully develops a key moment in the storyline, and also surreptitiously foreshadows coming events and references previous ones within itself.[57]

In developing his use of montage, Eisenstein was influenced by Vygotsky's belief that human consciousness differs from that of an animal in that only the former is mediated by cultural tools, a primary one being language, and that this is also true of our emotional responses. But while this was a general influence, a more specific one was the substantial amount of work Vygotsky devoted to the psychology of different artforms. Vygotsky argued that 'art is the social technique of feelings', and Eisenstein developed this idea in a practical direction by using film-editing techniques to organize and control viewers' responses in order to optimally produce aesthetic effects in his audience.[58] He proposed that a piece of art's effectiveness works simultaneously at the level of the structural organization of that artform, and that of its emotional and broader intellectual and psychological effects on the viewer.

Eisenstein believed that a central issue for any work of art is the paradoxical coexistence of opposing dimensions in the viewer: logical and sensuous, cognitive and emotional, rational and irrational, conscious and unconscious. He argued that 'while the conventional film directs the emotions, this suggests an opportunity to encourage and direct the whole thought process as well.'[59] Building on Vygotsky's views about the importance of inner speech in the thought process, Eisenstein noted that in contrast to outer verbal speech, inner speech is closer to image-based thinking; even though it shares fundamental similarities with language, it can involve not only words but also pictures, sounds, and even writing. As such, this offers possibilities for a film to manipulate thought in a multi-modal fashion.

One phenomenon that particularly interested Eisenstein in this respect was a medical condition called 'synaesthesia'. Whereas normally we have five clearly defined sensory classes—touch, sight, hearing, smell, and taste—in people with this condition there is a blurring of the divisions between these classes. For instance in his work as a clinical neuropsychologist Alexander Luria treated a patient called Solomon Shereshevsky, who claimed to 'see sounds as colours, and colours as sounds'.[60] Becoming familiar with this case through his friendship with Luria, Eisenstein proposed that while synaesthesia in its natural form is an abnormality affecting only a few individuals, film could seek to provoke such a blurring of the senses in an audience in order to heighten the aesthetic response.

One reason why Eisenstein viewed such blurring as particularly effective was that he saw the aesthetic experiences that this provokes as taking 'one back to the early evolutionary phases of sensuous thought, where no differentiation of perception yet existed.'[61] In other words, despite our species being unique in having a delineation of perception that is a product of rational, conceptual thought, for a deeper experience we may benefit from temporarily losing some of this delineation. And Eisenstein sought to create such an impression by manipulation of visual images and sound in his later films, for instance in *Alexander Nevsky*, which featured music by Sergei Provokiev, in particular during its famous 'Battle on the Ice' sequence when Russian defenders repel an invading German army.[62] Released in 1938, there were clear parallels with a contemporary situation in which the Soviet Union was about to fight a war with Nazi Germany. But this film can also be viewed as one of the first truly audiovisual cinematic experiences, and as such it had a major impact on later film-making.

So much for Eisenstein's impact on film-making in the past, but perhaps of even greater importance is the way that his and Vygotsky's views about the nature of aesthetic experiences might help us to understand, and perhaps shape, the visual arts in the future. In particular, the digital revolution that first came to prominence in the 1990s and increasingly defines our lives today continues to open up new avenues for artistic production by increasing and enriching the range of available technologies of production and modes of delivery through which moving images and sound can operate.[63] All this has implications for our understanding of human consciousness, because such technology offers new means of portraying the fragmented and often unconscious nature of a lot of human thought in ways that may both further such understanding and potentially produce great future art. And while this poses challenges for film-making, it also seems clear that this revolution could have implications for the development of other forms of art, which includes literary fiction.

CHAPTER 20

FACT AND FICTION

———◦◦◦———

Like so many others, one of my main influences in life from a young age has been the fictional world as expressed in stories and novels. At primary school, I progressed rapidly through the 'Peter and Jane' books that were a key component of teaching literacy at the time,[1] partly because of a desire to read the more interesting stories further on in the series. Perhaps appropriately for someone who was an amateur cave explorer in their youth before beginning a scientific career, I was later drawn to factual accounts of the lives of famous scientists and explorers, and to adventure stories and science fiction. However, by my early teenage years, I was already beginning to thirst for more challenging novelistic fiction, and this thirst was satisfied and further fuelled by the books I found in my local library.

I can still recall the musty smell of the small library that was just a mile from my home in the Bradford suburb of Idle—famous for its 'Idle Working Men's Club'.[2] My quest for reading matter was aided by the librarian, who was the sister of novelist John Braine, author of *Room at the Top*. This novel's gritty story of working-class life and aspirations captured the frustration and anger at the lack of opportunity in 1950s Britain where people, especially those from the back streets of northern industrial towns, 'knew their place'.[3] The novel was also frank in its description of sex at a time when people were not supposed to talk about such things in public. Maybe that is why John Braine's sister never prevented me borrowing 'adult' novels, while only in my early teens, including Vladimir Nabokov's *Lolita*. I cannot fully remember what my first impressions were of this novel whose central character is a paedophile, but something that did strike me even at that age was the way that a great writer can use language to manipulate the thoughts and emotions of the reader, in this case drawing us into the mind of, and even making us empathize with, a twisted individual.[4] I also developed a love for the novels of William Golding, particularly his *Pincher Martin*, ostensibly about a shipwrecked sailor's struggle for survival on a rock in the Atlantic, but actually a profound discussion about the human condition, what it means to live but also to die, and how even a highly unlikeable human being may possess qualities with which we can empathize, and wish for him to achieve some kind of redemption.[5]

How much can novels tell us about the various mechanisms underlying human consciousness? Some might say, very little, and argue that only scientific study can uncover such mechanisms. I do not agree with this viewpoint; my reasoning is that because language plays such a key role in shaping human consciousness, the fictional explorations of the human condition that we find in novelistic literature can greatly add to our scientific understanding by concretizing that condition in its diverse forms. And it seems that I am not alone in this belief, for as influential a thinker as Noam Chomsky has said that, 'It is quite possible... that we will always learn more about human life and personality from novels than from scientific psychology.'[6] Personally I believe these two approaches to be equally valuable, each providing insights that the other cannot, in ways I will explore in more detail in the rest of this chapter. And I will also explore a related question: how much do novels draw on new insights about the nature of consciousness, so increasing their ability to inform us about the human condition, and its relationship to changing forms of society?

In pondering such questions, a feature of novels worth noting is how they differ from a form of artistic expression such as a play which relies not only on the written text of the playwright, but also on how this is interpreted by the play's director—with the possibility of leaving out or altering parts of the text—the actors' own interpretations, and the interactions between actors, and with the audience. Unfortunately a consideration of staged drama, or for that matter that other key area of literature, poetry, is outside the scope of this book. But my point here is that unlike staged drama, a novel is a fixed piece of text with a readership that has no necessary contact with the author, or with other readers. And given what I have said previously about viewing human consciousness as a kind of dialogue, this might suggest that novels are less dialogic, and therefore less representative of consciousness, than plays.

Yet such a view neglects two important features of great novels. One is the fact that as novelist David Lodge has noted, 'the silence and privacy of the reading experience afforded by books mimic the silent privacy of individual consciousness.'[7] And given that we have seen that such a consciousness is dialogic, this gives ample capacity for a dialogue to form between the reader and the words that they are reading on a page or screen. Interestingly, both the neuroscientist Antonio Damasio and the literary critic Ian Watt have argued that the very idea of an individual consciousness is a recent concept, dating back only a few hundred years, and linked this to the rise of the novel as a form of mass entertainment. So according to Watt, the idea of society being composed of a 'developing but unplanned aggregate of particular individuals having particular experiences at particular times and in particular places' echoes the fact that unlike earlier narrative fiction, which tended to recycle familiar stories, 'novelists were the first storytellers to pretend that their stories had never been told before.'[8]

A second important feature of great novels is the capacity for characters—and interactions between them—to take on a life of their own that can go way beyond the original intentions of the novelist. This was expressed by the novelist Somerset Maugham when he wrote that 'a character in a writer's head, unwritten, remains a possession; his thoughts recur to it constantly, and while his imagination gradually enriches it, he enjoys the singular pleasure of feeling that there, in his mind, someone is living a varied and tremulous life, obedient to his fancy and yet in a queer, wilful way independent of him.'[9] Of course, it is impossible for an invented character to literally have a life and thoughts of their own, independent of the author. So how do novelists achieve such an illusion? Here it is important to recognize that both conscious and unconscious processes may be at work in the creation and development of fictional characters in a novel.

One way in which characters can take on an apparent life of their own is through their interaction with other characters. It is through such interaction that both dialogue and actions can develop beyond the bounds of the author's will. Mikhail Bakhtin, whose ideas about the dynamics of language and thought I have mentioned previously, believed that the novelist Fyodor Dostoevsky was particularly skilful in this approach to fiction, which he called 'polyphonic' (Figure 28). Bakhtin counterposed this to 'monophonic' fiction, in which the author's opinions and world viewpoint tend to dominate the utterances of the different characters. In contrast, he argued that Dostoevsky played a pioneering role by creating characters that were 'not voiceless slaves...but free people, capable of standing alongside their creator, capable of not agreeing with him and even of rebelling against him.'[10]

For Bakhtin, an essential feature of Dostoevsky's novels is that no one individual's ideas dominate; instead, it is how the words of each character relate to the other characters that gives these novels their power. For instance, in *The Brothers Karamazov*, Ivan Karamazov is distinguished not by his philosophy as such, but by the way that philosophy is shaped in the face of objections and additions from other characters, each capable of mounting powerful arguments of their own. As Bakhtin put it, 'every thought of Dostoevsky's heroes...senses itself to be from the very beginning a rejoinder in an unfinalized dialogue. Such thought is not impelled

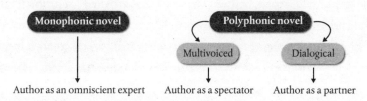

Figure 28 Bakhtin's view of the difference between monophonic and polyphonic novels.

towards a well-rounded, finalized, systematically monologic whole. It lives a tense life on the borders of someone else's thought, someone else's consciousness.'[11] Not surprisingly given the close links between Bakhtin and Valentin Voloshinov, this echoes what I have noted previously about Voloshinov's idea of individual human consciousness as a boundary phenomenon. Importantly though, it shows that great novels can have a role to play in depicting that phenomenon, giving it concrete life, and thereby giving us insights into our own minds.

Another intriguing possibility given our focus in this book on brain mechanisms is that Dostoevsky's great skill in creating a polyphonic world might have been linked to the fact that he suffered from epilepsy.[12] Reflecting his experience of his condition, a number of characters in Dostoevsky's novels are epileptic, the most famous being Prince Myshkin in *The Idiot*. As we have seen, epilepsy can be an extremely debilitating condition. However, Dostoevsky identified a more positive aspect to the disorder when he felt 'a happiness unthinkable in the normal state and unimaginable for anyone who hasn't experienced it...I am then in perfect harmony with myself and the entire universe.'[13] This feeling preceded a full-blown epileptic seizure. Could this feeling of harmony with the rest of the world have somehow helped Dostoevsky to more easily imagine the thoughts and desires of others?

One potential response to Bakhtin's view of Dostoevsky's fiction is that if his characters are as independent as the former suggests, then surely the novelist was being somehow passive or irresponsible in his role as author. In fact this was not Bakhtin's opinion, for he argued against the idea that Dostoevsky 'only assembles others' viewpoints, completely rejecting his own viewpoint'. Instead, he saw the novelist as creating 'a completely new and special interrelation between his truth and that of others'.

To highlight the significance of this shift, Bakhtin compared it to the Copernican revolution in physics. Much as Copernicus shifted the focus from the Earth to the Sun as the centre of planetary movements in the Solar System, so Bakhtin believed that Dostoevsky effected a 'small Copernican revolution' in removing himself, the author, from the centre of the textual world, to make way for a dynamic in which characters have more capacity to exert their own forces.[14] Given that such characters are fictional, this raises the question of what drives them.

One possibility is that important social issues of the time—for instance, as heard in conversation, debate, and argument by the author of the novel—inform the dialogue that is, after all, still placed in the mouths of different characters by the author. From this perspective it is surely no coincidence that Dostoevsky's novels have been described as teeming with 'thwarted characters on the verge of exploding—Raskolnikov in *Crime and Punishment*, Ivan Karamazov, the radicals of *The Possessed*'[15]—given that when he was writing, basic democratic rights and freedom of expression were suppressed by the Tsarist state. And cultural critic Marshall Berman has argued that Dostoevsky was driven to produce what are arguably the first truly modernist works of literature because of the stark contradiction in late-nineteenth-century St Petersburg between the façade of modernity as expressed in the city's new architecture and fashions, which offered the prospect of an exciting freedom from the shackles of the past, and the actual reality of an autocratic society still deeply moored in the feudal era.[16] However, it would be mistaken to assume that there is any simple, direct link between social circumstances and the dynamic of a good novel; indeed, it seems fair to say that if there is such an obvious link, the novel is likely to be the worse for it, becoming crude social commentary or thinly veiled propaganda rather than great art.

Instead, I believe that the best novels have a complexity and ambiguity of meaning—and this might be said to be increasingly a feature of literature in the modern era—that itself reflects the many contradictions in society and the individual psyche within that society. Importantly, this means there can be multiple readings of great novels, with different readers interpreting them in various ways, some of which may be distinct from—or even go way beyond—what the original author might have intended. And in what follows we will examine several novels that have multiple interpretations and which I also believe illuminate and enhance our understanding of different aspects of consciousness.

One example of a novel with many different possible interpretations is Emily Brontë's *Wuthering Heights*. Although it was published in 1847, this novel is surprisingly modern in both its outlook and structure. At its centre is the passionate, doomed love affair between Cathy Earnshaw and the foundling Heathcliff, but the novel is also noteworthy for its themes of revenge, betrayal, and instances of sadistic cruelty, all set against the savage beauty of the Yorkshire moors. Such themes help explain why the novel received many negative reviews when it was first published in 1847, with some critics finding it 'coarse', 'disagreeable', and 'a compound of vulgar depravity and unnatural horrors'.[17] Used to fiction in which characters fell clearly into categories of good versus evil, what seems to have particularly offended such reviewers was the moral ambiguity at the novel's heart. Yet for successive generations of readers this ambiguity has been part of the book's enduring appeal.

The ambiguity in *Wuthering Heights* is characterized by some of the novel's central characters. It would take a hardened soul not to be moved by the intensity of the passion of Cathy and Heathcliff for each other and also for the wild untamed moor, yet in other respects this pair are cruel and manipulating, and Heathcliff is capable of acts of the utmost brutality. But this ambiguity is also true of some more peripheral characters. In a novel with multiple layers of storytelling, it was surely a stroke of genius to have such tales of passion, violence, and unearthly goings-on relayed by the insipid, emasculated narrator Mr. Lockwood. At first, helped by the fact that this is his own account, we primarily view Lockwood as a victim, since when he first visits Wuthering Heights, he is treated inhospitably, attacked by Heathcliff's dogs, has icy water splashed down his neck, sleeps in a haunted chamber, and sinks up to his neck in snow.[18] Yet it is Lockwood who, during a dream or an actual supernatural visitation in which Cathy's ghost begs to be let in at the window of his bedroom, tries to drive the childlike apparition away by rubbing her wrist over the shards of broken window glass until the blood flows freely and stains the bedclothes.

The ambiguity at the heart of *Wuthering Heights* has complicated attempts to understand its underlying meaning. One interpretation seeks to locate it in the social and political tumult of the time. The time between 1801, when the novel begins, and 1847, when it was actually published, was after all a period of British history when the Industrial Revolution was in full swing, but so was the Chartists' campaign for universal (male) suffrage—the first mass working-class political movement in history—that increasingly found itself in a direct, often violent, confrontation with the forces of the state.[19] Meanwhile, this period saw the abolition of slavery, but also an influx of Irish immigrants fleeing famine, who were also often radicalized by the experience of living in an occupied nation.[20]

Heathcliff is living in Liverpool, centre of the slave trade and the port of entry for Irish immigrants, when he is discovered as a child by the elder Mr Earnshaw and brought to live at Wuthering Heights.[21] Because of this, some critics have suggested that Heathcliff represents the oppressed Irish, or black, masses, and his struggle against Hindley Earnshaw a symbol of rebellion against such oppression. For instance, literary critic Arnold Kettle has argued that the basic motive force of the novel is social in origin, and located the source of Cathy and Heathcliff's passionate attraction for each other in the rebellion forced on them by the injustice of Hindley and his wife Frances. For Kettle, Heathcliff is 'the outcast slummy [that] turns to the lively, spirited, fearless girl who alone offers him human under-standing and comradeship. And she, born into the world of Wuthering Heights, senses that to achieve a full humanity, to be true to herself as a human being, she must associate herself totally with him in his rebellion against the tyranny of the Earnshaws and all that tyranny involves.'[22]

There is a problem though with viewing Heathcliff primarily as a rebel against oppression. First, this would be quite at variance with Emily Brontë's own often quite reactionary political views,[23] and, secondly, on his return to Wuthering Heights—having found wealth in an unspecified way—Heathcliff becomes an oppressor himself in many respects. This is true not only in his expropriation of Hindley, which might be justified as revenge for past ills, but also in the abusive and exploitative manner with which he treats Isabella Linton and Hareton Earnshaw.[24] Heathcliff may well express the social tensions of his time, but he surely cannot be reduced to them, and this is surely part of his fascination.

A quite different reading of *Wuthering Heights*, relevant to what I said previously about the potential link between creativity and mental distress, sees evidence of neurosis in the structure and content of the novel. Here it is worth noting Freud's view that works of literature represent the external expression of the author's unconscious mind and a key motivation for writing a novel is to satisfy repressed wishes that might have developed during the writer's infancy and were instantly suppressed and dumped in the unconscious. Notably, Freud believed that the creative writer, like a child at play, creates a world of fantasy that is the fulfilment of unconscious wishes expressed in a disguised form.

Freud surely overstates his case in trying to explain everything according to his theory of the repressed unconscious, yet it is also true that Emily Brontë had many reasons to create a fantasy world as a buffer against reality. After his wife died aged only 38, Patrick Brontë sent his girls to the Clergy Daughters' School in Cowan Bridge in the Yorkshire Dales, which 'was run on spartan lines, designed to encourage resignation and humility'.[25] The terrible conditions at the school—vividly portrayed in Charlotte Brontë's *Jane Eyre*—almost certainly contributed to the deaths of Emily's sisters Elizabeth and Maria, from tuberculosis in 1825.[26] Later, in 1848, Emily's only brother, Branwell, also died from the disease, in his case as a consequence of his body being weakened by his addiction to alcohol and opiates.[27]

That Emily Brontë's fiction may have been driven by neurosis is based upon the claim that her writing became a compulsive ritual that allowed her to offset her anxiety about the misfortunes of her family. In support of this claim, critic Moussa Pourya Asl has pointed to a repetition in *Wuthering Heights* involving violent men, a cycle of names, a system of masters and servants, and reincarnation of character types.[28] Orphanhood and abandonment also characterize the novel and may reflect the losses Emily experienced in her life. Certainly, many of the characters—including

Cathy, Edgar Linton, Hareton, Isabella, and the younger Catherine—share with her the traumatic experience of their mother's early death.

There is also a repeating theme of sadism in the novel, which might be one way of unconsciously channelling desires of revenge as a response to misfortune. So at one point we learn that Hareton has hanged 'a litter of puppies from a chair-back', while at another, Heathcliff hangs Isabella's 'little pet springer' dog.[29] Meanwhile, Edgar Linton, for sport, tortures cats to death and Lockwood finds 'a heap of dead rabbits' in the Heathcliff household.[30] And of course, there is the scene already mentioned, in which Lockwood drags the wrist of Cathy's childlike apparition against a broken window to escape her presence.

Yet focusing only on the 'monstrous symbolic conflicts' in the novel ignores another central element, namely the air of homely domesticity that forms the backdrop both to the telling of the story and to Wuthering Heights itself. Of key importance here is servant Nelly Dean, actually the principal narrator, Lockwood's own account acting as the key that unlocks this interior narrative. As the only person with first-hand know-ledge of all three generations of the linked Earnshaw and Linton families, through serving both at Wuthering Heights and Thrushcross Grange, Nelly is well placed to tell the families' tragic tale. Yet what is most striking about Nelly's account is her 'matter-of-fact' telling of this fantastical, and at times unearthly, story. As literary critic David Daiches has noted, 'at no point does Nelly throw up her hands and exclaim: "for God's sake, what is going on here? What kind of people are they?"'[31] Instead the story is related quite casually as she goes about her daily business.

The importance of domestic routine is an often underappreciated fea-ture of Wuthering Heights itself. Despite the intensely passionate and possibly supernatural nature of the events that we learn have taken place at this isolated farmhouse on the bleak and windswept moors, our first impression of the sitting room at Wuthering Heights as related by Lockwood is of a 'huge, warm, cheerful apartment', which 'glowed delightfully in the radiance of an immense fire'.[32] This counterposition of the homely and the fantastical plays an important role in the believability and power of *Wuthering Heights* as an illustration of the human condition. As Daiches puts it, 'in fact, throughout the novel, the homely and the familiar and the wild and extravagant go together, the former providing a setting for the latter, with the result that the simplest domestic detail can, in virtue of being made the scene of such monstrous conflicts of passion, become symbolic. Conversely, passion set in such scenes becomes credible.'[33]

This juxtaposition of the ordinary and the fantastical, the matter-of-fact and the supernatural, is sustained until the famous closing lines of the novel. Significantly, it is the naïve and superficial narrator Lockwood who, happening upon the adjacent graves of Cathy, Heathcliff, and Edgar Linton, tells how he 'lingered around them, under that benign sky; watched the moths fluttering among the heath and hare-bells; listened to the soft wind breathing through the grass; and wondered how anyone could imagine unquiet slumbers, for the sleepers in that quiet earth.'[34] Such lines encapsulate the ambiguity at the heart of *Wuthering Heights*, for while this peaceful ending may be viewed as fitting for the spirit of reconciliation at the end of the novel—particularly personified by the union of Hareton Earnshaw and Catherine Linton—there seem plenty of reasons, given the unearthly passions described in the story, to imagine things will not be quite so peaceful and straightforward.

If ambiguity is a feature of *Wuthering Heights*, then it could be said to be the defining characteristic of *The Turn of the Screw* by Henry James. Published in 1898, the novel's unnamed narrator is a young woman, a parson's daughter, who is engaged as governess to two angelic children—ten-year-old Miles and eight-year-old Flora—at Bly, a remote English country house.[35] However, what initially seems to be a pastoral idyll soon turns harrowing, as the governess becomes convinced that the children are consorting with a pair of malevolent spirits. These are the ghosts of former employees at Bly: Peter Quint, a valet, and Miss Jessel, a previous governess. In life, scandalously, the pair had been illicit lovers and both died young—Quint after falling and hitting his head while leaving the pub, and Miss Jessel possibly by committing suicide after discovering that she was pregnant with Quint's child. Their phantom visitations with the children hint at Satanism and possible sexual abuse.

The governess becomes convinced that her primary mission is to save the souls of the children; yet in trying to shield them from the ghosts, she ends up traumatizing Flora and possibly contributing to the death of Miles from a heart attack. The novel has perplexed readers and critics ever since it was first published, with the key unsolved question being whether the ghosts are real or merely delusions of the governess's mind.[36] In fact, literary critic Brad Leithauser has argued that 'all such attempts to "solve" the book, however admiringly tendered, unwittingly work toward its diminution; its profoundest pleasure lies in the beautifully fussed over way in which James refuses to come down on either side...the book becomes a modest monument to the bold pursuit of ambiguity.'[37] However, this does not mean that we cannot learn a lot both about the novel's underlying power and about the workings of the human mind, by exploring different interpretations of the story, for the ambiguity of the story itself reflects some of the ambiguous nature of thought processes.

From the point of view of seeing the unconscious as a crucial part of human consciousness, the idea that the ghosts are products of the governess's deluded mind is both plausible and potentially highly suggestive about the workings of that unconscious. For instance, one interpretation of the novel views the governess as driven by both sexual and economic motives, being attracted sexually to the children's aloof, absent guardian uncle on their initial meeting and entertaining thoughts that if he were to reciprocate such feelings, she might end up as mistress of a grand estate.[38] However, because such feelings clash with her strict religious upbringing, which has taught her that both sexual attraction and coveting material wealth are sinful, she represses such feelings, instead projecting them on to the 'ghosts' conjured up by her unconscious. This possibility is suggested by scenes such as one in which the governess accidentally sinks down on a staircase in the same spot where she had seen Miss Jessel's ghost earlier, and then in the schoolroom, where both women seem to have a certain right to be present; indeed, the governess even feels like the intruder at one point.

Fear of the lower classes may also be a factor in the governess's unconscious feelings.[39] It is significant that the novel was published in 1898, only a decade after the mass unionization of London's East End that began with the match girls' strike in 1888 which I mentioned previously, and also a decade after the horrific murders of prostitutes committed by the serial killer 'Jack the Ripper' in East London, which occurred in the same year as that strike.[40] For many middle-class individuals, fears about the political radicalism of the working class could be combined with prejudiced beliefs about the supposed sexual licentiousness and violent nature of this class. For both the fictional governess and many middle-class readers of the novel, Miss Jessel's crime would have been not only the illicit nature of her sexual relationship with Peter Quint, but also the very fact that she was consorting with such a lower-class individual.[41]

As persuasive as the argument is that the ghosts are constructions of the governess's own repressed desires, a powerful aspect of the story is that an equally persuasive case can be made for the opposite view. In particular, the possibility with which the governess becomes obsessed, namely that both children can see the ghosts as clearly as she can and are complicit in their corruption by the latter, is kept alive right to the end of the novel by both their behaviour and the ambiguity in their language. For instance, when Miles asks the governess to send him back to boarding school, stating that he is 'a fellow, don't you see? who's—well, getting on,' is he showing a guileless desire to be with boys his own age, or the precocity of the abused child? And when Flora says to the governess, 'I don't like to frighten you,' is she confirming why she is universally admired for good behaviour, or making a covert threat?[42]

Importantly, given how much I have placed language at the centre of human conscious awareness, a good deal of the power of *The Turn of the Screw* stems from the way it uses language to unsettle us. The governess herself gives us a sense of the nature of what is to follow at the start of her account when she says, 'I remember the whole beginning as a succession of flights and drops, a little see-saw of the right throbs and the wrong.'[43] And as literary critic William Penny has noted, the account continues as a 'series of equivocations, provisos, and meanderings that further cloud our understanding of the young woman's motivations.'[44] Moreover, in her interactions with the elderly housekeeper Mrs Grose—her only adult companion in the house—the governess's sentences often go unfinished, and the whole text is littered with the presence of hyphens and dashes. It is almost as if the absence of language is as important as words in conveying the ambiguity at the heart of this novel.

This ambiguity is so powerful that Brad Leithauser believes it 'provides an unrivalled opportunity to read in a bifurcated fashion, to operate paragraph by paragraph on two levels. Logically, the effect of this ought to be expansive. James is trafficking in openness; readers can shift, at whim, from ghostly tale to character study.'[45] Yet as Leithauser goes on to say, paradoxically the ultimate effect is the opposite of openness, turning it 'into the most claustrophobic book I've ever read. Yes, you're free to shift constantly from one interpretation to the next, and yet, as you progress deeper into the story, each interpretation begins to seem more horrible than the other…If the governess *is* mad, she has unwittingly killed a bright and beautiful little boy; this is a tragedy, but a local one. If the ghosts are genuine, however, there are jagged cracks in the firmament above us all, and nobody is safe.'[46] The chink of doubt as to whether such spirits might actually exist, exposing our rational explanations as illusions themselves, helps to give this story its enduring power.

The idea that reality itself might be just an illusion of human consciousness in certain circumstances is dealt with in William Golding's *Pincher Martin*. Published in 1956—the year of the Suez Crisis and the Hungarians' revolution against their Stalinist regime—this novel takes us inside the head of a sailor battling to survive after his ship is blown apart by a German torpedo during World War II.[47] We follow the sailor, Christopher Martin, as he first tries to keep afloat in the freezing, turbulent seas of the north Atlantic, and then, after miraculously finding a rocky island, his battle to stay alive on it, while hoping to be rescued.

However, in the very last line of the novel, we realize that Martin drowned almost immediately after his ship was sunk and it now becomes clear that the whole battle to survive on the island was merely an illusion of his dying moments. In fact, clues are scattered throughout the course of the novel that all was not as it seemed. For instance, we learn that the 'island' bears an uncanny resemblance to a tooth that Martin once had extracted.[48] Meanwhile, increasingly odd events begin to occur as the novel progresses that make us begin to question Martin's sanity and therefore the reality that he seems to be experiencing.

Pincher Martin is interesting in several respects in terms of the ideas about human consciousness that we have been considering in this book. One is the difference between sanity and madness.[49] We expect a sane person to have a more coherent consciousness than someone who is mentally disturbed, but the closer that Martin gets to realizing his true predicament, the more confused become not only his thoughts, but also the whole 'reality' of the island. We realize that the carefully constructed battle to survive that we have been experiencing with Martin is really just his conscious mind refusing to accept what is actually happening—that he is drowning.[50] And it is the unconscious that gradually begins to make itself known, with increasing force, as a way of conveying the reality of the situation.

Another important aspect of *Pincher Martin* is that it powerfully illustrates how great literature can allow us to identify with the thoughts and desires of another human being, even one who should be highly unlikeable. In creating Christopher Martin, Golding deliberately went out of his way to create a deeply objectionable person.[51] In flashbacks that occur with increasing frequency as the novel progresses, we learn that Martin was selfish, greedy, and a potential rapist and murderer.[52] Yet it is very difficult for the reader not to feel deeply caught up in his struggle to survive, and to suffer as he suffers, as we share in his plight.[53] One reason for this is that not only does the novel manipulate our sense of empathy for a fellow human being, but also Martin's attempts on the rock to fashion a shelter for himself, then acquire food and drink, and finally to try and order the world around him mimic how human beings have distinguished themselves from animals. In particular, the way that Martin seeks to make sense of his rock by naming things on it plays on the importance of language in our lives.[54]

Previously in this book, I looked at the role that dreams play in human consciousness, particularly their link with creativity. Dreams have many curious aspects, one being that although they can allow us to indulge in hidden fantasies that we might not admit to when awake, our desires can also often be thwarted in dreams. Another aspect is the way that objects, people, and situations that would not normally belong together can do so in a dream.[55] This reflects the fact that dreams do not have to respect the normal rules of time and space. The possibility that dreams may provide insights into the nature of human consciousness is explored in *The Unconsoled*, a novel by Kazuo Ishiguro published in 1995.

Ostensibly, the novel is about a famous pianist called Ryder who arrives in a nameless European city where he is to play an important concert.[56] Yet as the novel progresses we come to realize that this is a far from real environment. At times Ryder feels like the character 'K' in Franz Kafka's novel *The Castle*, caught up in a mysterious and vaguely sinister plot that surpasses his understanding.[57] His efforts to get to the concert hall where he is to play often seem like a parody of K's struggles, just as his attempts to cope with a host of dithering acquaintances recall the other character's frustrating encounters with an incompetent bureaucracy. Yet at other times, Ryder is more like Alice on a visit to Wonderland: he finds himself going in and out of strange little doors that apparently lead him in circles, and he meets an assortment of curious people who draw him into a series of intrigues that threaten to distract him from his appointed duties.[58] Indeed, practically everyone that Ryder meets has an important favour to ask him that steals more of his precious time. An elderly porter at the hotel asks him to mediate a family problem; a woman named Sophie begs him to talk to her troubled son; and the hotel manager asks for advice about his own son, an aspiring musician.

Yet as the novel progresses, Ryder experiences momentary flashbacks during which he realizes he does seem to know things about many of these people, including their most intimate thoughts.[59] The woman named Sophie appears to have once been Ryder's girlfriend or wife; her son, Boris, is his child or stepson, and the elderly porter at the hotel turns out to be Sophie's father; however, we also learn that he has not spoken to his daughter in years. At first glance, it seems counterintuitive that Ryder might be in such a predicament. He is an accomplished artist, renowned concert pianist, and respected public figure. Yet there is an increasing feeling that his life has come to nothing—at least in terms of personal relations. This sense of lack of achievement is enhanced by Ryder bizarrely bumping into a succession of characters from his youth in England—perhaps echoing the idea developed previously in this book that an individual consciousness is heavily shaped by the interactions we have with our peers as well as parents and teachers—who involve him in more distractions through their demands on his time, and rarely seem in awe of what he has achieved so far professionally.

In this sense, the novel reads as a catalogue of wasted opportunity, a failure to seize the moment, an inability to communicate, inconclusive wondering, and general time-wasting. Crucially, Ryder seems to continually miss his chance to connect as a human being. This becomes a book, as one critic has put it, about the 'destructive power of love without empathy'.[60] Ryder always falls short in comprehending the minds of those around him. He forever misinterprets, misunderstands, and misdirects. And so it is that Sophie, the mother of his child, says to him in the end, 'Leave us, you were always on the outside of our love.'[61]

In fact, it is not just Ryder himself, but everyone he encounters, who seems to be suffering from an inability to connect or communicate with a loved one.[62] Brodsky, a local drunk rehabilitated and cast in the role of resident artistic genius, has been estranged for years from the woman he loves. And the hotel manager, Hoffman, appears to be seriously out of touch with the feelings of his wife, and also with their son, Stephan, a talented young pianist.

Ishiguro himself has claimed that in writing The Unconsoled, he wanted to 'present a biography of a person, but instead of using memory or flashbacks, you have him wandering about in this dream world where he bumps into earlier, or later, versions of himself. They're not literally so; they are to some extent other people.'[63] In line with what I have said previously about the fluidity of meaning within the inner dialogue that constitutes an individual person's thoughts, Ishiguro believes that in dreams an individual consciousness can 'bend and twist the whole world into some big expression of his feelings and emotions.' This also fits with another idea put forward in this book: that individual consciousness is a boundary phenomenon, uniting both our inner thoughts and the influence of wider society.

By presenting the story in this way, Ishiguro is able to ask profound questions about our relationships with others, our responsibilities and obligations, and how we can progress through life without facing up to our real thoughts and desires. To some extent, for Ryder, any realizations he makes about the negative features of his life and interactions with others, and how he might overcome these in future, remain only partial. For displacement in dreams may be precisely there to shield us from unmediated confrontation with painful truth.[64] Yet in other respects, there is definitely a sense of resolution and moving forward in this novel.

It is true that the concert never happens, but we do get a firm conclusion about whether Ryder will make his family relations work. We also find out what has been happening in hotel-owner Hoffman's marriage, and a way into a fulfilling future for his son, while even Brodsky's troubled life hastens towards a conclusion. It is impressive, in fact, how many loose ends are tied up in the concluding pages. Even the series of bizarre encounters with figures from Ryder's past lead to a certain sense of closure in finally dealing with problematic issues from his youth. Because of this, the novel ends on a surprisingly uplifting note: that although life and relationships may often seem difficult and seemingly intractable, through perseverance and a willingness to tackle such difficulties and face up to our thoughts and desires—but also our responsibilities and obligations to those around us—we all have the possibility to progress in life and perhaps find real happiness in the process.

CHAPTER 21

SCIENCE AND TECHNOLOGY

———❦———

Perhaps more than many, I think I can say that I have been fairly consistent in my chosen career path. At the age of fourteen, I decided I wanted to be a biochemist—or more specifically a French biochemist, as I had just been on my first French exchange trip, and fallen in love with the country and its culture, food, wine, and general ambience. My goal at that time was therefore to end up working in France, ideally in a renowned scientific institution such as the Pasteur Institute in Paris. Reality conspires against our childhood hopes and desires in various ways, and I never made that step of moving abroad. But I did achieve my dream of becoming a biochemist, albeit in Oxford rather than Paris. By curious coincidence, my wife, who is Portuguese, also wanted at the age of fourteen to eventually become a biochemist in the Pasteur Institute. As a step to that goal she did a PhD at the Imperial Cancer Research Fund in London, where I also did my PhD, and where we met. As a consequence she decided to remain in Britain and build her career here rather than moving to Paris. Today, I like to tell my son and daughter that they owe their lives to cancer research!

My wife and I have had ups and downs in our research careers, but we have retained our passion for science; today it is what occupies a good deal of our working hours, and sometimes keeps us awake at night. Above all though, our research is also just the job that we do, and we both consider ourselves fairly ordinary people. Such a viewpoint is not shared by the majority of the population. A common stereotypical view of scientists is that they are very clever, but also rather odd—the typical absent-minded, socially inept, nerdy individual personified in popular culture through

TV shows such as *The Big Bang Theory*.[1] While scientists can be viewed as doing great good—such as discovering new treatments for cancer—they may also be seen as responsible for environmental catastrophes and consequently as irresponsible people, willing to risk all in their search for knowledge or lacking concern for the consequences of their actions.[2] Whether good or bad though, a common perception is that scientists as a group think and behave in very different ways from the rest of humanity.

In this chapter, I want to challenge that perception, and argue that while scientists may occupy a very specific niche in society in their role in the construction of new knowledge, they are prone to the same strengths and weaknesses in terms of their individual minds, and are affected by the same social prejudices, as the rest of humanity. Moreover, far from being an alien activity, the pursuit of scientific knowledge is something that is a key resource for humanity as a whole, and everyone with an interest in the origins and fundamental nature of human conscious awareness should have a strong interest in knowing more about the process of scientific discovery, and an input into debates about both the direction and application of scientific research. To understand why, it is worth looking at how science first developed.

Ever since our proto-human ancestors first began using and designing tools to manipulate the world around them, it seems fair to say that humankind has been practising a form of science and technology. And unlike other animals ranging from crows to chimps which have been observed to skilfully use natural objects as tools,[3] a distinguishing feature of our species is that we not only use tools, but also are continually improving on them and developing new kinds.[4] Our ability to develop different tools over time is closely bound up with our creativity and imagination, for in order to design a new tool, it is necessary to have some preconception of something that does not yet exist. Karl Marx recognized this in relation to creation of animal and human habitats when he noted that 'a spider conducts operations that resemble those of a weaver, and a bee puts to shame many an architect in the construction of her cells. But what distinguishes the worst architect from the best of bees is this, that the architect raises his structure in imagination before he erects it in reality.'[5]

We humans have another unique quality: our ability to discover new things about the world and how it functions. The development and use of novel tools may be described as technology, while finding out new things about the world can be considered as science. Yet it would be unwise to draw too narrow a dividing line between these two subjects. We can only develop new technologies by gaining greater insights about the world around us and how it works. At the same time, new technologies drive scientific discovery by making it possible to investigate the world in more varied and fundamental ways. Indeed, the molecular biologist Sydney Brenner once stated that scientific progress 'depends on the interplay of techniques, discoveries and new ideas, probably in that order of decreasing importance.'[6]

Scientific discovery generally requires observation but also experimentation. Crucially, the scientific method involves putting forward a hypothesis and then studying whether reality corresponds to our proposed view.[7] Humans and even our proto-human ancestors have probably been carrying out such observation and experimentation ever since we first began interacting with nature in novel ways, something made possible by the development of tools. Once we developed language, and an ability to think in a complex fashion about objects and people, cause and effect, time and place, and to communicate such ideas to other humans, we were able to frame questions about the world and test them.

Imagination and creativity are crucial features of humankind, whose links to the rest of consciousness I discussed previously. Yet when asked to name creative individuals, most people would tend to think of a musician, a novelist, or a painter—in other words, artistic individuals—not a scientist or a technologist.[8] Bill Wallace, a science teacher at Georgetown Day School in Washington, DC, thinks that one reason for this is the way many people are taught about science. According to him, 'a lot of kids think that science is a body of knowledge, a collection of facts they need to memorize.'[9] Wallace believes that allowing students to come up with their own solutions to open-ended questions can foster creativity in the classroom. For instance, he asked his students to design experiments to investigate how sensitive fruit flies are to alcohol. He notes that 'I had seven groups of students, and got seven different ways to measure inebriation. And that's what I would call creativity in a science class.'[10] Wallace thinks an approach to science that emphasizes only facts and concepts leaves little room for the creative thinking central to science. He believes that 'if instead, you teach science as a process of learning, of observing and of gathering information about the way that nature works, then there's more room for incorporating creativity.'[11]

One way of demonstrating the importance of creativity in science is to redefine what it means to be creative, in contrast to common perceptions. For Robert DeHaan, a biologist at Emory University in Atlanta, Georgia, 'creativity is a new idea that has value in solving a problem, or an object that is new or useful.'[12] That can mean composing a piece of music that is pleasing to the ear or painting a mural on a city street for pedestrians to admire, but it can also mean dreaming up a solution to a challenge encountered in the lab. Many scientists describe science not as a set of facts and vocabulary to memorize, or a lab report with one 'right' answer, but as an ongoing journey, a quest for knowledge about the natural world.

One theme that emerged from our earlier exploration of creativity and imagination is the importance of unconscious brain activities in the creative process. It is interesting then that recent thinking about creativity in science has emphasized the importance of 'associative thinking'—the process whereby the mind is free to wander, making possible connections between unrelated ideas. The process runs counter to what most people would expect to do when tackling a challenge. Most would probably think the best way to solve a problem would be to focus on it—to think analytically—and then to keep reworking the problem. Yet Robert DeHaan believes that 'the best time to come to a solution to a complex, high-level problem is to go for a hike in the woods or do something totally unrelated and let your mind wander.'[13]

An occasion where a walk in the woods did lead to a major scientific breakthrough happened on Christmas Day in 1938. An unsolved puzzle at this time was the mechanism by which unstable radioactive elements break down into smaller, more stable ones. Interest in this problem was first stimulated by Henri Becquerel's discovery in 1896 that uranium emits radiation, followed by Marie and Pierre Curie's identification of the even more radioactive elements polonium and radium, and finally by Ernest Rutherford's demonstration that these elements are intermediates in a chain of disintegration starting with uranium and ending with lead.[14] Such studies showed that the old alchemist's dream of transforming one element into another was not so far-fetched. And they also led to the realization that atoms are composed of even smaller particles—electrons, protons, and neutrons.[15]

One individual whose imagination was stimulated by such findings was the German physicist Lise Meitner, who decided to devote her life to trying to identify the mechanism by which one element could change into another. However, Meitner had to battle against many obstacles in her career. For instance, when Rutherford visited her he first exclaimed, 'Oh, I thought you were a man,' then went off to talk science with her colleague Otto Hahn, leaving Meitner to go shopping with his wife.[16] But far greater challenges faced Meitner, for she was a Jew, and therefore in great danger in Hitler's Germany. Fleeing the country to Sweden in July 1938, she was lucky to escape alive, but had lost her laboratory, house, life savings, and homeland. She did, however, continue communicating with Hahn in Berlin, who kept her up to date with his research findings. On 22 December 1938 he reported that having irradiated uranium with neutrons, he produced two much smaller elements—barium and krypton. This apparent division into two made no sense given the view of atomic structure at the time.[17]

However, while walking through a forest near the town where she was staying, on her way to visit friends who had invited her for Christmas dinner, Meitner suddenly realized the findings could be explained if the nucleus of the uranium atom was like a wobbly, unstable drop of water that could divide at the slightest provocation.[18] Sitting down on a tree stump with pencil and paper she realized through a series of equations that this 'splitting of the atom' was mathematically feasible and, according to Einstein's equation $E = mc^2$, would release huge amounts of energy. Meitner's revolutionary insight was a transformative moment in physics but it also opened the door to the development of the atomic bomb and the terrible loss of life that resulted when it was used on the Japanese cities of Hiroshima and Nagasaki. While some physicists helped develop the bomb because of fears about the Nazis getting there first, Meitner remained adamant that 'I will have nothing to do with a bomb!'[19]

The practical application of the discovery of nuclear fission for purposes of mass destruction demonstrates the double-edged character of science in modern society, on the one hand leading to great good through new medicines or novel ways of feeding humanity, and on the other, fuelling the development of new forms of warfare or environmental pollution. In fact, this double-edged character of science has been true ever since the birth of class society. It is important not to forget, however, that we owe the very development of science to this type of society. Previously, I pointed to some oppressive features of class society and that it brought with it not just class divisions, but also sexism and racism, war and other forms of destruction. However, it is also true that this form of society is responsible for the amazing knowledge that we now possess, not only about our own planet, but also about the whole Universe. At the same time, we have gained important insights into the factors that govern not just our bodies but also our minds, and finally about human society itself.

One way that class society led to the development of science and technology was through the birth of the city, and subsequent new types of production and trade between cities. Both writing and the first forms of mathematics were initially developed to allow cities to track food supplies and other goods that for the first time in history formed a surplus.[20] Meanwhile, the development of agriculture led to increased interest in predicting the weather and the seasons. This was particularly important in a society like ancient Egypt whose agriculture relied heavily on the seasonal flood of the river Nile to fertilize the land.[21]

One of the advantages of slave societies such as those of the ancient world was that—freed from having to do any menial work—the upper classes now had the leisure time to sit and think. This allowed the development of new ideas about the natural world. However, a negative feature of such societies was that being unaccustomed to getting their hands dirty, upper-class individuals rarely did experiments to test their ideas, and therefore the practice of science lagged far behind scientific theory. Having said that, certain areas of technology that were deemed particularly important for ancient society, like the architectural advances required to build the monuments that celebrated these societies' achievements, but also the war weapons required to defend the cities, did see much progress at this time.

There is often a tendency to distinguish between pure and applied science. But the distinction between these two forms has always been somewhat blurry. Today Archimedes is famed for his discoveries in geometry and, increasingly, for the realization that he came close to developing calculus, the mathematics of movement. Yet in his own era, Archimedes was better known for the war machines he designed to defend his home city of Syracuse in Sicily—then one of the world's greatest cities—against the Romans.[22] In fact, it may have been his development for this purpose of new types of catapults, and study of the dynamics of the ballistic missiles they fired, that led to Archimedes' interest in the science of moving objects.

In general though, the separation between the free artisans and slaves who made things and the leisured class who just sat and thought stifled the development of science and technology in ancient societies. This separation between the leisured class and the people who did the work was also a feature of feudal society. Another limitation of both societies was that the main point of producing a surplus in these societies was to feed, pamper, and entertain their rulers. This limited scientific development, because once ruling class needs had been satisfied there was little impetus for further development of the productive forces.

In contrast, the rise of capitalist society gave a new impetus to science and technology by giving birth to a society founded solely on commodity exchange. Since any object—including people—can be a commodity under capitalism, this led to a fantastic expansion in the types of objects produced, but also a drive to develop new means of production. There is a direct link here with science and technology, as shown by the way that development of steam engines for the production of textiles and new forms of transport such as the train and the steamship led to the birth of the science of thermodynamics. This not only had repercussions for the development of further technologies such as electricity, but also transformed our view of the material nature of the Universe, culminating in Einstein's theory of relativity.

Science does not develop independently of the society in which it arises. The impact of societal values on science can be seen, for example, in the scientific questions that are deemed worthy of asking, and the funding available to study such questions.

Scientific understanding itself is sometimes said to be neutral of society; however, this is only partially true. By putting forward hypotheses about the natural world and then testing them against reality, scientists gain a better understanding about the true nature of the world. Indeed, the philosopher Karl Popper claimed that science can only disprove, not prove anything.[23] In fact the biggest breakthroughs in science often occur when a hypothesis about the world is proved wrong. For while confirmation of a hypothesis by an experiment does not change the hypothesis, when the opposite happens this can force scientists to reconsider and put forward new hypotheses. Sometimes the realization that a view of nature is wrong can have revolutionary implications, such as when Einstein showed that Newton's mechanical view of the Universe only partially explained the behaviour of mass in space and time.

Although science ought to progress by new findings demonstrating the invalidity of a hypothesis, there is often a huge amount of resistance by scientists to giving up a false idea. Sometimes it takes several different pieces of evidence to result in what US philosopher Thomas Kuhn called a 'paradigm shift', in which a new way of looking at the world finally becomes acceptable.[24] However, just because a key part of scientific progress involves testing hypotheses against reality, this does not mean that society's values do not have an influence on scientific ideas themselves. Darwin's discovery that all species are a product of evolution by natural selection was a revolutionary step forward in our understanding of life, and indeed the origins of our own species. However, the emphasis on competition—the survival of the fittest—detracts from the fact that cooperation is as much a feature of the natural world and of the evolutionary process.[25] In this case, Darwin's objective view about the dynamics in nature was overly influenced by the values of free-market economics that tended to dominate all discussions at that time in Victorian England.

I mentioned previously that a key idea at the heart of the revolutions that took place in the Netherlands and England in the sixteenth and seventeenth centuries, and in France and America at the end of the eighteenth century, was the importance of the primacy of the individual. The view of society as being just a collection of individuals had an influence on perceptions about the physical world. For instance, it inspired Newton's view of the Universe as being akin to a piece of clockwork, that is, something reducible to its separate parts. Such an approach to understanding the natural world is known as reductionism. It was the guiding principle for the British renaissance in science that occurred in the seventeenth century following the revolution in 1649 that brought the capitalist class to power. As well as leading to Newton's great discoveries in physics, this period saw major steps forward in biology and medicine. In fact, this is not the only example where a social revolution has acted as a major catalyst for dramatic changes in the sciences. The period around the French revolution of 1789 was associated with major scientific discoveries and technologies that included the isolation of chemical elements, a better understanding of human metabolism, the physics of electricity, and the first theory of evolution.[26] Remarkable technical innovations at this time included the metric system, ballooning, the invention of canned food, and the semaphore telegraph.

While reductionism has been an immensely powerful approach for understanding the natural world, it also suffers from important limitations. Here we should distinguish reductionism as a tool for dissecting the natural world and making it more comprehensible, and reductionism as a philosophy about the structure of that world. As a molecular biologist, I use reductionist methods routinely. For instance, they allowed my colleagues and I to show that a human egg is kick-started to develop into an embryo by a chemical 'signal'—a rise in the concentration of calcium ions in the egg—induced by a protein, PLCzeta, produced by the sperm and injected into the egg at fertilization.[27] And this led to our subsequent discovery that certain types of infertility in men are caused by a defect in this molecular process.

Reductionism as a philosophical idea is allied to the reductionist method, but takes it further, in proposing that if we could only catalogue all the different parts of the system, then we would be able to fully describe its operation as a whole. In biology, this idea informs approaches such as the Human Genome Project, whose aim was to map and obtain the DNA sequence of all the genes in the genome, or the current Connectome Project,[28] which seeks to identify all the neurons in a human brain and also the synaptic connections between them.

The Human Genome Project has indeed become a vital resource for those studying the functional roles of genes in specific biological processes, and it is possible that the Connectome Project may also prove very valuable for those wanting to know more about the basic architecture of the brain. Yet increasingly, for complex processes like the development of a fertilized egg into the different cell types, tissues, and organs of the foetus, or the way that different organs interact in metabolic processes, or indeed how our brains work, the path to understanding is now seen by many to require going beyond simply identifying a complex system's parts.[29] Instead, there is growing awareness that the whole determines the parts as much as the parts determine the whole. This has particularly important implications for understanding human consciousness, and it is significant that Lev Vygotsky, whose ideas have influenced many of the proposals about consciousness developed in this book, also put forward some more general ideas about the best approach for studying complex systems.

Vygotsky believed that rather than seeking to explain the operation of a complex system as the sum of its most basic parts—which in the case of the body as a whole could mean the genes, or in the case of the human brain its individual neurons—instead, we should be seeking to identify a 'unit of analysis' that somehow makes it possible to represent the properties of a system at not only the microscopic but also macroscopic level.[30] In biology we can say that, to some extent, such a unit of analysis was identified by Charles Darwin through his theory of evolution by natural selection. Importantly, this principle governs both the macroscopic—the interrelationship of species in a complex ecosystem—and the microscopic—the way that the amino acid sequence of a particular protein changes over millions of years because of mutations in the gene that codes for the protein.

Similarly, the development of the periodic table by Dimitri Mendeleev transformed chemistry by identifying a pattern that explained both the macroscopic properties of different elements—for instance, why chlorine is a reactive gas, while gold is an inert metal—and later their differences at the level of atomic structure.[31] Meanwhile in physics, quantum mechanics can explain phenomena at not only the subatomic, but also higher levels. Computers and smartphones, lasers, GPS tracking devices, and MRI scanners are all technologies based on our scientific understanding of the quantum world.[32] Finally, in terms of understanding things at a cosmic level, Einstein's theory of relativity has played a revolutionary role.[33]

However, despite such successes, it would be mistaken to assume that there is nothing more to be done in terms of identifying more fully encompassing units of analysis for these different branches of science. In biology, the discovery of the gene as the basic unit of inheritance by Gregor Mendel, and the subsequent identification of DNA as the molecule that genes are composed of, played a highly important role in connecting the microscopic and macroscopic worlds.[34] In a sense, while natural selection still provides the overarching structure of explanation, these other discoveries, together with a more detailed understanding of cellular structure, has helped to 'flesh out' the unit of analysis in this case. This example shows how such a unit is not necessarily a fixed entity, but one that can expand in meaning.

This is important because one of the revelations of recent years—outlined previously in this book and in more detail in my book *The Deeper Genome*—is that there is far more to genes than simply stretches of DNA coding for proteins.[35] In particular, there is now increasing recognition that genomes can only be understood as 3D entities, and that chemical modifications to the DNA and the proteins associated with it also play key roles in gene expression, as do the newly discovered class of regulatory RNAs. Finally, tying all these discoveries together is the new field of epigenetics, which has revealed that both the cellular and bodily environment, working via the control mechanisms just mentioned, can influence not only genome function in an individual organism but also in its offspring. Importantly, this suggests that the twentieth-century 'modern synthesis' of Darwinism and traditional genetics needs modernizing if it is truly to explain the multiple levels of biology.[36]

Meanwhile, in physics, we currently face a curious schism in explanatory power. On the one hand, Einstein's theory of relativity explains the structure and dynamics of the Universe at a macroscopic level, accounting for gravity and all of the things it dominates: orbiting planets, colliding galaxies, the dynamics of the expanding Universe as a whole.[37] On the other hand, quantum theory, which handles the other three forces—electromagnetism and the two nuclear forces—is extremely adept at describing what happens when a uranium atom decays, or when individual particles of light hit a solar cell.[38] However, the problem is that relativity and quantum mechanics are fundamentally different theories with different formulations. It is not just a matter of scientific terminology; it is a clash of genuinely incompatible descriptions of reality.[39] Relativity gives nonsensical answers when scaled down to quantum size, eventually descending to infinite values in its description of gravity. And quantum mechanics runs into serious trouble when blown up to cosmic dimensions.

Here the problem is that two plausible units of analysis have been identified that each explain nature across a wide sphere of influence, but then somehow break down at a universe-wide level. Therefore, if a unit of analysis is to be found that encapsulates the microscopic and the macroscopic in the physical world, then it must somehow reconcile these two approaches, or identify a completely different way of understanding this world.

One of the goals of Lev Vygotsky was to identify an all-encompassing unit of analysis for the study of human consciousness. As we have seen, he did this by showing that it was the 'mediation' of the human brain by cultural tools—particularly language—that created the self-awareness unique to our species. In addition, Vygotsky saw this as a social process, in which 'higher mental functions appear on the inter-psychological plane before they appear on the intra-psychological plane'; that is, initially as social interactions between individuals before they become internalized as part of the formation of the individual psyche.[40] This led Vygotsky to propose that the unit of analysis for human consciousness should somehow reflect this dynamic interaction. In his view, 'word meaning' could represent such a unit since 'the word is related to consciousness as a miniature world is related to a large one, as a living cell is related to an organism, as an atom to a cosmos...The meaningful word is a microcosm of human consciousness.'[41] At first glance, this seems like an odd choice for a unit that is supposed to be so expansive. However, if we interpret word meaning to encompass not just the complex interactions that bind words together in a framework shared by interacting individuals, their relationship to the inner speech of thought, and their link to the wider values and tensions of a particular society, but also the unconscious, emotions, and creative impulses, then it may be easier to see what Vygotsky had in mind.

Yet word meaning as expressed in this expansive way must also somehow connect with specific neural processes and with the particular biology and architecture of the human brain. One important emerging insight, highlighted at various points in this book, is the recognition that while our brains may share the same basic molecular and cellular processes as those of our closest animal cousins such as chimpanzees, there is increasing evidence of radical restructuring of the human brain. This restructuring seems to have occurred not only at the level of novel neural connections between different regions of the brain, but also through important differences in the expression of genes in these different regions, compared to the situation in our primate cousins. This therefore could be seen as evidence that, while such differential expression of genes shapes the neural activity of the human brain, that activity—which reflects not just the basic biology of the brain but also social influences—in turn may affect brain function at its most basic level, that is, through gene expression.

Perhaps most importantly, as shown by recent evidence for the importance of brain waves in coordinating the activity of different regions of the brain, it is possible to see how language—and other forms of culture that we have mentioned such as music, art, and literature—might work through such means to coordinate our unique human conscious awareness. For electrical waves—as well as rapid changes in chemical messengers through the brain—provide a material route for language to mediate brain function in exactly the sort of manner that Vygotsky predicted with his idea of a culturally mediated consciousness.

Identifying such connections only constitutes the very beginning of a major research project. A key question for future research will be how such molecular and cellular differences relate to the distinct aspects of human consciousness, and an important challenge will be to find ways to study this question given the limits to which we can experiment on actual living human brains. These are issues that I will return to in the final chapters of this book, but before I do that, I want to consider two further aspects of human consciousness that are often posed as a diametric opposite to science—namely religion and spirituality.

PART VII

THE FUTURE OF MIND

CHAPTER 22

MIND AND MEANING

———◆———

Despite being an atheist since my mid-teens, when in a city I love to visit its old religious buildings. As a child growing up in Bradford, a visit to York with its minster was always a treat. Later in life Westminster Abbey and St Paul's in London, Winchester Cathedral, and then Notre Dame in Paris, and the great cathedrals of Milan and Florence in Italy were a revelation. In Spain I encountered the beautiful Islamic architecture of the Alhambra in Granada and the Mesquita at Cordoba. And like so many others before me, I was awed by the grandeur and beauty of such architecture, the planning that went into these monuments, the back-breaking toil involved in their construction, and the thought of all those who have entered them over the centuries, each with their own fears, desires, and hopes for the future.

If these are the positive feelings I experience towards places of worship, the negative ones stem from the fact that I also associate churches with funerals, and that particular awkwardness that an atheist can feel when paying their respects to a loved one's memory while at the same time being uncomfortable about the religious ceremony associated with the event. Such was definitely the case with my sister's funeral. For despite the best efforts of family and friends to celebrate the life of someone who had lived and loved life to the full, it was impossible to escape the fact that she had decided, aged 53, to end her life in the most horrific fashion. But what also struck me forcibly was that my sister had been characterized as mentally ill because of her psychotic delusions, and yet we were now being asked to pray to some invisible supernatural being for the salvation of her immortal soul. This brings me to the question of how religion might have originated and why it still maintains such a strong hold on the human mind, even amidst the science and technology of the twenty-first century.

Of course, a religious person could answer that their feelings stem from the existence of God. But since this book is concerned primarily with aspects of consciousness that can be examined scientifically, my approach here will be solely to ask what material reasons we can find to explain belief in the supernatural and link this to this book's theme, which is how culture has shaped the human mind. One suggestion is that such belief arose as a by-product of response mechanisms that evolved to protect our species in its first phase of existence. For instance, consider a scenario in which one of our prehistoric ancestors walking in the African savannah heard a rustling in the grass. While this might be just the wind, it could equally signify the presence of a predator, such as a lion. Kelly James Clark, of the Kaufman Interfaith Institute at Grand Valley State University in Michigan, has argued that those of our ancestors who recognized that other organisms have 'agency'—an ability to act of their own accord—could have been more likely to survive than those without this recognition.[1]

Another feature of human beings is what has been called our 'theory of mind'—the capacity to understand that other humans have thoughts and desires and to have some idea what these might be.[2] In particular, only people, through the way that language has uniquely transformed our brains, have a true sense of ourselves as individual rational beings but also possess the ability to decipher the thoughts of other people, either by observing their actions and words or through direct interrogation of others through a series of questions.

Yet while helping humans to survive, such attributes may have led to supernatural beliefs. For as well as attributing agency to predators, and rationality to each other, our ancestors may have started extending them to things with no agency or rationality at all. As Clark notes, 'you might think that raindrops aren't agents. They can't act of their own accord. They just fall. And clouds just form; they're not things that can act. But what human beings have done [in the past] is to think that clouds are agents.'[3] Similarly, as our ancestors could see that other human beings had rational minds, they might have started to wonder whether such rationality was more widespread in the world than just within our species. And such a misplaced perception of agency and rationality in nature could have given rise to the idea of conscious, powerful, and unruly invisible spirits that need placating by ritual ceremony.

Indeed, a characteristic feature of hunter-gatherer society—the type of society in which our species lived for the vast majority of its existence—is 'animism', the belief that objects, places, and animals all have a distinct spiritual essence.[4] To some extent, this belief may have helped to develop the sustainable manner in which hunter-gatherer societies interact with nature, by engendering feelings of respect for other organisms. And in a world in which floods, lightning, disease, and other natural disasters could strike out of the blue, it may also have helped our ancestors make sense of an unpredictable, often unkind, world in which loved ones might meet their end in the prime of life, or even in early childhood.

The associated idea that human beings might somehow survive in some form after death—what might be called the ultimate comfort—seems to have been a feature of our species from early in its existence. Indeed, recent finds have indicated that belief in an afterlife may even have been shared by Neanderthals. For instance, Enrique Baquedano and colleagues of the Regional Archaeological Museum of Madrid discovered that in a cave 93 kilometres north of Spain's capital city, the jaw and six teeth of a Neanderthal child had been buried, and surrounding the burial site were several blackened hearths, within which the researchers found the horn or antler of a herbivore, apparently carefully placed there.[5] Baquedano has argued that this arrangement 'may therefore have been of ritual or symbolic significance.'[6] He believes that the fires were lit as a sort of funeral ceremony.

Today, a belief in life after death is a central feature of most religions, although in some Eastern faiths, such as Hinduism or Buddhism, there is instead a belief in reincarnation on Earth until perfection of the soul results in release from the cycle of death and rebirth into some sort of Nirvana.[7] The fear of one's own death, and the pain of losing someone dear, can fuel hope that something of oneself and one's loved ones persists after death. Following my father's sudden death, my mother and sister, both formerly atheist or agnostic, became deeply religious. In my mother's case, this followed a visit to a spiritualist church where she claimed she had been presented with evidence that my father still existed in an afterlife.

This is not the place to consider the evidence that some spiritualists have deliberately sought to trap and deceive bereaved individuals desperately searching for some sign that a loved one still exists in some spiritual plane. Of course, not all mediums are necessarily frauds; some may actually believe they are channelling messages from the dead. What is clear is that during periods of great tragedy, such as World War I in which millions of young men lost their lives in battle, there has been an increase in the number of people seeking to communicate with the dead. So a British commentator wrote in 1919 about 'mothers and friends of fallen soldiers resorting to table-rapping, creakings, automatic writing through the medium of the planchette, Ouija, heliograph, etc., in the hope of once more communicating with their loved ones.'[8] A desire to believe in the supernatural can rely upon a human tendency to seize upon an unlikely chance occurrence as meaningful, rather than just coincidence. If a medium calls out to their audience, 'I have a message for someone called John', if there is someone in the audience of that name, they may believe that the message is for them, and the medium may subsequently respond to a bereaved person's questions in such an ambiguous fashion that they may really think they are communing with a dead loved one.

Of course, there will be people who will swear that the messages or signals they believe they have received from a dead loved one only make sense if there is an afterlife. I once had a highly unlikely encounter that certainly tested my sense of scientific rationality. As I related earlier, a week prior to my father's death from a fall down the pothole Gaping Ghyll, I had set off for a walking holiday in the Scottish Highlands with my then girlfriend, only learning of his death when I returned home. What I did not mention is that on our return to Bradford, my girlfriend and I broke our journey by staying a night in the Yorkshire Dales. While there, we climbed Ingleborough, the mountain on whose slopes Gaping Ghyll is located, and even passed the pothole during our ascent. While descending the mountain, an ambulance belonging to the Cave Rescue Organization passed us, and I remarked to my girlfriend that I hoped it was not attending a serious incident. It was only on arriving home and learning about my father's death that I realized his dead body had been in the ambulance.

Now clearly, the chance of my happening to have unwittingly made a detour to a mountain where my father had recently died seems a very unlikely one indeed, and certainly my mother saw this as a sign that my father was trying to communicate with us after his death. For myself, my rational viewpoint makes it impossible to see the event as anything other than an amazing coincidence, but it has made me more aware of the factors that might lead a person to believe—against all scientific evidence—in a world beyond the grave.

Of course, there is far more to religion than a belief that human beings survive in some form after their death. In particular, the idea of a single all-powerful supernatural being—a god—is a key feature of monotheistic religions such as Judaism, Christianity, and Islam, while Hindus for example worship multiple gods, though these are recognized as facets of a single all-pervading spirit. Multiple gods were also a key feature of the religions of ancient societies like those of Egypt, Greece, and Rome. It might therefore be expected that belief in an all-powerful supernatural being, or beings, has been a feature of human societies since their inception. Yet studies of hunter-gatherer societies have shown that although they typically believe in a variety of supernatural spirits, these have only limited powers, are generally not omniscient, and usually lack concern for morality and human affairs.[9]

Interestingly, the concept of sin as an offence against a supernatural being does not seem to be a feature of hunter-gatherer societies. Instead, individual human wrongdoing against other humans is considered to be something for humans themselves to deal with.[10] This suggests that a belief in all-powerful gods who make moral demands on humans is a relatively recent development within human society, and raises the question of its origins.

Assessed on their own terms, the great religions might be viewed as a highly positive aspect of human society. All stress the importance of living a 'good' life—meaning one that avoids negative behaviours such as boastfulness, greed, jealousy, anger, and laziness—and typically preach the importance of peaceful behaviour, caring for those who are hurt or in distress, and giving to those in need. Yet sadly throughout the history of civilization religion has also been used to justify the persecution, and sometimes execution, of those who transgress the moral code or are considered to hold heretical views, as well as outright warfare. While major religions may sometimes be pitted against each other, as in the Crusades, there has also been much strife within different denominations of a religion like Christianity, ranging from the Thirty Years War to the situation in Northern Ireland today. As the character Louis Cyphre—a thinly disguised Devil—cynically expressed it in the 1987 film *Angel Heart*, it seems that 'there's just enough religion in the world to make men hate each other, but not enough to make them love.' So what might explain this aspect of religion?

One explanation is that modern-day religions originated in the first class-based societies as a way to justify the inequality of such societies and keep the mass of people in their place, and to some extent they still play this role. And indeed, the religions of ancient class societies did largely replicate the hierarchical nature of such societies, but also their warlike nature. Moreover, the promise that good behaviour by a person in their lifetime would be rewarded by a blissful life after death, coupled with the threat that bad behaviour would mean a descent into eternal damnation within some form of hell, can be seen as a highly effective form of self-discipline.[11] For instance, ancient Egyptians believed that if a deceased person was found wanting after a series of tests in the underworld, they were deprived of their sense organs, and forced to walk on their heads and eat their own excrement.[12]

Today, we could laugh at such beliefs as a product of their time were it not that the notion of hell remains an important aspect of many modern religions, as shown by the outcry in the Catholic Church when a newspaper article in 2018 quoted Pope Francis as apparently saying that 'bad souls' are not punished. Instead the Pope was supposed to have said that 'those who repent obtain God's forgiveness and take their place among the ranks of those who contemplate him, but those who do not repent and cannot be forgiven disappear.'[13] Vatican officials quickly rebuked the newspaper, saying that it had misquoted the Pope, and affirmed the view that hell 'really exists and is eternal'.[14] Clearly, the prospect of sinning souls quietly 'disappearing' does not have the same impact as eternal suffering. Yet to only focus on religion as a ruling class form of control beyond the grave—with the reward of a blissful heaven balanced against the threat of eternal damnation—or for that matter, to solely concentrate on the irrational and superstitious aspects of religion does not do justice to the fact that religion has at times also been associated with much more progressive aims.

The idea that religion simply acts to keep ordinary people in their place is sometimes supported by a famous quote from Karl Marx, in which he spoke of religion being 'the opium of the people'. Yet focusing on this phrase in isolation is misleading, since it was part of a longer statement: 'Religion is...the expression of real suffering and a protest against real suffering. Religion is the sigh of the oppressed creature, the heart of a heartless world, and the soul of soulless conditions. It is the opium of the people.'[15] This double-edged character of religion means that while it may sometimes be used to stifle ideas of struggle against oppressive conditions in society, and even be used in highly reactionary ways, at other times it can act as a very powerful stimulus to revolts against earthly injustices.

One factor that has allowed religion to play such a dual role throughout history is the ambiguity at the heart of all major religions regarding their attitude to exploitation and oppression. An important reason for such ambiguity is that such religions' widespread appeal is based on being able to voice the interests of all classes in society, yet by definition this leads to potential contradictions, given that the interests of such classes may often be diametrically opposed. The Bible's New Testament is a good example of such contradictions.

Christianity was originally a movement of oppressed people: it first appeared as a religion of slaves and emancipated slaves, of poor people deprived of all rights, and of peoples subjugated or dispersed by Rome.[16] But once subsumed as the dominant religion of the Roman Empire, it underwent substantial revision, as reflected in the often contradictory messages of the four Gospels, which, far from being the work of Jesus's disciples, were written years after their death. Multiple authorship and editing, but also the transition from a religion of the oppressed to one backed by the state, may explain why in the Gospels Jesus can overturn the tables of moneylenders in the temple, and yet when asked what he thinks of Roman taxation, also reply, 'Give to Caesar what is Caesar's.'

In fact given what we have said about the fact that human consciousness can itself often be contradictory, this makes it easier for such mixed messages to be accepted by the human brain. And while the altruism that is supposed to be at the heart of every great religion builds upon the fact that social cooperation seems to be wired into the human brain in ways that we have already discussed, the unconscious bias against things considered to be out of the ordinary, mediated in particular by the amygdala region of the brain, means that believers may be easily swayed to see those labelled unbelievers or heretics as a threat to society.

In medieval Europe, all discussion about society was framed in religious terms, for that was the only philosophical framework available. The dominant view was that everyone had their place in society, ordained by God, and those accused of heresy faced often violent retribution. Yet feudal society also witnessed periodic revolts by the poor, for instance the Peasants Revolt in England in 1381. During this revolt one of its leading participants, the radical preacher John Ball, asked, 'when Adam delved and Eve span, who then was the gentleman?' He added that 'from the beginning all men by nature were created alike, and our bondage or servitude came in by the unjust oppression of naughty men.'[17] This reinterpretation of the Book of Genesis expressed the peasants' dreams of liberty and equality. During the armed peasants' march on London, Ball preached that 'things cannot go right in England and never will, until goods are held in common and there are no more villeins and gentlefolk, but we are all one and the same.'[18] What is interesting for the current discussion is how the rebellious peasants in this case developed an ideology that subverted the traditional reading of the Bible so that instead of upholding the existing rigid feudal order, the stories in this religious text were used as a plea for a much more progressive society.

The years following the English Revolution of the mid-seventeenth century saw a flourishing of new scientific theories and discoveries ranging from Newton's laws of motion to the discovery of the cell as the biological unit of life by Robert Hooke.[19] Yet it would be a mistake to separate the English Revolution from the religious fervour that inspired its activists. For although groups like the Levellers can appear remarkably modern, with their demands for a form of universal suffrage, their use of newspapers, petitions, mass demonstrations, and particularly the involvement of women in their movement, in other ways many of their followers were a clear product of their time, believing in magic, prophecies, and miracles.[20]

We can get a sense of the fervour of such activists from the work of a particularly gifted individual of this era—John Bunyan. A soldier in the Parliamentary army during the English Civil War, but then imprisoned for his beliefs during the religious clampdown that accompanied the monarchy's return in 1660, in jail Bunyan began writing *The Pilgrim's Progress*.[21] Portrayed as a dream, Bunyan's book became an instant classic with its story of Christian's pilgrimage to the Celestial City that takes him through many adventures, dangers, and false turns.[22] One particularly powerful aspect of Bunyan's allegory is the way he renders abstract moral dilemmas concrete through encounters with characters like Mr Worldly-Wiseman, Mr Facing-Both-Ways, Lord Time-Server, and so on.[23] Bunyan also sees an individual's direct communion with God as more important than obeying the rules of a repressive society if the two contradict each other, thereby providing a holy licence for the struggle for a better world. For this reason historian E. P. Thompson saw *The Pilgrim's Progress* as one of the 'foundation texts of the English working-class movement'.[24]

By default religion played a central role in human conscious awareness and communication between people in past times, but what about the situation today? Here we face an apparent conundrum, for while in the past, belief in the supernatural could be excused as ignorance of the fact that droughts, floods, earthquakes, volcanic eruptions, plagues, and other challenging phenomena have scientifically explainable natural causes and are not the work of an angry god, then surely today religious views should be just as untenable as the idea that the Earth is flat. Moreover, we now have no need for creation myths given the evidence that human beings, and all other species on Earth, are products of an evolutionary process that can be explained in a purely materialist manner. Yet today, religion continues to be a source of inspiration for billions. So how do we explain the persistence of religious beliefs? One obvious reason, if religion can indeed be 'an expression of real suffering and a protest against real suffering', is that such suffering is very much still alive in the world. And in such a situation, religion can not only offer solace, but also may inspire radical action. Such was certainly the case in the black civil rights movement of the 1960s.

Martin Luther King's speech at the end of the 'March on Washington' in 1963, which drew around a quarter of a million people, is rightly celebrated as one of the greatest speeches of the twentieth century.[25] It was centred on the unforgettable plea 'I have a dream.' The speech is notable for the way the rhythmic pattern of King's words becomes stronger with each repetition of the phrase, and also for the poetic character of his lines. For instance, he described the 'valley of despair' faced by black Americans, and the southern US states 'sweltering with the heat of oppression', but also a future in which 'one day on the red hills of Georgia, the sons of former slaves and the sons of former slave owners will be able to sit down together at the table of brotherhood.'[26] Perhaps most importantly, King looked forward to a world when religious divisions themselves would be a thing of the past, with his concluding sentence that when 'we let freedom ring, when we let it ring from every village and every hamlet, from every state and every city, we will be able to speed up that day when *all* of God's children, black men and white men, Jews and Gentiles, Protestants and Catholics, will be able to join hands and sing in the words of the old Negro spiritual: Free at last! Free at last! Thank God Almighty, we are free at last!'[27] In this respect it is interesting that King employed the language of religious fervour, but used it to make a highly rational plea for a unified form of humanity that was not divided by skin colour or particular faith.

From a consistently scientific point of view religion can only be seen ultimately as a form of illusion, being based on a belief in supernatural forces, yet it provides both meaning and sometimes a justification for action in ways that cannot simply be dismissed as irrational. But what about people who do not believe in the supernatural? Is there a way for non-believers to find deeper meaning in life while at the same time maintaining that everything in the Universe, including human consciousness, can be explained in purely material terms? For surely, finding such deeper meaning is a highly important aspect of human conscious awareness, if that awareness is to rise beyond the mundane reality of daily existence.

To some extent, merely recognizing the connection of human beings, with our conscious awareness, to the rest of nature can provide a sense of deeper meaning. Addressing critics who believed that his theory of evolution by natural selection diminished our sense of wonder about the world around us, Charles Darwin replied that 'there is grandeur in this view of life'.[28] Knowing that the intensity of what we feel when we fall in love or experience the birth of our first child is partially based on a surge of brain chemicals like dopamine or oxytocin should not undermine the experience, but rather can help us connect it to the rest of nature. Similarly, recognizing that great human cultural achievements—whether the pyramids of Giza, the music of Mozart, Michelangelo's art, or the writing of Emily Brontë—are ultimately the product of the actions of interconnected neurons and glial cells in different parts of the human brain should not lessen the importance of these artistic creations.

Yet while accepting that our minds are based on brain matter and nothing more, an important aspect of finding deeper meaning in the world also means recognizing that human consciousness has unique properties that distinguish it from the rest of nature. This position is different from that of Christof Koch, of the Allen Institute for Brain Science in Seattle, who has argued that 'consciousness is…probably present in most of metazoa, most animals, [and] it may even be present in a very simple system like a bacterium'.[29] In contrast, I have argued that a proper understanding of human consciousness requires not only an awareness of qualitative differences of the human brain, in its interconnections, internal divisions of labour, and even patterns of gene expression, but also identifying the material basis of this consciousness. This means moving beyond a crude, mechanical understanding of the material basis of human consciousness, without at the same type lapsing into an unscientific idealism.

In fact mechanical materialism and idealism are far more closely related than might at first be imagined. For a model of the human mind that fails to take into account the true complexity of the material basis of human consciousness inevitably ends up bringing in non-materialist explanations to compensate for this deficit. For instance, we saw previously how René Descartes failed to extend his mechanical model of the body to the mind, arguing instead that this was an unknowable supernatural soul. Meanwhile, John Locke and later behaviourists like B. F. Skinner simply ignored the material basis of human consciousness, treating it instead as a 'blank slate'.[30] But this not only failed to explain in material terms why human beings have a conscious awareness that other species lack, it also provided no explanation for why individual humans can differ so much in terms of personality, aptitude, motivation, and so on. Being unanchored in this way, it was an idealist vision.

This tendency to lurch from mechanical materialism into idealism is true even of some thinkers who specifically set out to provide a materialist view of the mind. Daniel Dennett of Tufts University in Massachusetts has strongly argued against the idea that consciousness can be viewed as if a 'homunculus' or little human being sits inside our skulls. A major problem with such a scenario, which also echoes what the philosopher Gilbert Ryle has called a 'ghost in the machine' view of consciousness, is that one might justifiably ask what is going on inside the mind of such a homunculus (Figure 29).[31] Instead Dennett proposes a 'bottom-up' approach, which views the human mind as the combined product of unconscious, evolved processes that somehow come together to provide the appearance of an individual 'I', but which in reality has no conscious entity at its core. While a valuable and considered approach to the problem of consciousness, one problem with his account is Dennett's espousal of the 'meme' theory of the evolution of human culture.[32]

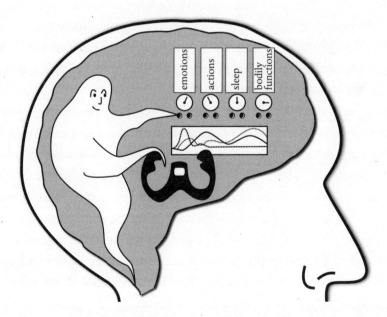

Figure 29 Cartesian dualism and the idea of a homunculus controlling the brain.

While today memes tends to mean those images or video clips—often humorous, cringeworthy, or carrying some life message—that spread rapidly on social media, the term was first introduced in the 1970s by Richard Dawkins in his book *The Selfish Gene*, where he used it to describe an idea, behaviour, or style that spreads from person to person within a culture—examples of memes being 'tunes, ideas, catch-phrases, clothes fashions, ways of making pots or of building arches'.[33] By analogy with biological inheritance, Dawkins argued that 'just as genes propagate themselves in the gene pool by leaping from body to body via sperm or eggs, so memes propagate themselves in the meme pool by leaping from brain to brain via a process which, in the broad sense, can be called imitation.'[34]

Now as a description of the way that popular images can spread rapidly across the Internet, almost like a viral infection, this was a remarkably prophetic and insightful vision. However, the potential idealism at the heart of Daniel Dennett's belief that meme theory provides a scientific explanation for human consciousness is exposed by science writer Jonnie Hughes' uncritical claim that this theory demonstrates that 'ideas are, to a very real extent, "alive" in their own right—surviving, reproducing, evolving, going extinct, just like living things,' and that we humans are 'little more than their hosts, their habitats'.[35]

Such a view turns reality on its head by seeing ideas themselves as free-floating entities that exist separate from human beings, and to some extent even dominate them, rather than being the creation of real, living human individuals, and the social interactions between them. In fact, the tendency to substitute 'the fantastic form of a relation between things' for actual relations between human beings is a feature of an 'alienated' view of the world.[36] And as I discussed earlier, this has developed within capitalist society because of the tendency of people within such a society to view the products of their labour not as their creations but as 'alien objects' standing above, and often in oppressive opposition to them. However, we also saw how such an alienated view of the world affects relations between all human beings in society, so that even something as fundamental to human nature as sex can be viewed as a commodity, rather than as a source of individual pleasure and fulfilment. And as shown by Hughes' comment above, alienation can mean that our very ideas end up being seen not as the creative products of our individual brains and interactions with other human individuals, but as objects over which we have little control and that can even dominate and oppress us.

Such a sense of alienation from our own minds can manifest itself in some odd ways. One is the idea that consciousness is not a unique product of human minds, but rather something distributed across nature as a whole. For instance, Christof Koch, whom I mentioned earlier, sees consciousness as a quantitative measure of an organism's awareness of itself and the world, with humans only at one end of this spectrum.[37] Not only does such 'panpsychism' not explain the qualitative shift that human self-awareness represents compared to other animals—demonstrated by our technology, our civilization, and our impact on the planet, something evident for no other species—but it can easily slide into a mystical view of the Universe. For if consciousness is something distributed across nature, it is not a great leap to see conscious awareness even in the basic fabric of the Universe. For instance, Gregory Matloff, a physicist at New York City College of Technology, has argued that a 'proto-consciousness field' may extend through all space, with stars being thinking entities that deliberately control their paths.[38] Put more bluntly, the entire cosmos may be self-aware. This is surely only a short step to belief in the idea that 'god is in everything'.

Comparing the different theories of human consciousness mentioned above, one general problem is that they tend to underestimate—and yet also overestimate—the task of explaining consciousness while maintaining a materialistic mode of explanation. To begin first with underestimation of the problem, theories of consciousness generally fail to properly explain the importance not only of size, but also structure, in distinguishing the human brain from those of other species and relating this to the distinctive features of human consciousness in such a way as I have been seeking to do throughout this book.

This underestimation of the role of structure also extends to the cell and the genome. Far too often theories of consciousness refer to the basic structural units of the brain that we share with other species—our neurons and glial cells—as if we can treat these as black boxes, whereas in fact our deepening understanding of the complex nature of these cell types and the way that they interact as a network in the brain as a whole through genetic and epigenetic mechanisms has an important impact on our understanding of how ideas form and change. And these structural considerations mean that the distinctive type of conscious awareness that I have often referred to in this book by the term 'human consciousness' cannot be something that we share with other animals and certainly not with inanimate objects.

In terms of overestimation of the problem of consciousness, one source of confusion in attempts to explain the human mind is a failure to properly identify the material basis of thought. It is this that leads to the belief that ideas and concepts are free-floating entities—memes. In both cases, there is a failure to relate such ideas and concepts to actual human brain structure and function and actual social interactions between human beings, and therefore the basis of ideas and concepts in such theories remains unknowable.

In contrast, I have argued here that words themselves are the material basis not only of the language that humans use to communicate with other human beings, but also of thought itself. In this respect given the initial focus of this chapter on religion, it is interesting that the Bible begins with the phrase 'In the beginning was the Word, and the Word was with God, and God was the Word.'[39] This is clearly an idealist vision, but interesting nevertheless in highlighting the centrality of words for our species. I have also argued that thought is a dialogical process, in which the individual mind represents a nexus between the brain and society. This has the consequence that the consciousness of each human individual is a social construction. However, lest this be taken to mean that human beings are merely word processors, the fact that thoughts are also formed in an organic brain—with a specific molecular and cellular structure, and precise interconnections between different cells—also gives consciousness its unique character, in terms of both the human species and differences between individuals. In particular, the fact that brain waves of different frequencies appear to play a key role in coordinating the activity of different regions of the brain is likely to be central to the construction in a human brain—through language—of the individual 'I'.

The unique interface between the organic brain of a human individual and the symbolic framework based on language that human consciousness represents is one reason why it is unlikely to be straightforward, or even possible, to create conscious awareness in a machine—an issue that I will explore in more detail in the next chapter. But how does this interface manifest itself in the specific personality of a human individual? And how does this relate to the question of how that individual finds meaning in the world?

The subjective nature of consciousness is key to what it feels like to be human. The aesthetics of an art work, sensations on eating a new type of food, and the responses that other people stimulate in us depending on whether they are stranger, lover, friend, or foe overlap, but also differ in important ways, between individuals. Philosophers use the term 'qualia' (from the Latin for 'what kind of') to describe individual experiences—the colour of a sunset, the smell of a rose, or the sound of a lover's voice. The term was coined by Harvard University philosopher Clarence Irving Lewis, who defined qualia as 'recognizable qualitative characters of the given, which may be repeated in different experiences, and are thus a sort of universals' but also 'directly intuited...purely subjective'.[40]

This subjective aspect of consciousness is one reason why some people believe the human mind to be ultimately inexpressible in scientific terms. In contrast, I would argue that both the universal and subjective aspects of consciousness are explainable by a view of our conscious awareness that sees this as primarily mediated by language. One consequence of this view is that both conversations between individuals and our innermost thoughts involve a dialogue, and also that the meaning of words themselves is contested and evolving, because of changing experiences and social interactions, leading to an indeterminacy in the inner speech we use to express the world around us to ourselves and others.

In addition, the 'fuzzy' nature of inner thought, with its fluidity in terms of both word meaning and grammatical structure, and a dynamically changing meaning that may be only partially acknowledged by the individual and which forms the unconscious part of our thoughts mean that it may be difficult to express our inner thoughts not only to others, but even to ourselves. Finally, since thoughts are formed in each of us within brains that differ in subtle ways in terms of structure, interconnections, and the precise variety and amounts of neurochemicals—with chemical interactions also going on with the rest of the body, not to speak of the drugs and other chemicals we ingest—all must play a significant role in the development of the individual subjective experience. Moreover, it is increasingly recognized that both the specific brain biology bequeathed to us by our parents via their genes and the epigenetic impact that their life experiences may have had on them might have been passed on to us, their offspring, and may impact our mental outlook and subjective feelings.

As I have already indicated, one crucial aspect of human consciousness is its capacity to be affected by experience, and this can alter both our thoughts and feelings. The novelist Henry James brilliantly evoked this process when he wrote that 'experience is never limited, and it is never complete; it is an immense sensibility, a kind of huge spider-web of the finest silken threads suspended in the chamber of consciousness, and catching every air-borne particle in its tissue.'[41] Such an effect of experience can occur at both an individual and societal level. For instance I can remember one instance when my own individual consciousness was challenged by experience. When my sister visited India and brought back a finely decorated jacket, I was perturbed to see embroidered on it rows of swastikas. Yet this merely reflected the fact that the swastika is an ancient symbol that can represent the four seasons, the sun's rays, or the circle of life, depending on which account you read.[42] The appropriation of the symbol by Hitler and the German Nazis reflected their rather mystical ideas about nationality and race, and their desire to propagate a 'thousand-year Reich'.[43]

To the person who embroidered my sister's jacket, the swastika symbol clearly held a very different cultural and religious relevance, so that they were not concerned by its more sinister connotations, assuming they were even aware of these. In contrast, the reason that I was shocked to see swastikas on my sister's jacket was because it is difficult today as a person in Europe not to associate this symbol with the Nazi movement. And it is testament to the influence of a changing society on the mind that a symbol can also powerfully change in meaning. For imagine, how perceptions of this particular symbol must have rapidly changed in Germany in the 1920s and 1930s, perhaps being viewed initially as an object of ridicule when the Nazis were only a fringe organization, giving way to unease as the organization began to dominate the political scene, first through violence on the streets then through increased influence in the German Reichstag or parliament, and finally fear, following the Nazis' rise to power in 1933, after which they rapidly moved to suppress opposition to their dictatorship.[44] For Nazi supporters, however, the swastika was probably an important unifying factor in all sorts of ways, combining as it did the colours of the previous German regime defeated in World War I and an ancient mystical symbol. Given that Hitler himself described mass demonstrations like the Nuremberg rallies as events that 'burn into the little man's soul the conviction that, though a little worm, he is part of a great dragon',[45] it would be a mistake to underestimate the potency of such a combination of different symbolic elements.

On a more personal note, the significance of one piece of music has changed dramatically for me over the past two decades because of the differing contexts in which it was heard. Like many people, I first heard the piano piece 'Comptine d'Un Autre Été' by Yann Tiersen while watching the film Amélie. This film, by French director Jean-Pierre Jeunet, is about a quirky young Parisian waitress Amélie, who decides to surreptitiously work to change the lives of those around her for the better, while struggling with her own isolation.[46]

As Amélie's theme music, 'Comptine d'Un Autre Été' captured perfectly the bittersweet nature of Amélie's experiences portrayed in the film. But the piece acquired a more personal relevance for me when my sister's son—my nephew Jack—learned to play it and often did so when we visited my sister's home, and it became a typical accompaniment to our family gatherings there. However, the music acquired a much deeper significance after my sister's suicide when Jack played the piece—one of my sister's favourites—at her funeral. Now, it is impossible for me to hear this piece of music without associating it not only with my sister's funeral, but also many past times—both happy and sad—spent together.

The subjective nature of experience is what makes being an individual unique. Humans have evolved to be social beings. This extreme sociability defines our species, yet possessing conscious awareness, we are ironically also the only species aware of our individuality. According to the view of the mind proposed in this book, this sense of individuality is to some extent an illusion, 'a tenant lodging in the social edifice of ideological signs',[47] as Voloshinov described the human mind. Yet that tenant is us, and however tenuous, our individuality is what defines us. Each of us finds meaning through our experiences and actions in a social world. My research, teaching, writing, political activities, hobbies, and leisure interests acquire meaning because of my interactions with other human beings, whether friends, family, or total strangers. But ultimately only we can know how it feels to be us, that 'I' who seems to operate from the skull sitting at the top of our bodies. We cling to this sense of being a unique individual through life, and it is the fear of ceasing to be that 'I' that can haunt us as we contemplate the time when we will be no more through death.

One of the key drivers of religious ideas is the hope—against all scientific evidence—that we can continue to exist after death. Meanwhile for those of us who do not believe in the supernatural, one of the biggest challenges we face is how we deal with our own transient existence, and those of our loved ones, while still retaining a deeper sense of the meaning of life. The question of how to deal with such a challenge is something I will return to in the conclusion of this book. But before I do that, it is worth contemplating whether in the future individual minds will necessarily die, or whether there will ever be the possibility of cheating the death of individual consciousness in a non-supernatural manner. It is this issue, and other future possible scenarios, that I will examine in the penultimate chapter of this book.

CHAPTER 23

MIND AND MACHINE

M aking predictions about the future is seldom easy. Such is the pace of modern life that predictions can greatly overestimate progress in the future, yet fail to anticipate major social transformations that we now take for granted. I was five years old in July 1969 when I watched awestruck as the flickering black and white TV screen in my home showed men walking on the Moon, and I assumed while growing up in the 1970s that by the time I was an adult our species would have colonized the Moon, Mars, and perhaps even the rest of the Solar System. My youth may have been a factor in such optimism, but I was far from being alone in such expectations at this time; the British 1970s TV programme *Space: 1999* was typical in predicting the existence of a human colony on the Moon before the turn of the millennium.[1]

In fact, it would be easy to produce a long list of technological predictions made by enthusiasts in the twentieth century that have still not been realized. Technologies ranging from teleportation to scanners that can non-invasively diagnose and cure disease, which were presented as commonplace in the futuristic series *Star Trek*, are little closer now to practical realization than they were in the 1960s when first conceived.[2] Yet other developments that we now take for granted, like the Internet, personal computers, smart phones, and precision gene editing, may have featured in science fiction, but also seemed likely to remain as fiction not so many years ago. So how might future technologies impact on human consciousness? In this chapter we will consider whether artificial intelligence might come to rival that of human beings and possible interfaces between human and machine intelligence.

First though, I want to discuss how new techniques are continuing to add to our understanding of the human mind. There are many exciting technologies available now to the neuroscientist, such as genomic analysis, optogenetics, gene editing, and brain organoids. To what extent could such technologies be used to investigate the model of human consciousness outlined in this book? Here, we face several problems. For this model views the human brain as having undergone a qualitative shift compared to those of other species, implying that there will be limitations to what we can learn from the current main tool for studying brain function—animal models. At the same time, the creation of experimental systems for studying the more human-like aspects of consciousness raises potential ethical issues.

On the positive side, the fact that much of the basic neurobiology of the human brain is shared with other species means we can continue to learn a lot from animal models. This is particularly true now that new gene-editing methods make it possible to disable or subtly alter the function of genes not only in mice, but also in primates.[3] This is important because even though I have been arguing that the human brain has undergone a qualitative shift compared to other primates, these primate species are much more similar to a human in brain size and structure and also share many similarities with humans in their social interactions, so studies of gene-edited primates may offer significant advantages compared to mice. However, there are also potential problems in both the creation and study of gene-edited primates.

The cost and time required to create a gene-edited primate is one factor to consider; another is that in many countries at the cutting edge of neuroscience, there is substantial opposition to primate research by animal rights activists. For this reason, much recent development of gene-edited primates has taken place in China, because of generous funding for biomedical research currently in this country, and also because traditionally there has been less ethical opposition to primate research.[4] But such studies are likely to face increasing ethical scrutiny, particularly as neuroscientists from the USA and other countries become involved in such research in China.[5] Potential ethical issues include the fact that creating models of conditions like severe forms of autism in a primate like a macaque may have more detrimental effects on the well-being of an individual of this species—precisely because its forms of social interaction are more similar to those of a human—than is the case for mice. Such sensitivity to possible consequences will be even more important if scientists use gene-edited primates to study the biological basis of human conscious awareness.

Showing the pace of research in this area, Bing Su, a geneticist at the Kunming Institute of Zoology in China, has already introduced a human version of a gene called SRGAP2, thought to endow the human brain with processing power by allowing the growth of connections between neurons, and of MCPH1, a gene related to brain size, into macaques and reported the consequences.[6] As mentioned previously, a distinguishing feature of human brain development is that it is delayed compared to the situation in other primates. Although as a consequence, human babies are far more helpless at birth, such a delay also means that our brains have a far greater learning capacity, for far longer, than those of other primates. As a baby's brain develops after birth, MCPH1 is expressed in abundance, but much less so in non-human primates, and it has been suggested to play a role in the human brain's developmental delay. And Su and his team's recent study showed that the brains of monkeys with the human version of MCPH1 took longer to develop, and the animals performed better in tests of short-term memory as well as reaction time compared to normal monkeys.[7]

Su claims that this is the 'first attempt to understand the evolution of human cognition using a transgenic monkey model'.[8] However, Jacqueline Glover, a University of Colorado bioethicist, believes the study is likely to raise concerns about 'Planet of the Apes'-type scenarios 'in the popular imagination' and thinks that 'to humanise [the monkeys] is to cause harm. Where would they live and what would they do? Do not create a being that can't have a meaningful life in any context.'[9] But Larry Baum, a researcher at Hong Kong University's Centre for Genomic Sciences, has downplayed sci-fi comparisons, stating that 'the genome of rhesus monkeys differs from ours by a few percent. That's millions of individual DNA bases differing between humans and monkeys. This study changed a few of those in just one of about 20,000 genes. You can decide for yourself whether there is anything to worry about.'[10]

In experiments of this kind, it is likely that there will always be a conflict between the natural urge of scientists to probe the limits of knowledge and possible adverse consequences—in this case, an animal that may gain a much higher degree of self-awareness and which might both therefore recognize its plight as an experimental animal and begin to question its rights in this situation.[11] While it seems unlikely that a single gene change might give rise to these situations, as our understanding grows about the genetic differences that distinguish humans from apes, and given the capacity for gene editing to introduce many genetic changes simultaneously, it is a possible scenario that needs to be taken seriously.

Another area of research that is becoming ethically charged is the use of 'organoids' to study human brain function and the biological basis of mental disorders.[12] However, brain organoids suffer from various limitations as objects for studying human brain function, some of which I have already discussed. Perhaps the biggest obstacle to organoids being used to study adult brain function is their lack of a blood supply, for without this such organoids are unable to grow beyond embryonic size. However, Ben Waldau, a vascular neurosurgeon at the University of California, Davis Medical Center, has recently begun to tackle this problem. Using brain cells removed from a patient during surgery, Waldau and his colleagues coaxed some of these into iPS cells and others into the endothelial cells that line the insides of blood vessels.[13] The stem cells grew into brain organoids, which were incubated in a matrix coated with the endothelial cells, then transplanted into a mouse's brain. Two weeks later the organoid was alive and healthy—and capillaries had grown into its inner layers, with these blood vessels that had grown into the brain organoid being of human, not mouse, origin.

While many significant obstacles remain to be overcome in brain organoid studies, let us assume that at some point it will be possible to take cells from a person and derive an object of the same size as a normal human brain, with a blood supply, and a structure and interconnections identical to that of the latter. Such an object would be a great resource for neuroscience, but would its creation ever be justifiable on ethical grounds?

One important ethical question is whether an artificial human brain derived from stem cells would have conscious awareness. If so, would it have a sense of itself as an experimental preparation? And could a brain in that situation be said to have any rights? A complicating factor is that it might only be possible to answer such questions by allowing an artificial brain to become aware of its circumstances. For the conscious awareness of a brain in a living person is—as I have argued in this book—partly a product of that brain's biology, but also partly of the interactions of that person with the outside world. So an artificial brain in a vat might only become aware of itself if connected to incoming signals that allowed it to see, hear, smell, and possibly touch its surroundings. Most importantly, given the key role that language plays in the development of human consciousness, a brain in a vat would surely need some language input from the outside world to be able to develop conscious thoughts. Yet would such an input on its own be sufficient to lead to conscious awareness?

During normal human development, as children we learn the meaning of words partly through our parents and other adults talking to us, but also by how such words relate to the world around us. So any attempt to communicate with an artificial brain would presumably need to also use various sensory inputs to recreate the sense of a world outside the vat in which the brain was suspended. But how much of an experience would this be? To some extent this question was anticipated by philosophers long before ways to create brain organoids were developed. For instance, in the 1980s, Harvard University philosopher Hilary Putnam posed the question of how any of us would know that we are not just a brain in a vat of bubbling chemicals, hooked up to a virtual-reality simulator.[14] One answer to this question is to restate the fact that it is through our interactions with the real world—through our senses, but also our relationships with other human beings—that we find evidence that we exist in such a real world. But how can we be sure then that such interactions are real?

Such a conundrum was explored in the 1999 science-fiction film *The Matrix*, in which a computer hacker called Neo comes to realize that he, and the rest of humanity, are actually bodies in suspended animation, enslaved by sentient machines who feed on their life essence, and 'reality' is just an illusion fed to the brain.[15] Only a few humans have recognized their plight, and the rest of the film concerns their battle to liberate humanity. The main premise of the film is that virtual reality (VR) technology has become so advanced that the human race—bar these few enlightened individuals—is not aware it is merely part of a simulation.[16] Yet although there have been significant advances in VR technology since the film's release, we are still not at the point where anyone could mistake such a simulated world for the real thing—unless somehow the film's premise is true, and all human experience really is an illusion!

Returning to our scenario of an artificial human brain in a vat, given that such a brain would never have experienced being in an actual human body, it might be less of an issue if the simulated world were not an accurate reconstruction of human life. But we need to address how feasible it would be to recreate the senses for such an artificial brain. For the more we learn about the biology of both vision and hearing, but also smell, taste, and touch, the clearer it is becoming that the way we process inputs from the outside world through such senses is complex not only in terms of the brain's role, but also with regard to the biology of the eye, the nose, the tongue, the skin, and so on. Because of this, the artificial senses that we might create for a brain in a vat would surely be far inferior to our own biological means for experiencing the world. However, for an artificial brain that had never experienced anything else, perhaps such an experience would not be seen as impoverished.

An important difference between the evil intelligent machines in *The Matrix* and a conscientious and ethically minded human scientist who had created an artificial human brain would surely be that such a scientist would need to reveal to this brain its predicament. Unfortunately, such a course of action could itself create more ethical dilemmas. Such a brain—despite being the product of an evolutionary process in which living human brains generally only exist in the context of a body—would then become aware that it was actually only an object in a vat linked to various electronic inputs and outputs. Having gained conscious awareness of its situation, should such an artificial brain be viewed merely as an experimental preparation, or as equivalent to a human being with its own personality and possibly also its own rights?

Looking even further into possible future scenarios, our growing ability to develop functioning robots raises the question of whether an artificial human brain might be used to control such a robot, creating in effect a cyborg. However, the creation of such an entity could make a big difference in terms of an artificial brain's sense of identity in the world, as well as its rights. Such a cyborg might be much more powerful than a normal human being, and also potentially live far longer, since only its brain would be organic. To some extent this is the subject of the 2017 film *Ghost in the Shell*, in which Mira Killian, an anti-terrorism agent for a Japanese government of the future, believes she was created as a cyborg following a terrorist attack in which only her brain survived.[17] However, we eventually learn that Killian's brain was taken from a young political activist, Motoko Kusanagi, whose memories were suppressed to create the Killian cyborg, by an unscrupulous biotech company. Despite attempts by the company to kill Killian, she is able to use her amazing powers to expose it.

Clearly in this film, part of the company's crime is its use of an unwilling living person's brain to create their cyborg. But if such a cyborg were created using an artificial human brain generated from stem cells, would this also be seen as an infringement of human rights? What would surely be the case is that such a cyborg, possessing as it would a human brain—and presumably thereby the ability to think and reason—would need to have a say in the rights or wrongs of its creation, and its position as a conscious individual in society.

The only organ of human origin in a cyborg like the one described would be its brain. Yet such is the importance of the brain for our conception of what it means to be human that many people might be able to accept the idea that such a cyborg could be considered part of humanity. What, then, of a form of conscious awareness that was purely machine? How would this be considered, and how likely is it that such an entity could ever exist?

Certainly, there has been no shortage of predictions in recent years that an 'artificial intelligence' (AI) based on a computer will soon have not only reached the same mental capabilities as a human being, but also surpassed them. Upon reaching such a point, often referred to as a technological 'singularity', AI is then predicted to undergo an exponential technological growth that will transform life as we know it. For instance, Vernor Vinge, a mathematician at San Diego State University, who first coined this term in 1993, believes that 'we will soon create intelligences greater than our own. When this happens, human history will have reached a kind of singularity, an intellectual transition as impenetrable as the knotted space-time at the centre of a black hole, and the world will pass far beyond our understanding.'[18] Such predictions have helped to trigger a debate in recent years about whether such an all-powerful AI will save humanity or instead hasten our extinction.

Representing the first point of view, Frank Lansink, European CEO at IPsoft, believes that 'AI...has an integral role to play in the workplaces of the future. It will help unleash creativity, create new job activities and new occupations which combine ground-breaking technology with our most human skills.'[19] However, others are less convinced about such a rosy scenario. For instance, in 2017 the biotech entrepreneur Elon Musk stated that 'I have exposure to the very cutting-edge AI, and I think people should be really concerned about it...I keep sounding the alarm bell. But until people see robots going down the street killing people, they don't know how to react, because it seems so ethereal.'[20] And the physicist Stephen Hawking told a BBC interviewer in 2014 that 'the development of full artificial intelligence could spell the end of the human race.'[21] Expressing how such a scenario could arise, Kilian Weinberger, a computer scientist at Cornell University, has recently said that 'if super-intelligent AI—more intelligent than us—becomes conscious, it could treat us like lower beings, like we treat monkeys. That would certainly be undesirable.'[22]

Not surprisingly, such statements can leave many people feeling highly uneasy about the development of ever-more sophisticated computer and robotic systems. But is the rise of an all-powerful AI really such a likely scenario? One reason for caution is that such predictions are not new. Notably, Alan Turing, the mathematician who was one of the first pioneers of computer science, predicted in 1951 that machines would 'outstrip our feeble powers' and 'take control' within the next human generation.[23] In 1965, Turing's colleague Irving Good argued that 'ultra-intelligent' AI systems would soon design even more intelligent ones, ad infinitum: 'Thus the first ultra-intelligent machine is the *last* invention that man need ever make, provided that the machine is docile enough to tell us how to keep it under control.'[24] Clearly, such predictions were hopelessly misguided in their timescale, but how about their basic substance? Given that one commentator has described AI as 'all the things that computers still can't do', could the singularity be creeping up on us, and may computers already be close to reaching that point of all-powerfulness?

Certainly, computing power has made possible some amazing advances in recent years. Computers are already proficient at picking financial stocks, translating speech, and identifying cancer cells in a biopsy. And although the board game Go was thought to be so guided by intuition that it was unsusceptible to mastery by a computer, in 2016, AlphaGo, a 'deep learning' programme designed by a team at Google's DeepMind AI subsidiary in London, was able to dethrone Lee Sedol, the world's leading player of the game.[25]

Despite such successes, is it true that computers are approaching the point of matching, or even surpassing, human consciousness? To address this question, we need to take a closer look at the mode of action of programmes like AlphaGo. Such programs are sometimes referred to as 'neural networks', by analogy to a biological nervous system.[26] For each turn of a game, the AlphaGo network looks at the positions of the pieces on the Go board and calculates which moves might be made next and the probability of them leading to a win. After each game, it updates its neural network, making it a stronger player for the next turn.[27] All of this is highly impressive, but how does it compare with human consciousness?

In some respects, the computational activities of a deep learning programme may be quite similar to some of what goes on in a human brain. When we see an object, smell an odour, listen to someone speak, assess a work of art, or play a game of chess, each of these responses involves an act of computation. And since AlphaGo can carry out its computations far more rapidly than a human brain, this can lead to the idea that the programme is well on its way to a form of intelligence that compares to, or even surpasses, that of a human.

Such an idea has been bolstered by statements from AlphaGo's designers. According to Demis Hassabis, DeepMind's CEO, 'for us, AlphaGo wasn't just about winning the game of Go. It was also a big step for us towards building these general-purpose algorithms.'[28] What he means is that most AIs are described as 'narrow' because they perform only a single task, such as translating languages or recognizing faces; in contrast, general-purpose AIs could potentially outperform humans at many different tasks. And certainly this kind of adaptability may prove very useful; for instance, a new version of AlphaGo is currently being used to identify 3D structures of proteins involved in key bodily processes, which could be important for identifying new molecular targets for drug discovery.[29]

Yet such statements can also feed the idea that a computer program like this is comparable to the generalized thinking capacity of a human. I believe that this is a mistaken conclusion for a number of reasons. First, although AlphaGo is versatile in its ability to become expert at various different tasks, it can only do so one at a time.[30] Compare that to a human being, who can drive a car, conduct a conversation with someone in the car, listen to the car radio, and be also thinking about what to cook for dinner or the argument they had earlier with their spouse (although there are limits—and risks—to the number of mental tasks on which a person can safely be engaged, especially if in charge of a fast-moving vehicle).

Not only can human beings multitask in this way, but we also have a memory system that allows us to acquire skills in diverse other tasks, all of which can be called into operation as required as we go through life, even if we have not used such skills in years. In contrast, for a program like AlphaGo that has become expert at Go to learn a new task—such as modelling 3D structures of proteins—it has to forget everything it has learned about Go.

Second, human brains are far more than just computational devices. In particular, our brains are first of all organic structures, which has important implications in comparing their action to that of a computer. Indeed, there are huge differences between even an animal brain—say that of a monkey—and a computer program like AlphaGo. While the name 'neural network' is deliberately meant to suggest similarities between such a network and an organic brain—with even the individual units of the electronic circuit named 'neurons'—in reality there is a huge difference between these two entities.[31] For as we have seen, real neurons are far from being simple digital on/off switches like the units of a computer circuit; instead, they have a complex structure, a genome, but also a pattern of gene expression distributed across the different parts of the cell—and neurons also come in all shapes and sizes. In addition, as well as individual neurons being connected in local circuits via synapses, there is increasing evidence that different brain regions are connected via brain waves into a larger whole.

Moreover, both the local circuitry and these longer-range connections may be quite different in a human compared to a chimp brain, because of the role of language and other cultural factors mediating consciousness in our species. As well as electrical communication, chemical signals—including complex informational molecules like regulatory RNAs—are, as we have seen, now recognized as playing an important role in facilitating communication between neurons, but also between neurons and glial cells. Such signals also allow the brain to communicate with the body and vice versa. And all of this structure and electrical and chemical communication is the product of four billion years of evolution.

The third difference between human consciousness and a computer program like AlphaGo is that only the former sees meaning in what it does.[32] While AlphaGo is superb at number crunching as a way to get a successful result, there is no evidence that it has any awareness at all of what it is doing. For a grandmaster Go player like Lee Sedol to beat another human grandmaster at this game means more than just the result itself, because he has invested a huge amount of his life in getting to such a level, with all of the fame and recognition that this can bring him as an individual expert. It must be somewhat soul-destroying then to be beaten so easily by a machine, and yet for AlphaGo's part, all this program has done is unconsciously carry out an algorithm. At least in their current form, it is far from clear that any AI will acquire conscious awareness, despite being able to carry out an increasing variety of tasks that include not just playing games like Go, but also activities ranging from medical diagnosis to self-driving cars.

The different levels of structure and communication in the human brain, which range from synapses between individual neurons to waves that coordinate the different brain regions, also make unlikely another prediction—that it might soon be possible to 'upload' an individual human consciousness into a computer. In 2016 physicist Stephen Hawking stated that it should be possible to 'copy the brain onto a computer and so provide a form of life after death.'[33] Not only that, but a biotech company, Netcome, has been set up by MIT graduate Robert McIntyre to begin work in this very area. McIntyre believes that if we can decipher the 'connectome'—all the synaptic connections—of an individual human brain, a project which I mentioned earlier, it should then be possible to upload this information into a computer.[34] However, there are two major flaws in such proposals.

First, while it may be possible to map all the synapses in a human brain, there is no obvious way at the moment to reproduce the informational content of each neuron or glial cell, or the electrical or chemical long-range signals. Second, given the differences between computers and brains discussed above, it is not clear how the 'information' in a brain could be uploaded into a computer. For this reason a prediction in 2014 by another physicist, Michio Kaku of the City University of New York, that soon 'if I have a CD-ROM with all the [brain's neural] connections on a disk, I can put that on a laser beam, and I can shoot that into outer space at the speed of light...and then at the other end there's a relay station which absorbs the laser beam and puts all these memories into a robot, and so you can then begin to feel, and live on another star system,'[35] is also highly unlikely to be realizable.

Indeed, given that it is also currently impossible to cryopreserve a living person, perhaps the only way that our species might be able to colonize habitable planets that are light years away would be through advances not in neuroscience, but in reproductive biology. For recent advances in this field have brought a step forward the prospect of being able to maintain human foetuses outside the mother's body in an artificial womb. Applied to humans, such technology could help women unable to bear children in the normal way and support babies born extremely prematurely. But it could also make possible colonization of habitable planets too far away for a person to travel to in their lifetime. So a spaceship could be loaded up with frozen human embryos, food supplies for the future human off-spring, the means to generate different animal and plant species, and a number of robots. When the ship arrived at the destination planet, the robots—which would need to be highly versatile, but not consciously aware—would be programmed to resuscitate the human embryos and implant them into the artificial wombs, then nurture them through to adulthood. The robots would play a similar role in nurturing the embryos of the other animal and plant species on the ship.

While this scenario is currently unrealizable, with advances in repro-ductive, computer, and space science technology, it does not seem totally implausible that it could be a viable project in the next few decades. However, from an ethical point of view such a scenario throws up many issues. A key one, assuming that everything went to plan, and our proto-human colony did not end up dying because of unforeseen adverse events, would be the fact that at least the first generation of this space colony would be brought up and educated solely by robots that, while highly skilled, had no conscious awareness of their own.

What type of human beings would emerge from such an experience? How might human minds and consciousness develop under such conditions? Would they be a joyful new extension of humanity, or would our ambitious plan leave them emotionally scarred for life? We would probably only find out as a species back on Earth once our extra-planetary offspring started sending back messages to their home planet. Now obviously such a scenario is still science fiction but the fact that it is not that far-fetched given recent technological progress is surely testament to the power of the human mind and the society it has produced.

EPILOGUE

A Twenty-First-Century Mind

————◇————

A ll books must end and it is time to draw this one to a close. But can any account of so vast a topic as human consciousness ever have a conclusive ending? For even if this book has managed to provide some important new ideas about this subject, it will still only have scratched the surface of what is really going on inside our heads. Nevertheless, drawing things to a conclusion is what I must do, and I will do so by returning to the various alternative views about human consciousness first mentioned at the start of this book and assess how my account compares to, and hopefully builds on, these other viewpoints.

René Descartes surely makes a suitable starting point. He revolutionized our understanding of how the human body works by proposing that it can be understood as a machine, yet fell short in this project by refusing to extend his mechanistic view to the mind. Instead, Descartes argued that 'higher' mental functions can only be explained by reference to an immortal soul, opaque to scientific investigation. This introduced a dualism that persists to this day in discussions about human consciousness.[1] So we need to consider how much the view of the human mind developed here both recognizes and overcomes this dualism.

The difficulties in developing a view of human consciousness that does not end up as Cartesian dualism arise from two main directions. The first comes from the difficulty in accepting that something as personal as the individual 'I' is merely based on the unconscious actions of cells in the brain, particularly given the apparent complexity of human thought that at times seems to verge on the transcendental. The second difficulty arises from the problem of how to construct a view of the mind that can explain our sense of being an individual consciousness, without lapsing into a scenario in which that consciousness appears to be due to a tiny self-aware homunculus sitting inside our heads directing our sense of self.

I have argued that all previous attempts to explain human consciousness have failed to provide a properly materialistic account that shows how mind is connected to brain. Either the issue is ducked completely by ignoring the role of the organic brain and assuming that the mind can be viewed as a 'black box' or a complex form of computer, but without relating this to actual brain structure, as in different forms of behaviourism.[2] Alternatively, the human mind is assumed to be something that 'emerges' from neural connections in the brain, but without explaining how. Instead, such a 'bottom-up' approach ends up reaching for what Daniel Dennett has called 'skyhooks'—explanations of complexity that do not build on lower, simpler layers.[3] Ironically, Dennett himself has employed one such skyhook by making the concept of 'memes'—the view that ideas exist as free-floating entities with a life of their own—central to his view of consciousness, despite this being an idealist notion.

In contrast, the view of human consciousness developed in this book can explain the uniqueness of our species' conscious awareness, but in an entirely materialistic fashion. This approach views language—the system of abstract symbols linked in a grammatical structure but also one that connects the individual to the world outside via word meaning—and other forms of human culture like music, art, and literature, as a material force that has reshaped human brain functions at every level. This has led to a qualitative shift in such functions, compared to that of every other species, including our closest animal cousins, the great apes.

Unlike a purely 'bottom-up' approach to human brain function, this view sees language, as well as other mediators of human culture, as imposing both structural and dynamic changes in our brains. Structurally, it sees the different brain regions, as well as their interconnections, as altered in humans. Throughout this book, I have identified examples of such changes. For instance we have evidence that the cerebellum—formerly thought to be only involved in repetitive motor skills—has both increased in size in humans and increased its connections with the cerebral cortex, and this has allowed it to play unique roles in human imagination and creativity. Other important changes include the expression of certain genes in different brain regions becoming altered in humans. For instance, the gene TH, involved in secretion of the neurotransmitter dopamine, is expressed not only in the striatum—a region involved in movement—as in other species, but also in the human brain in the neocortex, pointing to distinctive roles for dopamine in 'higher' human brain functions.

Other important differences include human neurons becoming covered with a fatty myelin sheath much later in life than those of chimps. Although this sheath increases the speed with which neurons can conduct electrical signals, it decreases the capacity for change or 'plasticity' of such neurons. Therefore this change is likely to have played an important role in our species' ability to learn and modify our ideas. In general, the human brain is much more plastic than those of other primates; recall, for instance, the observation that the folded structure of the brain cortex was found to be far more similar between two chimp brothers than is the case for two human brothers, including those who are identical twins.

Such biological differences explain why attempts to teach chimpanzees and gorillas language have largely failed. Instead, while individuals of these species can clearly possess a remarkable capacity to learn to associate words with objects and possibly also with emotions, there has been no real indication that they can learn to use words in a structured manner, to express concepts of time, place, self versus other, and all the other ways that humans employ language to build up a conceptual framework of the world that includes the individual as a conscious being aware of his or herself and their relationship to the world around them.

So a unique aspect of brain function in humans is therefore likely to be the way that language, as well as other aspects of human culture, combines with our biological differences to shape our developing brains. I have highlighted how brain waves play an important role in coordinating different brain regions in particular activities that include working memory. Although such waves appear to mediate global brain activity in many other species, the 'culturally mediated' view of the brain developed here predicts a link between human brain wave activity and language in a way that is not found in these other species. Investigating how language affects the coordinating role of different brain waves associated with various aspects of human consciousness is likely to be a fertile area for future studies.

An important feature of brain waves is that they display what is known as 'non-linear dynamics'. Such dynamics occur throughout nature and can even be observed in a simple chemistry experiment—the famous Belousov–Zhabotinsky (BZ) reaction.[4] Named after its founders, the Russian chemists Boris Belousov and Anatol Zhabotinsky, this reaction is initiated by mixing together chemicals that include two forms of cerium ion (Ce^{3+} and Ce^{4+}) which vary in colour, in a plastic dish. The ratio of Ce^{3+} and Ce^{4+} subsequently begins to oscillate, causing the colour of the solution to fluctuate between brown and colourless. But far from being a simple colour switch, what next occurs is the production of amazing, rotating spiral waves, until the surface of the dish is a mass of swirling, fractal forms.[5] And there is evidence that brain waves also possess the same oscillating and fractal character.[6]

This is important given what I said about the potential role of such waves as mediators of consciousness, since this non-linear dynamic behaviour would mean that there is no need for consciousness to be centred in any specific part of the brain, conjuring up visions of a controlling homunculus or soul. Instead, consciousness would be a property of the whole organ. In fact, it seems entirely feasible that such order out of chaos also operates in an animal brain—say that of your pet dog or cat—explaining how its attention is suddenly taken up by something novel in its environment, whether a squirrel, a tasty snack, or just the sight of you. However, in humans, superimposed on this mechanism for switching attention from one thing to another, language also surely plays a unique and higher guiding role—acting through brain waves, but doing so in a structured way that can only occur in our species because of the word meaning and grammatical structure that language alone can provide.

It is increasingly recognized that not only electrical communication but also that mediated by chemicals such as neurotransmitters and by newly discovered 'signalling' molecules like regulatory RNAs play important roles in the brain. It will therefore be important in future to study how this other form of communication might be involved in reconfiguring the human brain in response to language and other cultural mediators.

In fact the examples mentioned above still only provide tentative indications that the human brain is radically different from those of other species, including the great apes. As the basic units of brain structure— neurons and glial cells—and the molecular mechanisms underlying brain function are very similar between humans and these other species, it will rather be the organization of brain structure, interconnections between brain regions, and brain dynamics that are likely to subtly differ in human beings. And this is where the practical and ethical obstacles to studying the dynamics of human brains have limited us.

To some extent, studies of molecular and cellular mechanisms in human brain slices from recently deceased individuals who have donated their bodies to science might be of value, although given how much I have stressed the importance of treating the brain as a whole entity, it is not clear how much would be gained from studying physiological responses in such human brain slices, as opposed to those taken from another primate, or even a rodent. However, given that I have also argued that the mediation of brain function by cultural factors—language being primary in this respect—is key to understanding human consciousness, it would be interesting to see what insights might be gained from molecular analyses of different brain regions taken from aborted foetuses at different stages of development and from corpses of human individuals of different ages and suffering from various mental disorders, all samples of course donated through ethical consent agreements.

In addition, as advances are made in our ability to generate increasingly complex brain organoids from stem cells derived from specific human individuals, scientists are already identifying ways to investigate not only basic human brain functions, but the effects of various brain disorders, in such experimental systems.[7] But as I have noted, there are many practical obstacles to overcome before such organoids can even approximate the complexity of human brain structure and function, while attempts to do so, for instance by linking such organoids to a blood supply, allowing them to reach adult size, would raise so many ethical issues that this may always remain an unacceptable scenario.

While 'humanized' animal models of brain function and fully functioning adult-sized human brains grown in a vat remain science fiction for the present—and some people may believe that we should keep things this way—important insights into human brain function as well as brain disorders can also come from analysis of the brains of living human beings. Such studies could be carried out in two main ways. First, a variety of sophisticated different forms of brain imaging analysis can be performed in a non-invasive fashion on individuals ranging from babyhood to old age.[8] Indeed, as I mentioned previously, it is now even possible to carry out such analysis on foetuses in the womb. Such type of analysis is allowing us to study the dynamics of brain activity, and how different regions of the brain are involved in a particular task. We have already noted how this type of approach has been used to study the dynamics of brain activity in musicians playing improvisational jazz, and this has provided clues about how different brain regions interact during such a creative performance.

Second, patients undergoing surgery can volunteer to have the activity of neurons in different regions of their brain analysed while they are still conscious. We saw previously how such an approach can identify which neurons become activated when a volunteer is shown an image of a celebrity such as Halle Berry or a landmark like the Eiffel Tower, providing clues both about how memories are stored and accessed and about how associations between people and places are formed in the brain. In future studies, both non-invasive imaging and this more direct—but invasive—analysis of specific neurons might be used to explore in more detail the key claims of this book, namely that language and other cultural mediators transform even the most basic brain functions in humans and that there has been a qualitative shift during human evolution in how different brain regions interact that has opened the human brain to such cultural mediation.

Of particular interest would be to carry out studies of the developing brains of children and adolescents. Imaging analysis of the brains of babies learning language is already happening, and therefore it would be very interesting to revisit some of the studies carried out by Lev Vygotsky and his colleagues in the 1920s and 1930s of children employing inner speech while performing different tasks or learning to group objects in a conceptual fashion, with a view to replicating these, but with imaging analysis of participants' brains.

A key reason for studying brain function, beyond intellectual curiosity, is to better understand psychiatric disorders like depression, schizophrenia, and bipolar disorder, as well as conditions like autism spectrum disorder. So what impact, if any, might the approach to understanding human consciousness that I have been developing in this book have upon diagnosis and treatment of such disorders? We should first acknowledge how complex a phenomenon mental disorder is, in both biological and social terms. There are unlikely to be easy answers for those seeking to reduce human mental disorders to either simple genetics or purely societal pressures. However, this does not mean that the view of human consciousness advanced here has nothing to offer in terms of better diagnosis and treatment.

For a start, the view advanced here that human brain function is regulated by top-down as well as bottom-up mechanisms, with brain waves of different frequencies and other global signalling mechanisms potentially playing an important role in coordinating the activities of different brain regions, suggests that mental disorders may occur due to a breakdown of such coordination, but in a way that distinguishes one disorder from another. This could explain why multiple regions of the genome have been linked to a condition like schizophrenia, because although mutations in very different genomic regions may affect different parts of the brain, the end result could be a breakdown in overall brain function that manifests itself in the symptoms by which we tend to define a schizophrenic.

Recall Lev Vygotsky's idea that if schizophrenia is due to a higher mental function being 'switched off', the mind may try to compensate by resorting to a developmentally more archaic function, but finding itself without the higher levels of control, this produces abnormal forms of behaviour and thought.[9] If schizophrenia is due to a breakdown of coordination between brain regions, then this could result in each region continuing to function, but in isolation. And if the abstract symbolism of language is one of the key coordinating elements in a human brain that provides it with the conceptual framework central to the development of conscious awareness, this could explain both the breakdown in capacity for conceptual thinking that Vygotsky noted in schizophrenics, as well as the loss of conscious awareness that can be one of the features of the disorder.[10]

Such an interpretation also allows us to explain a well-known symptom of schizophrenia—the tendency to hear 'voices'. For a major claim of the socially mediated view of human consciousness is that our innermost thoughts find expression through a form of 'inner speech'. Individual thought is seen as a kind of internal dialogue, with the different voices within that dialogue being drawn from our social interactions, and individuality itself as a boundary phenomenon that is a product of both such social interactions and the biology of the individual's brain. To some extent this leads to a conundrum, namely that individuality itself is partly an illusion, expressed by Charles Fernyhough by the statement that 'we are all fragmented. There is no unitary self. We are all in pieces, struggling to create the illusion of a coherent "me" from moment to moment.'[11] And yet, it is clear that most people exist from day to day with the feeling that they are a unified self. This is despite the fact that every time we go to sleep at night, we lose our waking conscious state, but still regain our sense of a unified consciousness when we wake up in the morning. However, for some schizophrenics, a breakdown of conscious awareness may express itself through an inability to recognize that the voices that make up the inner speech within our heads are merely expressions of our individuality, and not another person somehow existing inside our heads but external to us.[12]

How might such an explanation of the schizophrenic state lead to improvements in the diagnosis and treatment of this disorder? In practical terms, brain scans of schizophrenic individuals engaged upon some problem-solving activity that involves conceptualization—for instance, the block-sorting exercises pioneered by Lev Vygotsky and his colleagues—might make it possible to see whether coordination of different brain regions during such exercises is different compared to the situation in unaffected individuals. Such studies might identify distinctive dynamic features of the brains of schizophrenic individuals and also highlight differences between such individuals. After all, while a particular range of symptoms define a schizophrenic—symptoms which can include hearing internal voices, experiencing hallucinations, assigning unusual significance or meaning to normal events, and experiencing delusions—two people can be diagnosed as suffering from this condition while having a completely different set of symptoms in this range.[13] In other words, what we define as schizophrenia may be a spectrum of loosely related conditions, and brain imaging might help define how this relates to different brain dynamics of individuals in such a spectrum.

In addition to being used to study schizophrenics, such an approach might be used to investigate the brain dynamics of those with other mental disorders, such as clinical depression. Some types of depression can also be associated with delusions—as, for instance, was the case with my sister's condition—but recent studies have indicated that depression is not associated with the breakdown of conscious awareness found with severe schizophrenia.[14] Therefore, it may be that delusions in a depressive individual have a different root, possibly linked to the feelings of worthlessness characteristic of depression, and such differences might therefore be identified by imaging. In addition, such analyses might help define different types of depression, since the condition is increasingly recognized as being as heterogeneous as schizophrenia in its symptoms and perhaps in its biological mechanisms, as well as in the societal pressures that may push an individual into this state.

Not only is there evidence for a key role for an individual's social environment as well as their biology in the genesis of their schizophrenia or depression, but also an increasing recognition that novel psychotherapeutic approaches are as essential as the development of more effective drugs in the treatment of such disorders. I have previously mentioned the Open Dialogue approach, whose aim is to establish a long-term dialogue between the patient and their therapists, as well as the patient's family in the therapy. In fact this is only one of a number of innovative new methods for the treatment of mental disorder.

Another approach, reflecting the fact that enjoying as well as creating music, art, and literature is a key aspect of what it means to be human, uses such creative activities to boost mental health. So Laura Marshall-Andrews, a British GP based in Brighton, believes doctors should be able to prescribe music, arts, and writing courses to individuals with depression. According to her, 'GPs need the freedom and the flexibility to try out different therapies and use different interventions.'[15] Marshall-Andrews' belief in the potential healing power of creative pursuits reflects her experience of the Brighton-based Healing Expressive and Recovery Arts (HERA) programme which has helped patients make new friends, combat loneliness, and generally resulted in a huge increase in well-being among participants.[16]

One common factor in these two approaches is that they both involve sustained, high-quality interactions with other human beings. And given how much I have stressed the importance of language in the formation of human conscious awareness, and its role in our unique capacities of creativity and imagination, it is maybe not surprising that such enhanced social interactions can have a therapeutic effect. Yet as we have seen, in Britain, the USA, and many other countries across the world, provision of care for those with mental disorders is getting worse at the same time as societal pressures—job insecurity, stagnating salaries, and attacks on pensions—are leading to a dramatic rise in the incidence of such disorders.

Such societal pressures may be a factor in something else to which I have drawn attention: the highly polarized opinions within the health-care system about the causes of mental disorder—and therefore the best way to treat it. So psychiatrists tend to subscribe to a 'medical model', which sees mental disorders as primarily due to an alteration of biological structure and functioning that is best treated with drugs, while clinical psychologists, social workers, and related professionals are more likely to reject the idea of a biological basis for what they call 'mental distress', and see psychotherapy as the best treatment.[17]

This state of affairs means that currently the two main groups of pro-fessionals that deal with mental health issues fundamentally disagree about both the causes of mental disorders and the best way to treat them, which surely cannot be a situation that benefits the patient. So why is it that the viewpoints of these two groups have become so polarized?

To answer this question, we need to consider both ideological and social factors. Ideologically, psychiatrists are more inclined to biological reductionism because of the influence of this view of bodily functions—including those of the brain—on the medical profession as a whole. However, medics are also deeply intertwined with the pharmaceutical companies. Such companies can influence the views of psychiatrists in various ways. For instance, it is common in the USA for pharmaceutical companies to pay psychiatrists money to recommend to their colleagues particular drugs for treating mental disorders in meetings at which a company representative is also present.[18] If doubt or criticism of the drug is expressed, the psychiatrist's subsidy ends. Another common practice is that a company's drug is praised in a scholarly article written by company personnel, and a psychiatrist only makes a few additions, but their name appears on the article and they receive payment.[19]

However, such ideological and social influences are far from affecting only psychiatrists. Psychologists and social workers, who in my experi-ence are often politically more left wing, can tend to a social reductionism that emphasizes the importance of social influences, while largely disre-garding biological factors. Yet as we have seen, this 'blank slate' view of the mind is actually quite far removed from the view of human consciousness developed by Lev Vygotsky in the years following the Russian Revolution that saw conscious awareness as a sophisticated interplay of both biological and social factors and that has been an inspiration for this book. Instead, the blank slate view owes far more to the ideology of the Stalinist regime that swept away the influence of thinkers like Vygotsky and Voloshinov. But such has been the powerful influence of Stalinism among left-wing circles for much of the twentieth century that it continues to play a far more pervasive role in progressive thought—even that nominally opposed to Stalinism—than is often realized.[20]

Such are the potential ideological differences between these two sets of professionals. But there are also social factors at work. Psychiatry is often viewed as medicine's 'poor cousin', and there are good reasons for this.[21] During the past century, our understanding of bodily mechanisms has advanced through scientific understanding on a whole number of different fronts and led to important new drug treatments in areas as diverse as cardiology, cancer therapy, and treatment of bacterial and viral infections. Yet as we have seen throughout this book, our understanding of how the human brain works as a whole still remains far from clear, and as a consequence, drug treatments for mental disorders have often been far from inspiring, with even the mechanisms of action of those drugs that do seem to have a positive impact on mental health remaining obscure. This lack of clarity is due to the brain being so much more complex than any other organ in the body, but it also reflects a central theme of this book, namely that human conscious awareness is as much a social as a biological entity, and therefore human mental disorders have a major social input.

Faced with such a lack of clarity about both the mechanisms of human consciousness and how to intervene when it becomes disordered, one strategy for health professionals under pressure as cuts bite into the health service—coupled with rising levels of reported mental disorder—is to retreat into a 'two camps' mentality. So psychiatrists can end up arguing that the limited available funds should go into the development and provision of drug-based treatments for mental disorders, while psychologists and social workers may counterpose that it is 'talking therapies' and community care programmes that should be receiving such funds. The tension between these two groups of professionals is accentuated by the higher status that psychiatrists, being doctors, tend to have in the health system—a difference reflected not only in pay, but also in who ultimately decides the primary form of treatment.[22] And as pressures on national health services continue to increase, so this tension is likely to rise too.

This is only one of the effects of economic crisis on both the diagnosis and treatment of mental disorder. As the crisis continues—and with it not only social problems but also decreasing funds for public healthcare and other social services—there is increasing pressure on psychiatrists to prescribe drugs as a kind of 'sticking plaster',[23] when really what is required is a proper consideration of biological, but also social factors that may have triggered a particular mental disorder in a specific individual. But clinical psychologists and social workers are also affected by such pressures. In particular, while clearly these two groups of professionals play a critical role in supporting and trying to help many of the most vulnerable and disturbed individuals in the community, as economic and social pressures increase so can there also be a pressure on such professionals to 'police' individuals of the often more lower-class layers of society with whom they tend to work.[24]

In such a situation, while the focus on the social causes of mental disorders can identify new ways to diagnose and treat them, there is also a danger of overly rigid interpretations of causes. For instance, the evidence that being abused as a child seems to increase the chance of someone becoming schizophrenic later in life might lead to a view that all cases of the disorder ultimately have this as a primary cause. This could result in a 'blame the victim' viewpoint, though in this case it is not the sufferer that is blamed, but their parents.[25] And given that clinical psychologists and social workers typically have a university education, while many of the patients they deal with do not, this can lead to a feeling that the lifestyles of such lower-class individuals are the primary problem, rather than seriously trying to get to the root of why a particular individual—of whatever social class but certainly not ignoring their social background—has succumbed to a specific mental disorder.

While it is possible to find both ideological and social reasons for the current divergence among health professionals about the best way to treat mental disorders, it should be clear from the view of human consciousness presented in this book that such a polarization of opinion does not reflect the complexity of this topic and is ultimately damaging to attempts to better diagnose and treat mental disorders. For what should be evident from our survey of the mechanisms underlying the human mind is how mistaken it is to try and separate social and biological factors in this way. Instead we have seen how social pressures contribute to the development of practically all mental disorders except perhaps 'single-gene' conditions like Huntington's disease, and yet at the same time we have also noted that such pressures are increasingly recognized as manifesting their effects through biological mechanisms such as inflammation in the brain and epigenetic changes to the genome.

A particularly problematic aspect of a view that sees either defective biology or social pressures as the sole cause of mental disorder is that the relative importance of either of these types of effects is likely to differ greatly between individuals. Indeed, there is a growing awareness that the names given to conditions such as depression, schizophrenia, and bipolar disorder mask the fact that these are probably quite heterogeneous in both their symptoms and underlying causes. As such, it makes little sense to say that a person suffering from a particular mental condition should be treated only with drugs or only by psychotherapy. Instead careful consideration is needed of the individual and their type of disorder but also where it might fit on a spectrum of heterogeneity. Of course, assessment of patients on a case-by-case basis is a key principle of medicine in general. However, it needs particular emphasis in an area of clinical practice where patients have often been lumped together on the basis of fairly arbitrary methods of classification driven by superficial appearances, in the absence of clarity about the underlying basis of disorders that can affect the human mind.

Ultimately, we need to be able to better understand the mechanisms of action of drugs that target mental disorders. This could then provide a platform for development of more specific drugs that are more effective, with fewer side effects. But this will require a proper understanding of how drugs affect the detailed structure and functional responses of the human brain, not vague suggestions based on crude views of the brain as a vat of chemicals. And in particular this understanding will best come from a recognition that human consciousness is mediated by cultural factors, primarily but not only language, and that while the human brain may share similarities with those of other animals, it also has unique differences that affect its function, and how that is influenced by drugs. If such drugs can be developed, they will have much to offer, since even if a mental disorder has a strong environmental component and is rooted in the particular social interactions of an individual, drugs can still have an important role to play in their treatment. Importantly, drugs should not be seen as an alternative to psychotherapy, but rather as just one tool in a programme in which multiple approaches can be used in a tailored fashion to meet an individual's needs.

Such are the ways in which a better understanding of the human brain and consciousness might contribute to the diagnosis and treatment of mental disorders in the future. But it would be an erroneous scientific approach that did not also seek to identify the specific social causes of mental disorders and aim to address these, both in terms of dealing with a particular individual and as more general preventative measures for the population as a whole. Indeed it would be an odd kind of doctor that did not apply such an approach to disorders of the body as well as of the mind. For instance, while some individuals appear to be biologically more susceptible to diabetes type two, the chance of such individuals succumbing to this disorder is greatly increased if they are obese, which explains why the global incidence of diabetes is increasing dramatically as waistlines across the world expand.[26] Similarly, while some people have a genetic susceptibility to lung cancer, the chances of getting this condition are greatly increased by smoking, which is why campaigns have encouraged people to quit the habit since this link was established in the 1950s by the British epidemiologist Richard Doll.[27] These disorders, and I could name many more, clearly have highly significant environmental causes, and treating them is as much a social issue as a medical one. Nevertheless, the human mind seems particularly susceptible to social pressures.

Some believe that the human mind is particularly vulnerable to mental disorder as a by-product of our evolution. For instance, Ole Andreassen and colleagues at the University of Oslo claim to have found evidence that evolved genetic differences that distinguish our species from Neanderthals may have enhanced our powers of creativity and imagination, but also made us more vulnerable to schizophrenia.[28] So as well as showing that genetic differences linked to this disorder were only found in the genomes of modern humans and not in those of Neanderthals, their findings indicate that the brain regions in which the genes are expressed are linked to our human ability to think and understand in complex ways.[29]

But while vulnerability to mental disorder may be the unfortunate price we pay for our conscious awareness, it would be a major oversight not to also seek explanations for the current epidemic of mental disorder in existing society. This society has given us immense power, as our cities and technologies increasingly reach into every corner of the globe, and indeed even outer space no longer seems quite the 'final frontier' that it would have appeared to be to past generations. Proof of our species' ability to sustain itself is demonstrated by the fact that the human population is now almost eight billion and has grown more since the eighteenth century than in the entire history of humanity prior to that moment.[30] Human individuals are also living to a far older age, with global average life expectancy at birth in 2020 being estimated to be 73.2 years.[31] Yet while current society has brought us civilization and provided great material prosperity and wealth for humankind as a whole, it is also characterized by an increasing feeling of instability that can be destabilizing to individual human minds and which may ultimately threaten the future of human civilization itself, with all the potential implications this may have for the mental health of our species in the future.

One of the most tangible senses of such instability is the economic crisis that began in 2008 and has subsequently become a defining feature of recent years, leading to the sense of a society out of control. It was not someone on the far left of politics, but Sir Vince Cable, then leader of the Liberal Democrats, who generally occupy the middle ground in British politics, who said in June 2018 that 'scarcely a day goes by without a scandal erupting around bullying bosses, pilfered pension funds, business tax-dodging, chaotic private train operators, rewards for failure, bankers' bonuses, price gouging, or worker exploitation.'[32] A key problem since the beginning of the crisis has been that economic policies ranging from historically low interest rates to 'quantitative easing' have failed to rescue the world economy from what International Monetary Fund managing director Christine Lagarde has termed the 'new mediocre', and former US Treasury Secretary Larry Summers referred to as an era of 'secular stagnation', meaning a condition of negligible or no economic growth.[33]

This economic situation forms the backdrop to a recent rise in reported cases of mental disorders such as anxiety, stress, and depression. Worryingly, this rise is particularly concentrated among young people, particularly women. Data from the British NHS from 2017 revealed that the number of times a girl aged 17 or under was admitted to hospital because of self-harm jumped from 10,500 to more than 17,500 a year over the previous decade—a rise of 68 per cent.[34] The jump among boys was lower, but still 26 per cent. For Bernadka Dubicka, chair of the child and adolescent faculty at the Royal College of Psychiatrists, this demonstrates 'a growing crisis in children and young people's mental health, and in particular a gathering crisis in mental distress and depression among girls and young women.'[35] One factor argued to play a role in this rise in mental disorder is social media, with Dubicka stating that media forms like Snapchat and Instagram 'can be damaging and even destructive' to girls' mental health and that 'there's a pressure for young people to be involved 24/7 and keep up with their peer group or...be left out and socially excluded.'[36]

Such signs of mental disorder in young people may be merely the latest manifestation—perhaps exacerbated by the latest technologies—of a general problem of alienation within modern society, despite all its amazing material achievements. This means not only that most people on the planet are estranged from the fruits of their labours and typically see work not as an act of creativity but something burdensome and oppressive to them as individuals, but also that life itself becomes seen as a competition—at school, in the job market, for friends and sexual partners—with all the mental stresses that this can produce. But perhaps the most worrying aspect of society today is that alongside all these problems we are hurtling towards environmental catastrophe.

For all the great concentration of wealth and power in current society, it is not clear that any of the people supposedly in control of the different countries of the world are doing anything tangible to halt, let alone reverse, the continuing flow of greenhouse gases into the atmosphere which, if unchecked, will surely spell disaster. One of the reasons for this paralysis is that no one really seems in control of our society; instead it seems to be the market and the drive for profit that ultimately makes this type of society tick. As such, there are substantial vested interests who do not want to risk any threat to their possibility of short-term gain, even if it risks the long-term viability of current society and possibly that of civilization itself.[37] And the problem for those influential individuals who can see beyond the short-term—for instance Bill Gates, one of the world's richest men, who has recently pointed to the limitations of a free-market economic system for solving a problem as big as global warming[38]—is how to persuade companies and governments to move away from putting the drive for profit first, in favour of sustainable technologies that may not yield such an immediate profit but are ultimately better for the planet and humanity, given the tendency of capitalism to punish those not competitive enough to survive in the free market.

Given such major anxieties and societal pressures, it is not surprising that many individuals today feel a sense of despair for the future—not just in terms of their own immediate futures and those of their families and friends, but also about the long-term prospects for humanity.[39] This feeling was only exacerbated by the global COVID-19 pandemic caused by the coronavirus SARS-CoV-2 that swept the world in 2020 and led to not only many deaths but also a lockdown that brought the global economy to a juddering halt and confined individuals to their homes, cut off from physical contact with many family members and friends. And politically it has been accompanied by a trend described as the 'collapse of the political centre'—that is, a rise in populist movements and increasing political polarization.[40]

All of which sounds like grounds for a great deal of pessimism—at least for people who see themselves as progressive in their political views—and could feed into further rises in mental disorders such as depression. Perhaps given the tragic history of my family I might be seen as someone particularly likely to succumb to such a feeling. After all, for all the insights I hope this book may have provided about the potential mechanisms underlying mental disorders, on a personal level it is still hard for me to comprehend just how successive members in my family that I held so dear, and who seemed vibrant, sociable, and happy individuals, could end up dying in such horrific ways. And my progressive political ideals might seem to be thwarted by current circumstances. However, I personally do not feel pessimistic, and to end this book, let me explain why I have a sense of optimism about what lies in store in the years ahead. To some extent my sense of optimism may be an aspect of my personality. Given that biology plays a role in personality, it seems not implausible to me that certain people are born with a tendency to be optimistic. But as we are shaped as much by our environment and experiences as our biology, even the sunniest personalities will surely have that sense of sunshine knocked out of them by a series of adverse circumstances, and here the influence of others matters. In my own case, I owe thanks to a variety of people—friends, colleagues, but most of all my family—for helping me keep an optimistic state of mind.

It might seem strange that I have derived any optimism from my family, given that two of my grandparents, my father, and my sister all died in miserable circumstances, all linked, or probably linked, to mental disorder. And I should add that my mother, who nursed my sister throughout her descent into depression, herself died, aged 79, only a few months after my sister's death, of a heart attack that was possibly exacerbated by the physical and mental strain of looking after her daughter, and by the latter's eventual death in such terrible circumstances. Yet it was particularly my mother who influenced my approach to adversity, and this is partly due to the loving environment that she provided while I was growing up, but also because of her response to the many tragedies, which included losing her parents at an early age, and then her husband in the prime of life. Yet somehow, despite all these traumatic experiences, I remember my mother as one of the most positive people I have known.

That some individuals might have a biological disposition to optimism is the opposite of what would be expected according to a 'blank slate' view of the mind. Yet this possibility is bolstered by a recent report about a retired Scottish teacher called Jo Cameron, who has mutations in two genes that as a consequence appear to make her almost impervious to pain.[41] Intriguingly, given what I said previously about recent evidence that physical and emotional pain seem to trigger responses in similar regions of the brain, Cameron also scored zero on tests assessing levels of anxiety and depression and has said that she always feels happy and upbeat, whatever the circumstance. This extreme resilience to negative events was demonstrated when a van driver ran her off the road, resulting in Cameron's car landing upside down in a ditch. Far from being distressed or angry, Cameron climbed out of her wrecked vehicle and went to comfort the distraught young man who caused the accident.[42]

So what is the cause of Jo Cameron's resistance to physical pain and her perpetually sunny outlook on life? One of the genes mutated in her genome, named FAAH, is involved in the breakdown of anandamide, a bodily chemical central to pain sensation, mood, and memory.[43] Anandamide works in a similar way to the active ingredients of cannabis. The less it is broken down, the more its effects are felt. The other mutation is in a gene that boosts the action of FAAH, and the combination of the two mutations dramatically boosts levels of anandamide in the body. Somewhat appropriately, given that Cameron seems to exist in a state of a permanent natural 'high', this second gene is named FAAH-OUT.[44]

So should we all aspire to have our genomes engineered to make us like Jo Cameron? In fact our pain response plays a vital role in protecting us from physical injury and death. And indeed over the seven decades of Cameron's life she has accumulated a variety of injuries that include a degenerated hip and thumbs deformed by osteoarthritis, which only came to the attention of medical authorities when Cameron became unable to walk and handle items, not because she reported any pain from these injuries.[45]

While we might not want to emulate Cameron's imperviousness to pain, how about finding a way to provide the rest of us with her sunny disposition on life? In fact here too there are some negative features. One contributing factor to Cameron's positive outlook is that she quickly forgets negative incidents in her life. However, she is also more generally highly forgetful.[46] This provides another reason why it would not be a good idea to genetically engineer everyone to be more like Cameron, unless we want a planet of people who feel no pain and are perpetually happy, yet who are prone to serious injury because they have no painful warning signals, and cannot remember what you told them last week.

Still, by studying how Cameron's two gene mutations result in her resistance to both physical and emotional pain, it might be possible to develop new types of painkillers and drugs and other therapies for the treatment of depression. More generally, while Cameron is clearly a highly unusual individual, the existence of such an individual suggests that there may be many other people who have a less extreme tendency to be resistant to pain and have a more positive outlook on life, but who nevertheless have a genetic basis for this tendency.

At the same time, it would seem mistaken given the emphasis in this book on the ways that culture mediates our minds to only focus on biological reasons for someone adopting a positive outlook on life.

Importantly, there is evidence that changes that we make in our lives can also help us cope with, and see a way beyond, even the greatest misfortunes that the world throws at us. One story that inspired me following the deaths of my sister and mother was that of a 100-year-old black woman called Ida Keeling from the Bronx in New York, who is a record-breaking sprinter, having decided to take up running at the age of 60.[47] This was not just the whim of someone approaching old age but a person trying to find some meaning in life after undergoing the deepest personal tragedy. In her earlier years Ida—a single mother after her husband died of a heart attack aged only 42—was active in the black civil rights movements, shuttling her children to Malcolm X speeches and boarding a predawn bus for the 1963 March on Washington in which Martin Luther King made his famous 'I have a dream' speech. Ida's daughter Shelley Keeling has said, 'I always understood from mother that you die on your feet rather than live on your knees.'[48]

However in the late 1970s, Ida's resilience was tested to the limit by the descent of her two sons into serious drug addiction and their subsequent deaths from drug-related violence—one son being hanged and the other beaten to death with a baseball bat.[49] But as Ida fell into a deep depression, Shelley Keeling—a track-and-field athlete—suggested that her mother take up running as one way of coming to terms with her grief. And after some coaxing, Ida registered for a five-kilometre race through Brooklyn. It was a seminal moment for her. 'Good Lord, I thought that race was never going to end, but afterwards I felt free,' she noted. 'I just threw off all of the bad memories, the aggravation, the stress.'[50] My own running routine—a mere one-mile jog around my local park every morning—pales into insignificance besides Ida Keeling's efforts, but it was directly inspired by reading her story and realizing that daily running could be a way for me to deal positively with my grief.

Ida Keeling's story raises the question of why running has proved so beneficial as a way of maintaining a positive outlook on life, and how this differs from the way that genetic mutations have had a similar impact on Jo Cameron. In fact these two individuals may have more in common than might first appear, since there is increasing evidence that exercise can stimulate the production of the bodily chemical anandamide—the same cannabis-like substance that exists at such high levels in Cameron's body because of a defect in the proteins that control its breakdown. This explains why even moderate exercise—like a 15-minute jog in the park—can leave us feeling more satisfied with ourselves and at peace with the world.

Maintaining an optimistic outlook on life is important if we are to maintain a feeling of happiness, and sanity, despite the many adverse incidents that can affect us as individuals. But what if humanity itself seems to have lost its sanity? In the face of coming environmental catastrophe, a viral pandemic, and a political climate that seems increasingly reactionary, it is not surprising that some people believe that optimism about the future is misplaced.

However, I disagree with this view, firstly because it seems like an overly simplistic view of the current situation, and secondly because a survey of some key historical moments shows that ultimately ordinary people acting as a collective have made possible the great leaps forward that occasionally occur in human history. And those great leaps often occurred at moments in history that must have seemed, prior to such leaps, as the most depressing circumstances possible to those individuals who lived through them.

Take, for instance, the English Peasants' Revolt of 1381 that we have already mentioned. Today, with hindsight, we can view this revolt as a seminal moment in British history, which weakened the feudal system in England and may have been an important factor that facilitated the later overthrow of this system during the English Revolution of 1642–9.[51] Yet only a few decades before the Peasants' Revolt, a deadly bubonic plague strain known as the Black Death had swept across Europe, killing around 50 million people, equivalent to 60 per cent of its population.[52] But the resulting shortage of labour gave the peasants in southern England greater bargaining power, and this was one factor that led them to start having ideas 'above their station'.[53] Similar things could be said about the

French Revolution of 1789, which followed a severe winter in 1788 resulting in crop failure, famine, and bread riots in Paris.[54] In more modern times, the revolution that took place in my wife's home country of Portugal in 1974–5 and overthrew its dictatorship was triggered by the dissatisfaction of veterans returning from brutal wars in the Portuguese overseas colonies.[55]

All of which goes to show that major progress in society can come out of apparently desperate situations. But how does that help us today faced with an impending environmental catastrophe on several fronts and a world population that—far from learning from the mistakes of the past—appears increasingly to believe statements by right-wing demagogues who claim that the main problem in the world is not global warming, but the fact that someone with a different skin colour or religion, or who speaks in a different language, is out to steal their job or home? Despite such developments, I believe we can find good reasons to be positive about the current situation in terms of finding an audience for progressive ideals.

One positive development is the innovative approaches and technologies that young people in particular are bringing to social movements. Not that this should be surprising. After all, a key feature of our species is that the brain is especially plastic in human beings even into their late twenties, and as a consequence younger people are always more open to new ways of doing things. Yet as we have also learned recently, even after our formative years, human brains remain far more plastic than had been realized, meaning that each one of us has the potential for our thoughts and assumptions to change. At the same time, it is increasingly recognized that certain individuals whose brains are 'atypical' have much to offer to society in being able see beyond what is accepted as normal and inevitable, yet which may be detrimental to humanity's long-term well-being and the future of the planet.

So Greta Thunberg, who has become a symbol of the global movement to 'save the planet', has explicitly linked her farsightedness on this question to her autistic condition.[56] Individuals like her have a very important role to play in social movements, yet as we have also seen in this book, creativity and imagination are clearly key aspects of every person's consciousness, not just the preserve of a few gifted individuals. Finally, a characteristic that is unique to our species is our power to shape the world around us in a controlled, planned way.

Currently, we seem to have relinquished that control as human activities threaten to destroy the environment and our civilization with it. Yet the potential remains, and although there are daunting obstacles ahead, collectively human beings have the powers of creativity and imagination that could allow us to build a very different sort of society, one that is sustainable and works not against nature, but in harmony with it, and in which technology enriches our lives, rather than oppressing us. If we can develop such a type of civilization, then who knows what marvels we might expect to achieve in the future, not only on Earth, but also on more distant planetary destinations. If so, all such possible futures will have come from the human brain, which for an object weighing only a kilogram and a half, and with the appearance and consistency of cold porridge, is surely the most remarkable thing of all.

GLOSSARY

action potential Occurs when there is a significant increase in the electrical activity along the membrane of a neuron. Associated with neurons passing electrochemical messages down the axon, releasing neurotransmitters to neighbouring cells in the synapse.

amygdala Part of the brain's limbic system. This primitive brain structure lies deep in the centre of the brain and is involved in emotional reactions, such as anger or fear, as well as emotionally charged memories.

astrocyte Star-shaped glial cell that supports neurons, by helping to both feed and remove waste from the cell, and otherwise modulates the activity of the neuron. Astrocytes also play critical roles in brain development and the creation of synapses.

axon Long, single nerve fibre that transmits messages, via electrochemical impulses, from the body of the neuron to dendrites of other neurons, or directly to body tissues such as muscles.

brain imaging Various techniques, such as magnetic resonance imaging (MRI) and positron emission tomography (PET), that make it possible to capture images of brain tissue and structure and reveal which brain regions are associated with behaviours or activities.

brain waves Rhythmic patterns of neural activity in the central nervous system. These waves are also sometimes referred to as neural oscillations.

cell body Central part of the nerve cell that contains the nucleus of the neuron. The axon and dendrites connect to this part of the cell. It is also known as the soma.

cerebellum Brain structure located at the top of the brain stem that coordinates the brain's instructions for skilled, repetitive movements and helps maintain balance and posture. Recent research suggests it may also be involved in some emotional and cognitive processes.

cerebrum Largest brain structure in humans and positioned over and around most other brain structures. The cerebrum is divided into left and right hemispheres, as well as specific areas called lobes that are associated with specialized functions.

cortex Outer layer of the cerebrum. Sometimes referred to as the cerebral cortex.

dendrites Short nerve fibres that project from a neuron, which generally receive messages from the axons of other neurons and relay them to the cell's nucleus.

frontal lobe Front of the brain's cerebrum, beneath the forehead. It is the area of the brain associated with higher cognitive processes such as decision-making, reasoning, social cognition, and planning, as well as motor control.

glial cells Supporting cells of the central nervous system. They may contribute to the transmission of nerve impulses and play a critical role in protecting and nourishing neurons.

hippocampus Primitive brain structure located deep in the brain that is particularly important for memory and learning.

hypothalamus Small structure located at the base of the brain where signals from the brain and the body's hormonal system interact.

limbic system Group of evolutionarily older brain structures that encircle the top of the brain stem and play complex roles in emotions, instincts, and appetitive behaviours.

long-term potentiation (LTP) Persistent strengthening of a synapse with increased use, thought to underlie learning and memory.

microglia Small, specialized glial cell that operates as the first line of immune defence in the central nervous system.

midbrain Small part of the brain stem that plays an important role in movement as well as auditory and visual processing. It is also referred to as the mesencephalon,

myelin Fatty substance that encases most nerve cell axons. It helps to insulate and protect the nerve fibre and effectively speeds up the transmission of nerve impulses.

neurogenesis Production of new neurons by neural stem and progenitor cells. Rapid and widespread neurogenesis occurs in the foetal brain in humans and other animals, but this process also occurs in adult humans in the hippocampus and possibly other brain regions.

neuron Basic unit of the central nervous system. It is responsible for the transmission of nerve impulses. Unlike any other cell in the body, a neuron consists of a central cell body as well as several threadlike branches called axons and dendrites.

neurotransmitter Chemical that acts as a messenger between neurons and is released into the synaptic cleft when a nerve impulse reaches the end of an axon.

optogenetics Innovative neuroscientific technique that uses light to turn genetically modified neurons on and off at will in living animals.

plasticity Brain's capacity to change and adapt in response to developmental forces, learning processes, injury, or ageing.

prefrontal cortex Area of the cerebrum located in the forward part of the frontal lobe. It mediates many of the higher cognitive processes such as planning, reasoning, and the ability to assess social situations in light of previous experience and personal knowledge.

striatum Small group of subcortical structures, including the caudate nucleus, putamen, and nucleus accumbens, located in the midbrain. These regions are implicated in both movement and reward-related behaviours.

synapse Junction where an axon approaches another neuron's dendrite. It is the point at which nerve-to-nerve communication occurs. Nerve impulses traveling down the axon reach the synapse and release neurotransmitters into the synaptic cleft, the gap between neurons.

thalamus Brain structure located at the top of the brain stem. It acts as a two-way relay station, sorting, processing, and directing signals from the spinal cord and midbrain structures to the cerebrum, and from the cerebrum downwards.

ENDNOTES

Introduction

1. C. Baraniuk, 'It Took Centuries, But We Now Know the Size of the Universe', *BBC Earth* (13 June 2016), http://www.bbc.co.uk/earth/story/20160610-it-took-centuries-but-we-now-know-the-size-of-the-universe.

2. Editorial, 'The Human Brain Is the Most Complex Structure in the Universe. Let's Do All We Can to Unravel Its Mysteries', *The Independent* (2 April 2014), https://www.independent.co.uk/voices/editorials/the-human-brain-is-the-most-comple-structure-in-the-universe-let-s-do-all-we-can-to-unravel-its-9233125.html.

3. J. Amos, 'Voyager Probe "Leaves Solar System"', *BBC* (12 September 2013), https://www.bbc.co.uk/news/science-environment-24026153.

4. P. Bazalgette, 'We Have to Recognise the Huge Value of Arts and Culture to Society', *The Guardian* (27 April 2014), https://www.theguardian.com/culture/2014/apr/27/value-of-arts-and-culture-to-society-peter-bazalgette.

5. A. Kirk, P. Scott, and J. Wilson, 'World Mental Health Day: The Charts that Show that the UK is in the Midst of a Mental Health Awakening', *The Telegraph* (10 October 2017), https://www.telegraph.co.uk/health-fitness/mind/world-mental-health-day-charts-show-uk-midst-mental-health-awakening/.

6. A. MacDonald, 'New Insights into Treatment-Resistant Depression', *Harvard Medical School* (9 December 2010), https://www.health.harvard.edu/blog/new-insights-into-treatment-resistant-depression-20101209891.

7. A. Abbott, 'Colours Light up Brain Structure', *Nature* (31 October 2007), https://www.nature.com/news/2007/071031/full/news.2007.209.html.

8. M. Z. Donahue, 'New Clues to How Neanderthal Genes Affect Your Health', *National Geographic* (5 October 2017), https://news.nationalgeographic.com/2017/10/how-neanderthal-genes-affect-human-health-dna-science/.

9. C. Bergland, 'What Is the Human Connectome Project? Why Should You Care?', *Psychology Today* (27 November 2013), https://www.psychologytoday.com/gb/blog/the-athletes-way/201311/what-is-the-human-connectome-project-why-should-you-care.

10. K. Deisseroth, 'Optogenetics: Controlling the Brain with Light', *Scientific American* (20 October 2010), https://www.scientificamerican.com/article/optogenetics-controlling/.

11. J. Parrington, *Redesigning Life: How Genome Editing Will Transform the World* (Oxford University Press, 2016).

12. S. Jäkell and L. Dimou, 'Glial Cells and Their Function in the Adult Brain: A Journey through the History of Their Ablation', *Frontiers in Cellular Neuroscience*, 11:24 (2017).

13. J. Parrington, *The Deeper Genome: Why There is More to the Human Genome than Meets the Eye* (Oxford University Press, 2015).

14. R. R. Kanherkar, N. Bhatia-Dey, and A. B. Csoka, 'Epigenetics across the Human Lifespan', *Frontiers in Cell and Developmental Biology*, 2:49 (2014).

15. C. Fruhbeis, D. Frohlich, and E. M. Kramer-Albers, 'Emerging Roles of Exosomes in Neuron-Glia Communication', *Frontiers in Physiology*, 3:119 (2012).

16. N. Davis, 'Stress in Fathers May Alter Sperm and Affect Behaviour in Offspring', *The Guardian* (16 February 2018), https://www.theguardian.com/science/2018/feb/16/stress-fathers-behaviour-offspring-altering-sperm-micrornas.

17. I. M. Mansuy and S. Mohanna, 'Epigenetics and the Human Brain: Where Nurture Meets Nature', *Cerebrum*, 2011:8 (2011).

18. L. Jaroff, 'The Gene Hunt', *Time* (24 June 2001), http://content.time.com/time/magazine/article/0,9171,151430,00.html.

19. J. Crace, 'Double Helix Trouble', *The Guardian* (16 October 2007), https://www.theguardian.com/education/2007/oct/16/highereducation.research.

20. R. Lewontin and R. Levins, *Biology under the Influence: Dialectical Essays on Ecology, Agriculture, and Health* (Monthly Review Press, 2007), p. 253.

21. S. McLeod, 'Classical Conditioning', *Simply Psychology* (21 August 2018), https://www.simplypsychology.org/classical-conditioning.html.

22. J. Doward, 'Psychiatrists under Fire in Mental Health Battle', *The Guardian* (12 May 2013), https://www.theguardian.com/society/2013/may/12/psychiatrists-under-fire-mental-health.

23. H. Rose and S. Rose, *Genes, Cells and Brains* (Verso, 2012), p. 25.

24. T. Horwitz, K. Lam, Y. Chen, Y. Xia, and C. Liu, 'A Decade in Psychiatric GWAS Research', *Molecular Psychiatry*, 24:378–89 (2018).

25. Ibid.

26. B. J. Deacon, 'The Biomedical Model of Mental Disorder: A Critical Analysis of its Validity, Utility, and Effects on Psychotherapy Research', *Clinical Psychology Reviews*, 33:846–61 (2013).

27. C. Smyth, 'Drugs Remain an Enduring Medical Mystery', *The Times* (21 July 2018), https://www.thetimes.co.uk/article/drugs-remain-an-enduring-medical-mystery-b92rz3lmk.

28. J. Swanson, 'Unraveling the Mystery of How Antidepression Drugs Work', *Scientific American* (10 December 2013), https://www.scientificamerican.com/article/unraveling-the-mystery-of-ssris-depression/.

29. S. McLeod, 'Pavlov's Dogs', *Simply Psychology* (2018), https://www.simplypsychology.org/pavlov.html.

30. M. Meacham, 'Why the Brain Is Still a "'Black Box" and What to Do About It', *Association for Talent Development* (4 August 2017), https://www.td.org/insights/why-the-brain-is-still-a-black-box-and-what-to-do-about-it.

31. M. Atkisson, 'Behaviorism vs. Cognitivism', *Ways of Knowing* (12 October 2010), https://wok-nowing.wordpress.com/2010/10/12/behaviorism-vs-cognitivisim/.

32. K. A. Jamrog, 'Understanding Cognitive Behavioral Therapy', *New Hampshire Magazine* (20 July 2018), https://www.nhmagazine.com/August-2018/Understanding-Cognitive-Behavioral-Therapy/.

33. 'Expanding the Shrinks', *The Economist* (20 October 2014), https://www.economist.com/britain/2014/10/20/expanding-the-shrinks.

34. O. Burkeman, 'Why CBT is Falling out of Favour', *The Guardian* (3 July 2015), https://www.theguardian.com/lifeandstyle/2015/jul/03/why-cbt-is-falling-out-of-favour-oliver-burkeman.

35. P. McKenna and D. Kingdon, 'Has Cognitive Behavioural Therapy for Psychosis Been Oversold?', *British Medical Journal*, 348:g2295 (2014).

36. D. Campbell, 'Spike in Mental Health Patient Deaths Shows NHS "Struggling to Cope"', *The Guardian* (26 January 2016), https://www.theguardian.com/society/2016/jan/26/rise-mental-health-patient-deaths-nhs-struggling-to-cope.

37. Ibid.

38. Ibid.

39. J. Beattie, 'Cuts Blamed for "Appalling" Rise in Mental Health Deaths', *The Mirror* (26 January 2016), https://www.mirror.co.uk/news/uk-news/cuts-blamed-appalling-rise-mental-7248054.

40. K Griffiths, 'Depressed Bradford Woman Took Own Life in France', *The Telegraph and Argus* (10 August 2016), http://www.thetelegraphandargus.co.uk/news/14673774.Depressed_Bradford_woman_took_own_life_in_France/.

41. R. Spencer, 'America's Mental Health Care Crisis: Families Left to Fill the Void of a Broken System', *The Guardian* (27 May 2014), https://www.theguardian.com/world/2014/may/27/sp-americas-mental-health-care-system-crisis.

42. M. Maruthappu, R. Ologunde, and A. Gunarajasingam, 'Is Health Care a Right? Health Reforms in the USA and their Impact Upon the Concept of Care', *Annals of Medicine Surgery* 2:15–17 (2013).

43. Spencer, 'America's Mental Health Care Crisis'.

44. L. Szabo, 'Cost of Not Caring: Nowhere to Go', *USA Today* (12 January 2015), https://eu.usatoday.com/story/news/nation/2014/05/12/mental-health-system-crisis/7746535/.

45. L. Szabo, 'Psychiatric Beds Disappear Despite Growing Demand', *USA Today* (12 May 2014), https://eu.usatoday.com/story/news/nation/2014/05/12/disappearing-hospital-beds/9003677/.

46. D. Brown, 'Government Survey Finds that 5 Percent of Americans Suffer from a "Serious Mental Illness"', *Washingon Post* (19 January 2012), https://www.washingtonpost.com/national/health-science/government-survey-finds-that-5-percent-of-americans-suffer-from-a-serious-mental-illness/2012/01/18/gIQAjp5h9P_story.html?utm_term=.d3e60a4255d3.

47. J. Belluz, 'Anthony Bourdain's Death is One in a Growing Public Health Tragedy', *Vox* (8 June 2018), https://www.vox.com/science-and-health/2018/6/8/17441330/anthony-bourdain-suicide-rates-us-cdc.

48. K. Weir, 'Worrying Trends in U.S. Suicide Rates', *American Psychological Association* (March 2019), https://www.apa.org/monitor/2019/03/trends-suicide.

49. C. Koons, 'Latest Suicide Data Show the Depth of U.S. Mental Health Crisis', *Bloomberg* (20 June 2019), https://www.bloomberg.com/news/articles/2019-06-20/latest-suicide-data-show-the-depth-of-u-s-mental-health-crisis?fbclid=IwAR3wbuY3QJXOBm_gCS5fmGP_LdvHSk4Sps9pqTUviiWNUcdgqu2Xn24QeF8.

50. Ibid.

51. Royal College of Psychiatrists, Mental Health Network, NHS Confederation, and London School of Economics and Political Science, *Mental Health and the Economic Downturn* (2009), https://www.base-uk.org/sites/default/files/%5Buser-raw%5D/11-06/op70.pdf.

52. J. Wertsch, *Vygotsky and the Social Formation of Mind* (Harvard University Press, 1985), p. 186.

53. A. Kozulin, *Vygotsky's Psychology: A Biography of Ideas* (Harvester Wheatsheaf, 1990), p. 88.

54. Ibid.

55. Ibid.

56. P. Broks, 'The Mystery of Consciousness', *Prospect* (29 April 2007), https://www.prospectmagazine.co.uk/magazine/themysteryofconsciousness.

Chapter 1: Mind and Matter

1. C. Mesaroş, 'Aristotle and Animal Mind', *Procedia—Social and Behavioral Sciences*, 163:185–92 (2014).
2. D. Ribatti, 'William Harvey and the Discovery of the Circulation of the Blood', *Journal of Angiogenesis Research*, 1:3 (2009).
3. Ibid.
4. G. Gorham, 'Mind–Body Dualism and the Harvey–Descartes Controversy', *Journal of the History of Ideas*, 55:211–34 (1994).
5. J. Claes, 'Return of Descartes': the simulating mind, *The World as Computation* (23 May 2009), https://claesjohnsonmathscience.wordpress.com/article/return-of-descartes-yvfu3xg7d7wt-34/.
6. D. Cunning, 'Descartes on the Dubitability of the Existence of Self', *Philosophy and Phenomenological Research*, 74:111–31 (2007).
7. S. J. Wagner, 'Descartes's Arguments for Mind–Body Distinctness', *Philosophy and Phenomenological Research*, 43:499–517 (1983).
8. Ibid.
9. L. G. Schwoerer, 'Locke, Lockean Ideas, and the Glorious Revolution', *Journal of the History of Ideas*, 51:531–48 (1990).
10. J. Godfrey, 'John Locke', *Cathedral Blog, Christchurch, Oxford* (6 November 2019), https://www.chch.ox.ac.uk/blog/john-locke.
11. L. Jardine, 'Britain and the Rise of Science', *BBC History* (17 February 2011), http://www.bbc.co.uk/history/british/empire_seapower/jardineih_01.shtml.
12. C. Hewett, John Locke's Theory of Knowledge (*An Essay Concerning Human Understanding*), *The Great Debate*, (2006), http://www.thegreatdebate.org.uk/LockeEpistem.html.
13. M. S. Gazzaniga, *The Consciousness Instinct: Unraveling the Mystery of How the Brain Makes Mind* (Farrar, Straus and Giroux, 2018), p. 37.
14. Ibid., 40.
15. S. Ritchie, 'On Genetics Oliver James is on a Different Planet to the Rest of Us', *The Spectator* (8 March 2016), https://life.spectator.co.uk/articles/on-genetics-oliver-james-is-wrong-about-everything/.
16. S. McLeod, 'Cognitive Psychology', *Simply Psychology* (2020), https://www.simplypsychology.org/cognitive.html.
17. Ibid.
18. R. Dahm, 'Friedrich Miescher and the Discovery of DNA', *Developmental Biology*, 278:274–88 (2005).
19. M. Cobb, 'Sexism in Science: Did Watson and Crick Really Steal Rosalind Franklin's Data?', *The Guardian* (23 June 2015), https://www.theguardian.com/science/2015/jun/23/sexism-in-science-did-watson-and-crick-really-steal-rosalind-franklins-data.
20. L. Pray, 'Discovery of DNA Structure and Function: Watson and Crick', *Nature Education*, 1:100 (2008).
21. A. Ralston and K. Shaw, 'Reading the Genetic Code', *Nature Education*, 1:120 (2008).
22. T. Carvalho and T. Zhu, 'The Human Genome Project (1990–2003)', *The Embryo Project Encyclopedia* (6 May 2014), https://embryo.asu.edu/pages/human-genome-project-1990-2003.
23. 'What They Said: Genome in Quotes', *BBC* (26 June 2000), http://news.bbc.co.uk/1/hi/sci/tech/807126.stm.

24. I. Miko, 'Gregor Mendel and the Principles of Inheritance', *Nature Education*, 1:134 (2008).

25. R. A. Lewis, 'Genetic Checkup: Lessons from Huntington Disease and Cystic Fibrosis', *The Scientist* (19 October 2003), https://www.the-scientist.com/research/a-genetic-checkup-lessons-from-huntington-disease-and-cystic-fibrosis-50868.

26. M. Pagel, *Wired for Culture* (Penguin, 2012), p. 81.

27. H. Lacey, 'What Men and Women Do Best', *The Independent* (21 June 1998), https://www.independent.co.uk/life-style/what-men-and-women-do-best-1166324.html.

28. L. Hoopes, 'Introduction to the Gene Expression and Regulation Topic Room', *Nature Education*, 1:160 (2008).

29. F. Emmert-Streib, M. Dehmer, and B. Haibe-Kains, 'Gene Regulatory Networks and Their Applications: Understanding Biological and Medical Problems in Terms of Networks', *Frontiers in Cell and Developmental Biology*, 2:38 (2014).

30. B. Dennis, 'How Do You Tell a Kid He's Not Going to Grow Up?', *Washington Post* (6 May 2015), https://www.washingtonpost.com/news/to-your-health/wp/2015/05/06/how-do-you-tell-a-kid-hes-not-going-to-grow-up/?utm_term=.95ec512cf236.

31. P. Forbes, 'How Close Are We to a Cure for Huntington's?', *The Independent* (19 March 2018), https://www.independent.co.uk/news/long_reads/how-close-are-we-to-a-cure-for-huntington-s-a8235921.html.

32. I. Sample, 'Scientists Identify 40 Genes That Shed New Light on Biology of Intelligence', *The Guardian* (22 May 2017), https://www.theguardian.com/science/2017/may/22/scientists-uncover-40-genes-iq-einstein-genius.

33. N. Angier, 'Report Suggests Homosexuality Is Linked to Genes', *The New York Times* (16 July 1993), https://www.nytimes.com/1993/07/16/us/report-suggests-homosexuality-is-linked-to-genes.html.

34. K. O'Riordan, 'The Life of the Gay Gene: From Hypothetical Genetic Marker to Social Reality', *Journal of Sex Research*, 49:362–68 (2012).

35. S. Pinker, *How the Mind Works* (Norton, 2009).

36. S. Rose, '*How the Mind Works* by Steven Pinker', *New Scientist* (24 January 1998), https://www.newscientist.com/article/mg15721185-500-how-the-mind-works-by-steven-pinker/.

Chapter 2: Tool and Symbol

1. A. Kuhn, 'He Helped Discover Evolution, And Then Became Extinct (30 April 2013)', *NPR*, https://www.npr.org/2013/04/30/177781424/he-helped-discover-evolution-and-then-became-extinct?t=1532631979618.

2. D. Curnoe, 'When Humans Split from the Apes', *The Conversation* (22 February 2016), https://theconversation.com/when-humans-split-from-the-apes-55104.

3. K. Wong, 'Tiny Genetic Differences between Humans and Other Primates Pervade the Genome', *Scientific American* (1 September 2014), https://www.scientificamerican.com/article/tiny-genetic-differences-between-humans-and-other-primates-pervade-the-genome/.

4. J. N. Wilford, 'Almost Human, and Sometimes Smarter', *New York Times* (17 April 2007), https://www.nytimes.com/2007/04/17/science/17chimp.html.

5. J. Gabbatiss, 'Chimps and Orangutans among Species in Danger of Being Wiped out in Imminent Mass Extinction of Primates, Scientists Warn', *The Independent* (15 June 2018), https://www.independent.co.uk/environment/primates-mass-extinction-chimpanzees-gorillas-monkeys-scientist-warning-brazil-indonesia-a8400186.html.

6. 'Homo neanderthalensis', *Smithsonian National Museum of Natural History* (10 January 2020), http://humanorigins.si.edu/evidence/human-fossils/species/homo-neanderthalensis.

7. P. Madison, 'Neanderthals and Giant's Bones', *Fossil History* (30 August 2016), https://fossilhistory.wordpress.com/2016/08/30/neanderthals-and-giants/.

8. B. Trigger, 'Comment on Tobias, Piltdown, the Case against Keith', *Current Anthropology*, 33:274–5 (1992).

9. Ibid.

10. F. Engels, 'The Part Played by Labour in the Transition from Ape to Man', *Die Neue Zeit* (1895), https://www.marxists.org/archive/marx/works/1876/part-played-labour/.

11. Ibid.

12. R. Gray, 'The Mystery of Piltdown Man Is Solved: Charles Dawson, Who Found the Fake Human "Fossil", Was Probably behind the Hoax', *Daily Mail* (12 August 2016), http://www.dailymail.co.uk/sciencetech/article-3731455/The-mystery-Piltdown-Man-solved-Charles-Dawson-fake-human-fossil-probably-hoax.html.

13. Ibid.

14. C. K. Brain, 'The Day the Missing Link Turned up in the Post', *The Guardian* (18 December 2003), https://www.theguardian.com/science/2003/dec/18/science.research2.

15. K. Wong, 'Human Origins', *Scientific American* (24 November 2014), https://blogs.scientificamerican.com/observations/40-years-after-lucy-the-fossil-that-revolutionized-the-search-for-human-origins/.

16. J. Hawks, 'Human Evolution Is More a Muddy Delta Than a Branching Tree', *Aeon* (8 February 2016), https://aeon.co/ideas/human-evolution-is-more-a-muddy-delta-than-a-branching-tree.

17. C. Stringer, 'Human Evolution and Biological Adaptation in the Pleistocene', in R.A. Foley (ed.), *Hominid Evolution and Community Ecology* (Academic Press, 1984), p. 53.

18. M. Holloway, 'Mary Leakey: Unearthing History', *Scientific American* (16 December 1996), https://www.scientificamerican.com/article/mary-leakey-unearthing-hi/.

19. Z. Zorich, 'Which Came First, Humans or Tools?', *The New Yorker* (20 May 2015), https://www.newyorker.com/tech/elements/which-came-first-humans-or-tools.

20. C. Q. Choi, 'Human Evolution: The Origin of Tool Use', *Live Science* (11 November 2009), https://www.livescience.com/7968-human-evolution-origin-tool.html.

21. Ibid.

22. Ibid.

23. Ibid.

24. Ibid.

25. C. Harman, 'Engels and the Origins of Human Society', *International Socialism*, 2:65 (Winter 1994) http://pubs.socialistreviewindex.org.uk/isj65/harman.htm.

26. Ibid.

27. Y. Anwar, 'World's Oldest Butchery Tools Gave Evolutionary Edge to Human Communication', *Berkeley News* (13 January 2015), http://news.berkeley.edu/2015/01/13/stone-age-tools/.

28. Ibid.

29. M. Balter, 'Striking Patterns: Study Suggests Tool Use and Language Evolved Together', *Wired* (3 September 2013), https://www.wired.com/2013/09/tools-and-language/.

30. Ibid.

31. D. Guess, W. Sailor, G. Rutherford, and D. M. Baer, 'An Experimental Analysis of Linguistic Development: The Productive Use of the Plural Morpheme', *Journal of Applied Behavior Analysis*, 1:297–306 (1968).

32. B. Carey, 'Washoe, a Chimp of Many Words, Dies at 42', *New York Times* (1 November 2007), https://www.nytimes.com/2007/11/01/science/01chimp.html.

33. Ibid.

34. E. Wayman, 'Six Talking Apes', *Smithsonian Magazine* (11 August 2011) https://www.smithsonianmag.com/science-nature/six-talking-apes-48085302/.

35. S. Keegan, 'Gorilla Befriended by Robin Williams "Cried in Mourning" When She Learnt of Star's Death', *The Mirror* (13 August 2014), https://www.mirror.co.uk/3am/celebrity-news/gorilla-befriended-robin-williams-cried-4043840.

36. D. C. Palmer, 'On Chomsky's Appraisal of Skinner's Verbal Behavior: A Half Century of Misunderstanding', *The Behavior Analyst*, 29:253–67 (2006).

37. Ibid.

38. H. Terrace, 'Can Chimps Converse?: An Exchange', *New York Review of Books* (24 November 2011), https://www.nybooks.com/articles/2011/11/24/can-chimps-converse-exchange/.

39. J. C. Hu, 'The Troubling World of Koko the Gorilla and the Decline of Ape Language Research', *National Post* (21 August 2014), https://nationalpost.com/news/the-troubling-world-of-koko-the-gorilla-and-the-decline-of-ape-language-research.

40. Ibid.

41. Ibid.

42. Ibid.

43. V. Gill, 'Chimpanzee Language: Communication Gestures Translated', *BBC News* (4 July 2014), https://www.bbc.com/news/science-environment-28023630.

44. Ibid.

45. Ibid.

46. F. Y. Lin, 'A Refutation of Universal Grammar', *Lingua*, 193: 1–22 (2017).

Chapter 3: Individual and Society

1. M. Riddle, 'Survival of the Nicest: How Altruism Made Us Human & Why It Pays to Get Along', *Psych Central* (17 May 2016), https://psychcentral.com/lib/survival-of-the-nicest-how-altruism-made-us-human-why-it-pays-to-get-along/.

2. R. Christiansen, 'The Story behind the Hymn', *The Telegraph* (22 September 2007), https://www.telegraph.co.uk/culture/music/3668059/The-story-behind-the-hymn.html.

3. C. Miceli, 'Descartes' "I Think, Therefore I am"', *1000-Word Philosophy* (26 November 2018), https://1000wordphilosophy.com/2018/11/26/descartes-i-think-therefore-i-am/.

4. A. Deneys-Tunney, 'Rousseau Shows Us That There is a Way to Break the Chains—from Within', *The Guardian* (15 July 2012), https://www.theguardian.com/commentisfree/2012/jul/15/rousseau-shows-us-way-break-chains.

5. C. Moore, 'No Such Thing as Society: A Good Time to Ask What Margaret Thatcher Really Meant', *The Telegraph* (27 September 2010), https://www.telegraph.co.uk/comment/columnists/charlesmoore/8027552/No-Such-Thing-as-Society-a-good-time-to-ask-what-Margaret-Thatcher-really-meant.html.

6. R. M. Henig, 'Linked In', *New York Times* (1 November 2013), https://www.nytimes.com/2013/11/03/books/review/social-by-matthew-d-lieberman.html.

7. Ibid.

8. J. Moll, F. Krueger, R. Zahn, M. Pardini, R. de Oliveira-Souza, and J. Grafman, 'Human Fronto-mesolimbic Networks Guide Decisions about Charitable Donation', *Proceedings of the National Academy of Sciences USA*, 103:15623–8 (2006).

9. S. Wolpert, "'Tis better to give than to receive?', *UCLA Newsroom* (9 November 2011), https://www.uclahealth.org/tis-better-to-give-than-to-receive.

10. Ibid.

11. E. Esfahani Smith, 'Social Connection Makes a Better Brain', *The Atlantic* (29 October 2013), https://www.theatlantic.com/health/archive/2013/10/social-connection-makes-a-better-brain/280934/.

12. A. Woodward, 'With a Little Help from My Friends', *Scientific American* (1 May 2017), https://www.scientificamerican.com/article/with-a-little-help-from-my-friends/.

13. S. Toulmin, 'The Mozart of Psychology', *New York Review of Books* (28 September 1978), https://www.nybooks.com/articles/1978/09/28/the-mozart-of-psychology/.

14. J. Reader, 'What Made the Animal Human', *New Scientist*, 29 July:314 (1982).

15. L. S. Vygotsky, *Mind in Society: The Development of Higher Psychological Processes* (Harvard University Press, 1976), p. 127.

16. S. McLeod, 'Jean Piaget's Theory of Cognitive Development', *Simply Psychology* (2018), https://www.simplypsychology.org/piaget.html.

17. Ibid.

18. I. Sample, 'Comparing Piaget and Vygotsky', *University of Iowa* (28 October 2004), https://www2.education.uiowa.edu/html/eportfolio/tep/07p075folder/Piaget_Vygotsky.htm.

19. S. McLeod, 'Lev Vygotsky's Sociocultural Theory', *Simply Psychology* (2020), https://www.simplypsychology.org/vygotsky.html.

20. Sample, 'Comparing Piaget and Vygotsky'.

21. Ibid.

22. R. van der Veer and J. Valsiner, *Understanding Vygotsky: A Quest for Synthesis* (Blackwell, 1991), pp. 262–6.

23. J. Parrington, 'In Perspective: Valentin Voloshinov', *International Socialism*, 2:75 (July 1997), https://www.marxists.org/history/etol/newspape/isj2/1997/isj2-075/parrington.htm.

24. A. Kozulin, *Vygotsky's Psychology: A Biography of Ideas* (Harvester Wheatsheaf, 1990), pp. 158–9.

25. Ibid.

26. Van der Veer and Valsiner, *Understanding Vygotsky*.

27. A. Blunden, *Concepts: A Critical Approach* (Brill, 2012), p. 242.

28. Ibid.

29. Wertsch, J. V., *Vygotsky and the Social Formation of Mind* (Harvard University Press, 1985), p. 103.

30. Ibid.

31. F. Newman and L. Holzman, *Lev Vygotsky: Revolutionary Scientist* (Routledge, 1993), pp. 80–1.

32. Ibid.

33. Ibid.

34. M. Holborow, *The Politics of English: A Marxist View of Language* (Sage Publications, 1999), p. 24.

35. C. Brandist, in C. Brandist, D. Sheperherd, and G. Tihanov (eds), *The Bakhtin Circle* (Manchester University Press, 2004), p. 103.

36. V. N. Voloshinov, *Freudianism: A Critical Sketch* (Indiana University Press, 1976), p. 21.

37. V. N. Voloshinov, *Marxism and the Philosophy of Language* (Harvard University Press, 1986), p. 19.

38. Ibid., 13.

39. Ibid., 19.

40. Ibid., 81.

41. Ibid., 102.

42. J. V. Wertsch, *Voices of the Mind: A Sociocultural Approach to Mediated Action* (Harvester Wheatsheaf, 1991), pp. 53–6.

43. M. Epstein, 'Hyperauthorship in Mikhail Bakhtin: The Primary Author and Conceptual Personae', *Russian Journal of Communication* 1:280–90 (2008).

44. M. M. Bakhtin, *The Dialogic Imagination: Four Essays by M.M. Bakhtin* (University of Texas Press, 1981), p. 262.

45. M. M. Bakhtin, *Speech Genres and Other Late Essays* (Texas University Press, 1986), p. 78.

46. Ibid., 87.

47. C. L. Briggs and R. Bauman, 'Genre, Intertextuality, and Social Power', *Journal of Linguistic Anthropology*, 2:131–72 (1992).

48. Bakhtin, *The Dialogic Imagination*, pp. 293–4.

49. A. Larrain and A. Haye, 'The Discursive Nature of Inner Speech', *Theory and Psychology*, 22:3–22 (2012).

50. Ibid.

51. J. F. Ehrich, 'Vygotskian Inner Speech and the Reading Process', *Australian Journal of Educational and Developmental Psychology*, 6:12–25 (2006).

52. B. Alderson-Day and C. Fernyhough, 'Inner Speech: Development, Cognitive Functions, Phenomenology, and Neurobiology', *Psychological Bulletin*, 141: 931–65 (2015).

53. L. S. Vygotsky, *Thought and Language* (Massachusetts Institute of Technology, 1986), pp. 235–49.

54. J. Beck, 'The Running Conversation in Your Head', *The Atlantic* (23 November 2016), https://www.theatlantic.com/science/archive/2016/11/figuring-out-how-and-why-we-talk-to-ourselves/508487/.

55. Alderson-Day and Fernyhough, 'Inner Speech: Development'.

56. Beck, 'The Running Conversation in Your Head'.

57. A. Dana, R. R. Shirazi, and F. Z. Jalili, 'The Effect of Instruction and Motivational Self Talk on Performance and Retention of Discrete and Continuous Motor Tasks', *Australian Journal of Basic and Applied Sciences*, 5, 312–15 (2011).

58. S. Vickers, '*The Voices Within: The History and Science of How We Talk to Ourselves* by Charles Fernyhough—review', *The Guardian* (24 April 2016), https://www.theguardian.com/books/2016/apr/24/voices-within-history-science-how-we-talk-to-ourselves-charles-fernyhough-review.

59. Beck, 'The Running Conversation in Your Head'.

60. C. Fernyhough, 'Do Deaf People Hear an Inner Voice?', *Psychology Today* (24 January 2014), https://www.psychologytoday.com/intl/blog/the-voices-within/201401/do-deaf-people-hear-inner-voice?amp.

61. Ibid.

Chapter 4: Nerve and Brain

1. T. Newman, 'All You Need to Know about Neurons', *Medical News Today* (7 December 2017), https://www.medicalnewstoday.com/articles/320289.php.

2. F. Bezanilla, 'Ion Channels: from Conductance to Structure', *Neuron*, 60:456–68 (2008).

3. F. Jabr, 'Know Your Neurons: The Discovery and Naming of the Neuron', *Scientific American* (14 May 2012), https://blogs.scientificamerican.com/brainwaves/know-your-neurons-the-discovery-and-naming-of-the-neuron/.

4. J. DeFelipe, 'Cajal's Butterflies of the Soul', *OUPblog* (9 November 2013), https://blog.oup.com/2013/11/cajal-butterflies-of-the-soul-cerebral-cortex/.

5. K. Susuki, 'Myelin: A Specialized Membrane for Cell Communication', *Nature Education*, 3:59 (2010).

6. C. J. Schwiening, 'A Brief Historical Perspective: Hodgkin and Huxley', *Journal of Physiology*, 590:2571–5 (2012).

7. B. P. Bean, 'The Action Potential in Mammalian Central Neurons', *Nature Reviews in Neuroscience*, 8:451–65 (2007).

8. F. Jabr, 'Know Your Neurons: How to Classify Different Types of Neurons in the Brain's Forest', *Scientific American* (16 May 2012) https://blogs.scientificamerican.com/brainwaves/know-your-neurons-classifying-the-many-types-of-cells-in-the-neuron-forest/.

9. Ibid.

10. S. Acharya and S. Shukla, 'Mirror Neurons: Enigma of the Metaphysical Modular Brain', *Journal of Natural Science, Biology and Medicine*, 3:118–24 (2012).

11. Ibid.

12. M. Iacoboni, 'The Mirror Neuron Revolution: Explaining What Makes Humans Social', *Scientific American* (1 July 2008), https://www.scientificamerican.com/article/the-mirror-neuron-revolut/.

13. M. N. Rasband, 'Glial Contributions to Neural Function and Disease', *Molecular and Cellular Proteomics*, 15:355–61 (2016).

14. I. Sample and J. Hooper, 'Brain Cells Clue to Genius of Einstein', *The Guardian* (21 February 2007), https://www.theguardian.com/science/2007/feb/21/neuroscience.highereducation.

15. Rasband, 'Glial Contributions to Neural Function and Disease'.

16. W. S. Chung, N. J. Allen, and C. Eroglu, 'Astrocytes Control Synapse Formation, Function, and Elimination', *Cold Spring Harbor Perspectives in Biology*, 7:a020370 (2015).

17. K. Reemst, S. C. Noctor, P. J. Lucassen, and E. M. Hol, 'The Indispensable Roles of Microglia and Astrocytes during Brain Development', *Frontiers in Human Neuroscience*, 10:566 (2016).

18. Q. Li and B. A. Barres, 'Microglia and Macrophages in Brain Homeostasis and Disease', *Nature Reviews in Immunology*, 18:225–42 (2018).

19. Reemst, Noctor, Lucassen, and Hol, 'The Indispensable Roles of Microglia'.

20. R. Kwon, 'Rise of the Microglia', *Scientific American* (23 October 2015), https://www.scientificamerican.com/article/rise-of-the-microglia/.

21. Ibid.

22. Ibid.

23. M. Costandi, 'Human Brain Cells Boost Mouse Memory', *Science* (7 March 2013), http://www.sciencemag.org/news/2013/03/human-brain-cells-boost-mouse-memory.

24. Ibid.

25. C. Edmonson, M. N. Ziats, and O. M. Rennert, 'Altered Glial Marker Expression in Autistic Post-mortem Prefrontal Cortex and Cerebellum', *Molecular Autism*, 5:3 (2014).

26. M. W. Salter and B. Stevens, 'Microglia Emerge as Central Players in Brain Disease', *Nature Medicine*, 23:1018–27 (2017).

27. R. Bailey, 'Divisions of the Brain', *ThoughtCo.* (15 November 2018), https://www.thoughtco.com/divisions-of-the-brain-4032899.

28. J. Defelipe, 'The Evolution of the Brain, the Human Nature of Cortical Circuits, and Intellectual Creativity', *Frontiers in Neuroanatomy*, 5:29 (2011).

29. R. Bailey, 'The Limbic System of the Brain', *ThoughtCo.* (28 March 2018), https://www.thoughtco.com/limbic-system-anatomy-373200.

30. Bailey, 'Divisions of the Brain'.

31. Ibid.

32. Ibid.

33. L. Sanders, 'Brain Waves May Focus Attention and Keep Information Flowing', *Science News* (13 March 2018), https://www.sciencenews.org/article/brain-waves-may-focus-attention-and-keep-information-flowing.
34. Ibid.
35. Ibid.
36. Ibid.
37. M. A. Hofman, 'Evolution of the Human Brain: When Bigger is Better', *Frontiers in Neuroanatomy*, 8:15 (2014).
38. A. Hawks, 'How Has the Human Brain Evolved?', *Scientific American* (1 July 2013), https://www.scientificamerican.com/article/how-has-human-brain-evolved/.
39. C. C. Sherwood, F. Subiaul, and T. W. Zawidzki, 'A Natural History of the Human Mind: Tracing Evolutionary Changes in Brain and Cognition', *Journal of Anatomy*, 212:426–54 (2008).
40. Ibid.
41. Ibid.
42. Hawks, 'How Has the Human Brain Evolved?'
43. Sherwood, Subiaul, and Zawidzki, 'A Natural History of the Human Mind'.
44. Ibid.
45. A. Wnuk, 'Brain Evolution: Searching for What Makes Us Human', *Brain Facts* (9 April 2015), https://www.brainfacts.org/Brain-Anatomy-and-Function/Evolution/2015/Brain-Evolution-Searching-for-What-Makes-Us-Human.
46. J. Stromberg, 'Science Shows Why You're Smarter Than a Neanderthal', *Smithsonian Magazine* (12 March 2013), https://www.smithsonianmag.com/science-nature/science-shows-why-youre-smarter-than-a-neanderthal-1885827/.
47. Ibid.
48. C. Q. Choi, 'Brain Scans Reveal Difference between Neanderthals and Us', *Live Science* (8 November 2010), https://www.livescience.com/11078-brain-scans-reveal-difference-neanderthals.html.
49. S. Jacob, 'The Brains of Neanderthals and Modern Humans Developed Differently', *Max Planck Institute* (8 November 2010), https://www.mpg.de/623578/pressRelease201011021.
50. Stromberg, 'Science Shows Why You're Smarter Than a Neanderthal'.
51. J. L. Smit, 'Top 10 Remarkable Traits Neanderthals Have In Common with Modern Humans', *List Verse* (20 April 2017), https://listverse.com/2017/04/20/top-10-remarkable-traits-neanderthals-have-in-common-with-humans/.
52. Wnuk, 'Brain Evolution: Searching for What Makes Us Human'.
53. Ibid.
54. H. Glowacka, 'Babies, Birth and Brains', *Ask An Anthropologist* (2018), https://askananthropologist.asu.edu/stories/babies-birth-and-brains.
55. Wnuk, 'Brain Evolution: Searching for What Makes Us Human'.
56. Ibid.
57. T. Newman, 'White Matter: The Brain's Flexible But Underrated Superhighway', *Medical News Today* (16 August 2017), https://www.medicalnewstoday.com/articles/318966.php.
58. Ibid.
59. R. A. Hoglund and A. A. Maghazachi, 'Multiple Sclerosis and the Role of Immune Cells', *World Journal of Experimental Medicine*, 4:27–37 (2014).
60. N. Muhlert, V. Sethi, L. Cipolotti, H. Haroon, G. J. Parker, T. Yousry, C. Wheeler-Kingshott, D. Miller, M. Ron, and D. Chard, 'The Grey Matter Correlates of Impaired Decision-Making in Multiple Sclerosis', *Journal of Neurology, Neurosurgery and Psychiatry*, 86:530–6 (2015).

61. Y. Hakak, J. R. Walker, C. Li, W. H. Wong, K. L. Davis, J. D. Buxbaum, V. Haroutunian, and A. A. Fienberg, 'Genome-Wide Expression Analysis Reveals Dysregulation of Myelination-Related Genes in Chronic Schizophrenia', *Proceedings of the National Academy of Sciences USA*, 98:4746–51 (2001).

Chapter 5: Genome and Epigenome

1. N. Comfort, 'Genetics: We are the 98%', *Nature*, 520:615–16 (2015).
2. K. V. Morris and J. S. Mattick, 'The Rise of Regulatory RNA', *Nature Reviews in Genetics*, 15:423–37 (2014).
3. J. H. Gibcus and J. Dekker, 'The Hierarchy of the 3D Genome', *Molecular Cell*, 49:773–82 (2013).
4. C. Deans and K. A. Maggert, 'What Do You Mean, "Epigenetic"?', *Genetics*, 199:887–96 (2015).
5. E. V. Koonin and A. S. Novozhilov, 'Origin and Evolution of the Genetic Code: The Universal Enigma', *IUBMB Life*, 61:99–111 (2009).
6. S. A. Lambert, A. Jolma, L. F. Campitelli, P. K. Das, Y. Yin, M. Albu, X. Chen, J. Taipale, T. R. Hughes, and M. T. Weirauch, 'The Human Transcription Factors', *Cell*, 172:650–65 (2018).
7. I. Vorechovsky and A. R. Collins, 'How Much "Junk" is in our DNA?', *The Conversation* (22 February 2016), https://theconversation.com/how-much-junk-is-in-our-dna-53929.
8. D. Greenbaum and M. Gerstein, 'Illuminating the Genome's Dark Matter', *Cell*, 163:1047–8 (2015).
9. J. Perkel, 'Making Sense of our Variation', *Biotechniques*, 59:262–7 (2015).
10. J. Dekker and L. Mirny, 'The 3D Genome as Moderator of Chromosomal Communication', *Cell*, 164:1110–21 (2016).
11. S. Quinodoz and M. Guttman, 'Long Noncoding RNAs: An Emerging Link between Gene Regulation and Nuclear Organization', *Trends in Cell Biology*, 24:651–63 (2014).
12. X. Hua and L. Bromham, 'Darwinism for the Genomic Age: Connecting Mutation to Diversification', *Frontiers in Genetics*, 8:12 (2017).
13. K. A. Janssen, S. Sidoli, and B. A. Garcia, 'Recent Achievements in Characterizing the Histone Code and Approaches to Integrating Epigenomics and Systems Biology', *Methods in Enzymology*, 586:359–78 (2017).
14. Ibid.
15. Ibid.
16. I. Donkin and R. Barres, 'Sperm Epigenetics and Influence of Environmental Factors', *Molecular Metabolism*, 14:1–11 (2018).
17. E. B. Chuong, N. C. Elde, and C. Feschotte, 'Regulatory Activities of Transposable Elements: From Conflicts to Benefits', *Nature Reviews in Genetics*, 18:71–86 (2017).
18. J. Henson, G. Tischler, and Z. Ning, 'Next-Generation Sequencing and Large Genome Assemblies', *Pharmacogenomics*, 13:901–15 (2012).
19. Z. Dong and Y. Chen, 'Transcriptomics: Advances and Approaches', *Science China. Life Sciences*, 56:960–7 (2013).
20. K. Wong, 'Tiny Genetic Differences between Humans and Other Primates Pervade the Genome', *Scientific American* (1 September 2014), https://www.scientificamerican.com/article/tiny-genetic-differences-between-humans-and-other-primates-pervade-the-genome/.

21. T. Ghose, '"Big Brain" Gene Found in Humans, Not Chimps', *Live Science* (26 February 2015), https://www.livescience.com/49960-human-big-brain-gene-found.html.
22. Ibid.
23. Ibid.
24. K. Boes, 'Researchers Discover Human-Specific Gene for Building a Bigger Brain', *Sci Tech Daily* (17 January 2019), https://scitechdaily.com/researchers-discover-human-specific-gene-for-building-a-bigger-brain/.
25. R. Feltman, 'Scientists Pinpoint a Gene Regulator That Makes Human Brains Bigger', *Washington Post* (19 February 2015), https://www.washingtonpost.com/news/speaking-of-science/wp/2015/02/19/scientists-pinpoint-a-gene-regulator-that-makes-human-brains-bigger/?utm_term=.e8c44bbc577b.
26. Ibid.
27. K. Miller, 'How Our Ancient Brains Are Coping in the Age of Digital Distraction', *Discover Magazine* (20 April 2020), https://www.discovermagazine.com/mind/how-our-ancient-brains-are-coping-in-the-age-of-digital-distraction?fbclid=IwAR2lzqxOphb8UOwOr14NdZHAY_JdWEBzNBrTe7ozpTKiYUMoznCHqVbu5Mo.
28. D. Shultz, 'Humans Can Outlearn Chimps Thanks to More Flexible Brain Genetics', *Science* (16 November 2015), http://www.sciencemag.org/news/2015/11/humans-can-outlearn-chimps-thanks-more-flexible-brain-genetics.
29. Ibid.
30. Ibid.
31. 'Lifelong Learning is Made Possible by Recycling of Histones, Study Says', The *Rockefeller University* (1 July 2015), https://www.rockefeller.edu/news/9722-lifelong-learning-is-made-possible-by-recycling-of-histones-study-says/.
32. Ibid.
33. 'Transposons Stir in Germline, but Small RNAs Still Them', *GEN News* (3 November 2017), https://www.genengnews.com/gen-news-highlights/transposons-stir-in-germline-but-small-rnas-still-them/81255130.
34. K. Clancy, 'The Strangers in Your Brain', *The New Yorker* (17 October 2015), https://www.newyorker.com/tech/elements/the-strangers-in-your-brain.
35. Ibid.
36. Salk Institute, 'Scientists Discover Protein Factories Hidden in Human Jumping Genes', *Science Daily* (22 October 2015), https://www.sciencedaily.com/releases/2015/10/151022124518.htm.
37. Ibid.
38. S. Makin, 'Scientists Surprised to Find No Two Neurons Are Genetically Alike', *Scientific American* (3 May 2017), https://www.scientificamerican.com/article/scientists-surprised-to-find-no-two-neurons-are-genetically-alike/.
39. Cell Press, '"Brain-Only" Mutation Causes Epileptic Brain Size Disorder', *Science Daily* (11 April 2012), https://www.sciencedaily.com/releases/2012/04/120411132055.htm.
40. Makin, 'No Two Neurons Are Genetically Alike'.
41. 'Exosomes: The FedEx of the Nervous System?', *Alzforum* (12 December 2014), https://www.alzforum.org/news/conference-coverage/exosomes-fedex-nervous-system.
42. Ibid.
43. Ibid.
44. Ibid.

Chapter 6: Growth and Development

1. M. Caldwell, 'How Does a Single Cell Become a Whole Body?', *Discover Magazine* (1 November 1992), http://discovermagazine.com/1992/nov/howdoesasinglece146.

2. L. Wolpert and C. Vicente, 'An Interview with Lewis Wolpert', *Development*, 142:2547–8 (2015).

3. T. Gest, 'Nervous System', *University of Michigan Medical School* (14 August 2006), https://www.med.umich.edu/lrc/coursepages/m1/embryology/embryo/08nervoussystem.htm.

4. J. Lieff, 'Emotional Needs in Teens May Spur the Growth of New Brain Cells', *Scientific American* (26 March 2013), https://blogs.scientificamerican.com/mind-guest-blog/emotional-needs-in-teens-may-spur-the-growth-of-new-brain-cells/.

5. S. M. Pollard, A. Benchoua, and S. Lowell, 'Neural Stem Cells, Neurons, and Glia', *Methods in Enzymology*, 418:151–69 (2006).

6. Ibid.

7. S. Kean, 'The Cat Nobel Prize Part I', *Psychology Today* (14 April 2014), https://www.psychologytoday.com/us/blog/brain-food/201404/the-cat-nobel-prize-part-i.

8. Ibid.

9. Ibid.

10. Ibid.

11. L. Y. Ku, 'Cats and Vision: Is Vision Acquired or Innate?', *The Serious Computer Vision Blog* (1 June 2013), https://computervisionblog.wordpress.com/2013/06/01/cats-and-vision-is-vision-acquired-or-innate/.

12. Ibid.

13. J. Gallagher, 'Prof Colin Blakemore: Medical Research Defender Knighted', *BBC News* (13 June 2014), https://www.bbc.com/news/health-27834406.

14. 'Grand Challenge: Nature versus Nurture: How Does the Interplay of Biology and Experience Shape Our Brains and Make Us Who We Are?', *National Center for Biotechnology Information* (2008), https://www.ncbi.nlm.nih.gov/books/NBK50991/.

15. Ibid.

16. K. Eckart-Washington, 'Brain Differences in Blind People May Sharpen Hearing', *Futurity* (6 May 2019), https://www.futurity.org/blind-people-hearing-brains-2054422-2/.

17. N. Popovich, 'How the Deaf Brain Rewires Itself to "Hear" Touch and Sight', *The Atlantic* (11 July 2012), https://www.theatlantic.com/health/archive/2012/07/how-the-deaf-brain-rewires-itself-to-hear-touch-and-sight/259681/.

18. P. Schlegel, M. Costa, and G. S. Jefferis, 'Learning from Connectomics on the Fly', *Current Opinion in Insect Science*, 24:96–105 (2017).

19. D. Lee, T. H. Huang, A. De La Cruz, A. Callejas, and C. Lois, 'Methods to Investigate the Structure and Connectivity of the Nervous System', *Fly (Austin)*, 11:224–38 (2017).

20. S. R. Olsen and R.I. Wilson, 'Cracking Neural Circuits in a Tiny Brain: New Approaches for Understanding the Neural Circuitry of Drosophila', *Trends in Neuroscience*, 31:512–20 (2008).

21. M. Drozd, B. Bardoni, and M. Capovilla, 'Modeling Fragile X Syndrome in Drosophila', *Frontiers in Molecular Neuroscience*, 11:124 (2018).

22. Ibid.

23. K. J. Venken, J. H. Simpson, and H. J. Bellen, 'Genetic Manipulation of Genes and Cells in the Nervous System of the Fruit Fly', *Neuron*, 72:202–30 (2011).

24. A. M. Stewart, J. F. Ullmann, W. H. Norton, M. O. Parker, C. H. Brennan, R. Gerlai, and A. V. Kalueff, 'Molecular Psychiatry of Zebrafish', *Molecular Psychiatry*, 20:2–17 (2015).

25. Ibid.

26. M. Costandi, 'An Activity Map of the Whole Zebrafish Brain', *The Guardian* (18 March 2013), https://www.theguardian.com/science/neurophilosophy/2013/mar/18/an-activity-map-of-the-whole-zebrafish-brain.

27. Y. Tsuruwaka, E. Shimada, K. Tsutsui, and T. Ogawa, 'Ca(2+) Dynamics in Zebrafish Morphogenesis', *Peer Journal*, 5:e2894 (2017).

28. Ibid.

29. J. Liu, Y. Zhou, X. Qi, J. Chen, W. Chen, G. Qiu, Z. Wu, and N. Wu, 'CRISPR/Cas9 in Zebrafish: An Efficient Combination for Human Genetic Diseases Modeling', *Human Genetics*, 136:1–12 (2017).

30. A. Doyle, M. P. McGarry, N. A. Lee, and J. J. Lee, 'The Construction of Transgenic and Gene Knockout/Knockin Mouse Models of Human Disease', *Transgenic Research*, 21:327–49 (2012).

31. A. C. Komor, A. H. Badran, and D. R. Liu, 'CRISPR-Based Technologies for the Manipulation of Eukaryotic Genomes', *Cell*, 168:20–36 (2017).

32. J. Wright, 'Monkey Model Reveals New Role for Top Autism Gene', *Spectrum* (11 September 2017), https://www.spectrumnews.org/news/monkey-model-reveals-new-role-top-autism-gene/.

33. Ibid.

34. I. Johnstone, 'Artificial "Embryo" Created for First Time in Historic Breakthrough', *The Independent* (2 March 2017), https://www.independent.co.uk/news/science/embryo-lab-creation-scientists-life-cells-lab-breakthrough-a7608446.html.

35. Ibid.

36. I. Hyun, A. Wilkerson, and J. Johnston, 'Embryology Policy: Revisit the 14-Day Rule', *Nature*, 533:169–71 (2016).

37. Ibid.

38. S. Gregory, 'Human "Mini-brains" in Lab Mimic Development and Disease', *Bio News* (2 May 2017), https://www.bionews.org.uk/page_95978.

39. S. Fan, 'Bizarre Mini Brains Offer a Fascinating New Look at the Brain', *Singularity Hub* (16 May 2017), https://singularityhub.com/2017/05/16/bizarre-mini-brains-offer-a-fascinating-new-look-at-the-brain/.

40. B. Bae and C. A. Walsh, 'What Are Mini-Brains?', *Science*, 342:200–1 (2013).

41. J. Rennie, 'Mini-Brains Go Modular', *Quanta Magazine* (9 August 2017), https://www.quantamagazine.org/mini-brains-go-modular-20170809/.

42. Ibid.

43. Ibid.

44. J. Cepelewicz, 'The Oldest Mini-Brains Have Lifelike Young Cells', *Quanta Magazine* (29 August 2017), https://www.quantamagazine.org/the-oldest-mini-brains-have-lifelike-young-cells-20170829/.

45. Ibid.

46. S. Kane, 'Giant Heads Nearly Killed our Ancestors but Human Immaturity Saved us', *Business Insider* (11 April 2016), https://www.businessinsider.com/human-head-hips-evolution-2016-4?IR=T.

47. C. Barras, 'The Real Reasons Why Childbirth is So Painful and Dangerous', *BBC* (22 December 2016), http://www.bbc.com/earth/story/20161221-the-real-reasons-why-childbirth-is-so-painful-and-dangerous (2016).

48. B. Hrvoj-Mihic, T. Bienvenu, L. Stefanacci, A. R. Muotri, and K. Semendeferi, 'Evolution, Development, and Plasticity of the Human Brain: From Molecules to Bones', *Frontiers in Human Neuroscience*, 7:707 (2013).

49. L. S. Vygotsky, in *The Collected Works of L. S. Vygotsky*, Vol. 4, ed R. W. Rieber (Plenum Press, 1997), p. 223.

50. Ibid., 106.

51. L. S. Vygotsky, *Mind in Society: The Development of Higher Psychological Processes* (Harvard University Press, 1978), p. 39.

52. Ibid.

53. A. Kozulin, *Vygotsky's Psychology: A Biography of Ideas* (Harvester Wheatsheaf, 1990), pp. 138–9.

54. Ibid.

Chapter 7: Learning and Memory

1. N. J. T. Thomas, 'Plato and his Predecessors', *Stanford Encyclopedia of Philosophy* (2014), https://plato.stanford.edu/entries/mental-imagery/plato-predecessors.html.

2. R. Allen, 'David Hartley', *Stanford Encyclopedia of Philosophy* (2017), https://plato.stanford.edu/entries/hartley/.

3. S. A. Josselyn, S. Kohler, and P. W. Frankland, 'Heroes of the Engram', *Journal of Neuroscience*, 37:4647–57 (2017).

4. C. H. Bailey and E. R. Kandel, 'Synaptic Remodeling, Synaptic Growth and the Storage of Long-Term Memory in Aplysia', *Progress in Brain Research*, 169:179–98 (2008).

5. Ibid.

6. T. Lømo, 'Discovering Long-Term Potentiation (LTP)—Recollections and Reflections on What Came after', *Acta Physiologica (Oxford)*, 222: e12921 (2018).

7. Ibid.

8. R. Nauert, 'Probing the Molecular Basis of Long-Term Memory', *Psych Central* (8 August 2018), https://psychcentral.com/news/2011/03/21/probing-the-molecular-basis-of-long-term-memory/24517.html.

9. Ibid.

10. Massachusetts Institute of Technology, 'Neuroscientists Reveal How the Brain Can Enhance Connections', *Science Daily* (18 November 2015), https://www.sciencedaily.com/releases/2015/11/151118155301.htm.

11. Ibid.

12. J. Akst, 'Glial Cells aid Memory Formation', *The Scientist* (12 January 2010), https://www.the-scientist.com/the-nutshell/glial-cells-aid-memory-formation-43588.

13. T. Phillips, 'Transcription Factors and Transcriptional Control in Eukaryotic Cells', *Nature Education*, 1:119 (2008).

14. A. Iacoangeli, R. Bianchi, and H. Tiedge, 'Regulatory RNAs in Brain Function and Disorders', *Brain Research*, 1338:36–47 (2010).

15. E. McNeill and D. Van Vactor, 'MicroRNAs Shape the Neuronal Landscape', *Neuron*, 75:363–79 (2012).

16. A. Aksoy-Aksel, F. Zampa, and G. Schratt, 'MicroRNAs and Synaptic Plasticity—A Mutual Relationship', *Philosophical Transactions of the Royal Society of London B Biological Sciences*, 369:20130515 (2014).

17. Ibid.

18. Ibid.

19. M. R. Lyons and A. E. West, 'Mechanisms of Specificity in Neuronal Activity-Regulated Gene Transcription', *Progress in Neurobiology*, 94:259–95 (2011).

20. N. R. Smalheiser, 'The RNA-Centred View of the Synapse: Non-coding RNAs and Synaptic Plasticity', *Philosophical Transactions of the Royal Society of London B Biological Sciences*, 369:20130504 (2014).

21. E. G. Jones, 'Microcolumns in the Cerebral Cortex', *Proceedings of the National Academy of Sciences USA*, 97:5019–21 (2000).

22. Smalheiser, 'The RNA-Centred View of the Synapse'.

23. J. H. Ross, 'The Seahorse In Your Brain: Where Body Parts Got Their Names', *WBUR News* (16 December 2016), http://www.wbur.org/npr/505754756/the-seahorse-in-your-brain-where-body-parts-got-their-names.

24. S. Shapin, 'The Man Who Forgot Everything', *The New Yorker* (14 October 2013), https://www.newyorker.com/books/page-turner/the-man-who-forgot-everything.

25. Ibid.

26. Ibid.

27. Ibid.

28. D. Dieguez, 'Remembering Henry Molaison', *Brain Blogger* (3 March 2014), http://brainblogger.com/2014/03/03/remembering-henry-molaison/.

29. D. Zalewski, 'Life Lines', *The New Yorker* (23 March 2015), https://www.newyorker.com/magazine/2015/03/30/an-artist-with-amnesia.

30. M. D. Lemonick, 'Living in the now', *Aeon* (13 February 2017), https://aeon.co/essays/what-amnesiac-patients-can-tell-us-about-how-memories-are-made.

31. Ibid.

32. Ibid.

33. A. Abbott and E. Callaway, 'Nobel Prize for Decoding Brain's Sense of Place', *Nature* (6 October 2014), https://www.nature.com/news/nobel-prize-for-decoding-brain-s-sense-of-place-1.16093.

34. Ibid.

35. J. Goldhill, 'The "Jennifer Aniston Neuron" is the Foundation of Compelling New Memory Research', *Quartz* (23 July 2016), https://qz.com/740481/the-jennifer-aniston-neuron-is-the-foundation-of-compelling-new-memory-research/.

36. Ibid.

37. Ibid.

38. Ibid.

39. A. Trafton, 'Neuroscientists Identify Brain Circuit Necessary for Memory Formation', *MIT News* (6 April 2017), http://news.mit.edu/2017/neuroscientists-identify-brain-circuit-necessary-memory-formation-0406.

40. Ibid.

41. N. Ng, '"Heart of Brain" Breakthrough May Aid Treatment of Disorders, Hong Kong Scientists Say', *South China Morning Post* (18 September 2017), https://www.scmp.com/news/hong-kong/health-environment/article/2111678/heart-brain-breakthrough-may-aid-treatment.

42. Goldhill, 'The "Jennifer Aniston Neuron"'.

43. Ng, '"Heart of Brain"'.

44. B. Rasch and J. Born, 'About Sleep's Role in Memory', *Physiology Reviews*, 93:681–766 (2013).

45. Institute for Basic Science, 'Controlling Memory by Triggering Specific Brain Waves during Sleep', *Science Daily* (6 July 2017), https://www.sciencedaily.com/releases/2017/07/170706121204.htm.

46. Ibid.

47. D. Orenstein, 'Rhythmic Interactions between Cortical Layers Underlie Working Memory', *MIT News* (15 January 2018), http://news.mit.edu/2018/rhythmic-interactions-cortical-layers-control-working-memory-0115.
48. Ibid.
49. Ibid.
50. Y. Bhattacharjee, 'The First Year', *National Geographic* (January 2015), https://www.nationalgeographic.com/magazine/2015/01/baby-brains-development-first-year/.
51. Ibid.
52. Ibid.
53. Ibid.

Chapter 8: Thought and Language

1. F. Cowie, 'Innateness and Language', *Stanford Encyclopedia of Philosophy* (16 January 2008), https://plato.stanford.edu/entries/innateness-language/.
2. 'Scientists Identify a Language Gene', *Anthropologist in the Attic* (14 February 2010), http://anthropologistintheattic.blogspot.com/2010/02/scientists-identify-language-gene.html.
3. Ibid.
4. Ibid.
5. Ibid.
6. E. Callaway, ' "Language Gene" Speeds Learning', *Nature News* (18 November 2011), https://www.nature.com/news/language-gene-speeds-learning-1.9395.
7. N. Wade, 'A Human Language Gene Changes the Sound of Mouse Squeaks', *New York Times* (28 May 2009), https://www.nytimes.com/2009/05/29/science/29mouse.html?mtrref=www.google.co.uk&gwh=9FAD516E6AC529D9F47F48200C0DDCAD&gwt=pay.
8. E. Pennisi, ' "Language Gene" Has a Partner', *Science* (31 October 2013), http://www.sciencemag.org/news/2013/10/language-gene-has-partner.
9. S. Wolpert, 'Genes in Songbirds Hold Clues about Human Speech Disorders, UCLA Biologists Report', *UCLA Newsroom* (27 March 2018), http://newsroom.ucla.edu/releases/genes-in-songbirds-hold-clues-about-human-speech-disorders-ucla-biologists-report.
10. Ibid.
11. Ibid.
12. S. Samuel, 'Scientists Added Human Brain Genes to Monkeys. Yes it's as Scary as it Sounds', *Vox* (12 April 2019), https://www.vox.com/future-perfect/2019/4/12/18306867/china-genetics-monkey-brain-intelligence.
13. P. Deriziotis and S. E. Fisher, 'Neurogenomics of Speech and Language Disorders: The Road Ahead', *Genome Biology*, 14:204 (2013).
14. M. Konnikova, 'The Man Who Couldn't Speak and How He Revolutionized Psychology', *Scientific American* (8 February 2013), https://blogs.scientificamerican.com/literally-psyched/the-man-who-couldnt-speakand-how-he-revolutionized-psychology.
15. K. Cherry, 'Wernicke's Area Location and Function', *Very Well Mind* (28 October 2019), https://www.verywellmind.com/wernickes-area-2796017.
16. P. Tremblay and A. S. Dick, 'Broca and Wernicke Are Dead, or Moving Past the Classic Model of Language Neurobiology', *Brain and Language*, 162:60–71 (2016).
17. Ibid.
18. Ibid.

19. D. A. McLennan, 'The Concept of Co-option: Why Evolution Often Looks Miraculous', *Evolution: Education & Outreach*, 1:247–58 (2008).

20. M. C. Corballis, 'Mirror Neurons and the Evolution of Language', *Brain & Language*, 112:25–35 (2010).

21. Ibid.

22. G. F. Marcus and S. E. Fisher, 'FOXP2 in Focus: What Can Genes Tell Us about Speech and Language?', *Trends in Cognitive Science*, 7:257–62 (2003).

23. I. Sample, 'Neuroscientists Create "Atlas" Showing How Words Are Organised in the Brain', *The Guardian* (27 April 2016), https://www.theguardian.com/science/2016/apr/27/brain-atlas-showing-how-words-are-organised-neuroscience.

24. Ibid.

25. Ibid.

26. P. Klass, 'Language Lessons Start in the Womb', *New York Times* (21 February 2017), https://www.nytimes.com/2017/02/21/well/family/language-lessons-start-in-the-womb.html.

27. Ibid.

28. R. Nuzzo, 'Babies' Brains May Be Tuned to Language before Birth', *Nature* (25 February 2013), https://www.nature.com/news/babies-brains-may-be-tuned-to-language-before-birth-1.12489.

29. Ibid.

30. G. Miller, 'Pioneering Study Images Activity in Fetal Brains', *Science* (13 January 2017), http://science.sciencemag.org/content/355/6321/117.2.

31. Ibid.

32. J. Ducharme, 'Why It's So Hard to Learn Another Language after Childhood', *Time* (2 May 2018), http://time.com/5261446/language-critical-period-age/.

33. Ibid.

34. N. J. Ramirez, 'Why the Baby Brain Can Learn Two Languages at the Same Time', *The Conversation* (15 April 2016), https://theconversation.com/why-the-baby-brain-can-learn-two-languages-at-the-same-time-57470.

35. J. Palermo, 'Infant Language Learning Linked to Social Interaction', *Space Coast Daily* (16 July 2016), http://spacecoastdaily.com/2016/07/infant-language-learning-linked-to-social-interaction/.

36. T. Ghose, 'Baby Talk: Infants May Practice Words in Their Minds', *Live Science* (14 July 2014), https://www.livescience.com/46789-baby-brains-practice-language.html.

37. Y. Bhattacharjee, 'The First Year', *National Geographic* (January 2015), https://www.nationalgeographic.com/magazine/2015/01/baby-brains-development-first-year/.

38. Ibid.

39. Ibid.

40. Ibid.

41. Ibid.

42. Ibid.

43. O. M. Lourenço, 'Piaget and Vygotsky: Many Resemblances, and a Crucial Difference', *New Ideas in Psychology* 30:281–95 (2012).

44. C. Emerson, 'The Outer Word and Inner Speech: Bakhtin, Vygotsky, and the Internalization of Language', *Critical Inquiry*, 10:245–64 (1983).

45. J. V. Wertsch, *Vygotsky and the Social Formation of Mind* (Harvard University Press, 1985), p. 116.

46. L. Kohlberg, J. Yaeger, and E. Hjertholm, 'Private Speech: Four Studies and a Review of Theories', *Child Development*, 39:691–736 (1968).

47. J. Beck, 'The Running Conversation in Your Head', *The Atlantic* (23 November 2016), https://www.theatlantic.com/science/archive/2016/11/figuring-out-how-and-why-we-talk-to-ourselves/508487/.

48. C. Fernyhough, *The Voices Within: The History and Science of How We Talk to Ourselves* (Profile Books, 2016), p. 63.

49. V. N. Voloshinov, *Marxism and the Philosophy of Language* (Harvard University Press, 1973), p. 38.

50. F. Jabr, 'Catching Ourselves in the Act of Thinking', *Scientific American* (18 November 2013), https://blogs.scientificamerican.com/brainwaves/catching-ourselves-in-the-act-of-thinking/.

51. J. V. Wertsch and P. Tulviste, 'L. S. Vygotsky and Contemporary Developmental Psychology', *Developmental Psychology*, 28:548–57 (1992).

52. Voloshinov, *Marxism and the Philosophy of Language*, p. 39.

53. Ibid., 38.

54. Fernyhough, *The Voices Within*, p. 98.

55. Beck, 'The Running Conversation in Your Head'.

Chapter 9: Creativity and Imagination

1. M. Michalko, 'Scratch a Genius and You Surprise a Child', *The Creativity Post* (28 June 2012), https://www.creativitypost.com/article/scratch_a_genius_and_you_surprise_a_child.

2. W. J. T. Mitchell, 'What Is an Image?', *New Literary History*, 15:503–37 (1984).

3. W. James, 'The Principles of Psychology', *Classics in the History of Psychology* (1890), https://psychclassics.yorku.ca/James/Principles/prin18.htm.

4. A. Burmester, 'How Do Our Brains Reconstruct the Visual World?', *The Conversation* (5 November 2015), https://theconversation.com/how-do-our-brains-reconstruct-the-visual-world-49276.

5. E. Pelaprat and M. Cole, '"Minding the Gap": Imagination, Creativity and Human Cognition', *Integrative Physiological and Behavioral Science* (2011), http://lchc.ucsd.edu/mca/Paper/ETMCimagination.pdf.

6. L. Holzman, *Vygotsky at Work and Play* (Routledge, 2017), p. 94.

7. L. S. Vygotsky, 'Imagination and Creativity in Childhood', *Journal of Russian and East European Psychology*, 42:7–97 (2004).

8. Ibid.

9. L. S. Vygotsky, *Mind in Society: The Development of Higher Psychological Processes* (Harvard University Press, 1978), pp. 92–104.

10. M. Cole, *Revival: Soviet Developmental Psychology: An Anthology* (Routledge, 1977), p. 88.

11. J. E. Johnson, Eberle, T. S. Henricks, and D. Kuschner, *The Handbook of the Study of Play* (Rowman and Littlefield, 2015), p. 206.

12. B. M. Newman and P. R. Newman, *Development through Life: A Psychosocial Approach* (Cengage Learning, 2008), p. 206.

13. K. Oatley, 'A Feeling for Fiction', *Greater Good Magazine* (1 September 2005), https://greatergood.berkeley.edu/article/item/a_feeling_for_fiction.

14. C. Philby, 'Hollywood Ate My Novel: Novelists Reveal What It's Like to Have Their Book Turned into a Movie', *The Independent* (18 February 2012), https://www.independent.co.uk/arts-entertainment/books/features/hollywood-ate-my-novel-novelists-reveal-what-it-s-like-to-have-their-book-turned-into-a-movie-6940772.html.

15. P. Thagard, 'Empathy in Literature and Film', *Psychology Today* (18 May 2017), https://www.psychologytoday.com/gb/blog/hot-thought/201705/empathy-in-literature-and-film.

16. K. G. Klinge, 'Mapping Creativity in the Brain', *The Atlantic* (21 March 2016), https://www.theatlantic.com/science/archive/2016/03/the-driving-principles-behind-creativity/474621/.

17. Ibid.

18. Ibid.

19. C. Mackay, 'Researchers Identify Specific Neurons That Distinguish between Reality and Imagination', *Western University* (1 June 2017), http://mediarelations.uwo.ca/2017/06/01/researchers-identify-specific-neurons-distinguish-reality-imagination/.

20. Ibid.

21. A. García-Molina, 'Phineas Gage and the Enigma of the Prefrontal Cortex', *Neurología*, 27:370–5 (2012).

22. Ibid.

23. Ibid.

24. Ibid.

25. Z. Kotowicz, 'The Strange Case of Phineas Gage', *History of the Human Sciences*, 20:115–31 (2007).

26. C. Gregoire, 'Research Uncovers How and Where Imagination Occurs in the Brain', *Huffington Post* (18 September 2013), https://www.huffingtonpost.com/2013/09/17/imagination-brain_n_3922136.html.

27. Ibid.

28. A. Piore, 'Attention, Please: Earl Miller Wants to Make Us All Smarter', *Discover Magazine* (1 September 2016), http://discovermagazine.com/2016/oct/your-attention-please.

29. Ibid.

30. Gregoire, 'Research Uncovers How and Where Imagination Occurs in the Brain'.

31. T. Lewis, 'Human Brain: Facts, Functions & Anatomy', *Live Science* (28 September 2018), https://www.livescience.com/29365-human-brain.html.

32. L. Vandervert, 'The Prominent Role of the Cerebellum in the Learning, Origin and Advancement of Culture', *Cerebellum Ataxias* 3:10 (2016).

33. Ibid.

34. Stanford University Medical Center, 'Unexpected Brain Structures Tied to Creativity, and to Stifling It', *Science Daily* (28 May 2015), https://www.sciencedaily.com/releases/2015/05/150528084158.htm.

35. Ibid.

36. Ibid.

37. Ibid.

38. A. Lipsett, 'My Message: "Anybody Can Learn"', *The Guardian* (2 September 2008), https://www.theguardian.com/education/2008/sep/02/languages.schools.

39. Vandervert, 'The Prominent Role of the Cerebellum'.

40. C. Kalb, 'What Makes a Genius?', *National Geographic* (May 2017), https://www.nationalgeographic.com/magazine/2017/05/genius-genetics-intelligence-neuroscience-creativity-einstein/.

41. A. Cho, 'Gravitational Waves, Einstein's Ripples in Spacetime, Spotted for First Time', *Science* (11 February 2016), http://www.sciencemag.org/news/2016/02/gravitational-waves-einstein-s-ripples-spacetime-spotted-first-time.

42. H. Cotter, 'Michelangelo Is the Divine Star of the Must-See Show of the Season', *New York Times* (9 November 2017), https://www.nytimes.com/2017/11/09/arts/design/michelangelo-review-metropolitan-museum-of-art-carmen-bambach.html.

43. D. Goldsmith and M. Bartusiak, *E = Einstein: His Life, His Thought, and His Influence on Our Culture* (Barnes and Noble, 2008), p. 159.

44. W. Kremer, 'The Strange Afterlife of Einstein's Brain', *BBC News* (17 April 2015), https://www.bbc.com/news/magazine-32354300.

45. Ibid.

46. Ibid.

47. Ibid.

48. Ibid.

49. Ibid.

50. Ibid.

51. M. Neihart, 'Creativity, the Arts, and Madness', *Roeper Review*, 21:47–50 (1998).

52. G. Wilson, 'Delusions and Grandeur', *Times Higher Education* (1 November 2012), https://www.timeshighereducation.com/features/delusions-and-grandeur/421624.article.

53. A. Fink, M. Benedek, H. F. Unterrainer, I. Papousek, and E. M. Weiss, 'Creativity and Psychopathology: Are There Similar Mental Processes Involved in Creativity and in Psychosis-Proneness?', *Frontiers in Psychology*, 5:1211 (2014).

54. E. Jaafe, 'What Neuroscience Says about the Link between Creativity and Madness', *Fast Company* (14 November 2013), https://www.fastcompany.com/3021561/the-neuroscience-linking-creativity-and-mental-illness.

55. Ibid.

Chapter 10: Emotion and Reason

1. M. Rescorla, 'The Computational Theory of Mind', *Stanford Encyclopedia of Philosophy* (21 February 2020), https://plato.stanford.edu/entries/computational-mind/.

2. D. Robson, 'A Brief History of the Brain', *New Scientist* (21 September 2011), https://www.newscientist.com/article/mg21128311-800-a-brief-history-of-the-brain/.

3. T. Dixon, '"Emotion": The History of a Keyword in Crisis', *Emotion Review*, 4:338–44 (2012).

4. J. Johnson, 'What Does The Hypothalamus Do?', *Medical News Today* (22 August 2018), https://www.medicalnewstoday.com/articles/312628.php.

5. S. Barrie, 'Your Fat Cells Control Your Brain', *Huffington Post* (17 November 2011), https://www.huffingtonpost.com/stephen-barrie-nd/your-fat-cells-control-yo_b_732201.html.

6. B. Kotchoubey, 'Human Consciousness: Where Is It From and What Is It for', *Frontiers in Psychology*, 9:567 (2018).

7. A. Damasio, 'Why Your Biology Runs on Feelings', *Nautilus* (18 January 2018), http://nautil.us/issue/56/perspective/why-your-biology-runs-on-feelings.

8. Kotchoubey, 'Human Consciousness'.

9. N. Magon and S. Kalra, 'The Orgasmic History of Oxytocin: Love, Lust, and Labor', *Indian Journal of Endocrinology and Metabolism*, 15:S156–61 (2011).

10. S. K. Fineberg and D. A. Ross, 'Oxytocin and the Social Brain', *Biology and Psychiatry*, 81:e19–e21 (2017).

11. J. Cepelewicz, 'In Birds' Songs, Brains and Genes, He Finds Clues to Speech', *Quanta Magazine* (30 January 2018), https://www.quantamagazine.org/erich-jarvis-in-birds-songs-brains-and-genes-he-finds-clues-to-speech-20180130/.

12. Ibid.

13. Ibid.

14. L. Konkel, 'What Is Dopamine?', *Everyday Health* (10 August 2018), https://www.everydayhealth.com/dopamine/.
15. T. R. Mhyre, J. T. Boyd, R. W. Hamill, and K. A., Maguire-Zeiss, 'Parkinson's Disease', *Subcellular Biochemistry*, 65:389–455 (2012).
16. K. Kelland, 'Work on Brain's Reward System Wins Scientists a Million Euro Reward', *Reuters* (6 March 2017), https://www.reuters.com/article/us-science-brain-idUSKBN16D1TF.
17. Ibid.
18. Ibid.
19. C. Bergland, 'What Makes Us Human? Dopamine and the Cerebellum Hold Clues', *Psychology Today* (24 November 2017), https://www.psychologytoday.com/us/blog/the-athletes-way/201711/what-makes-us-human-dopamine-and-the-cerebellum-hold-clues.
20. Ibid.
21. C. Fernyhough, 'Getting Vygotskian about Theory of Mind: Mediation, Dialogue, and the Development of Social Understanding', *Developmental Review*, 28:225–62 (2008).
22. L. S. Vygotsky, *The Collected Works of L.S. Vygotsky* (Plenum Press, 1987), p. 347.
23. G. R. Mesquita, 'Vygotsky and the Theories of Emotions: In Search of a Possible Dialogue', *Psicologia: Reflexão e Crítica*, 25:809–16 (2012).
24. Ibid.
25. S. McLeod, 'The Zone of Proximal Development and Scaffolding', *Psychology Today* (2019), https://www.simplypsychology.org/Zone-of-Proximal-Development.html.
26. Ibid.
27. H. Grimmett, *The Practice of Teachers Professional Development: A Cultural-Historical Approach* (Springer, 2014), p. 13.
28. H. Mahn and V. John-Steiner, 'The Gift of Confidence: A Vygotskian View of Emotions', in G. Wells and G. Claxton (eds), *Learning for Life in the 21st Century: Sociocultural Perspectives on the Future of Education* (Wiley-Blackwell, 2002), pp. 46–58.
29. Ibid.
30. M. Nudelman and E. Brodwin, 'The 5 Most Addictive Substances on the Planet, Ranked', *Business Insider* (2 December 2018), http://uk.businessinsider.com/most-addictive-drugs-ranked-2016-10.
31. R. Gray, 'No Wonder They Called It the Stone Age! Ancient Humans Were Taking Drugs—Including Magic Mushrooms and Opium—up to 10,600 Years Ago', *Daily Mail* (4 February 2015), http://www.dailymail.co.uk/sciencetech/article-2939830/No-wonder-called-stone-age-Ancient-humans-taking-drugs-including-magic-mushrooms-opium-10-600-years-ago.html.
32. Ibid.
33. Ibid.
34. Ibid.
35. Ibid.
36. F. Smith, 'How Science Is Unlocking the Secrets of Addiction', *National Geographic* (September 2017), https://www.nationalgeographic.com/magazine/2017/09/the-addicted-brain/.
37. B. A. Mason, 'Slaves to Dopamine and the Hijacking of Our Brains', *Huffington Post* (29 December 2017), https://www.huffingtonpost.com/entry/slaves-to-the-rhythm-how-dopamine-hijacking-is-taking_us_5a453bd4e4b0d86c803c7583.
38. 'Biology of Addiction', *National Institutes of Health News in Health* (October 2015), https://newsinhealth.nih.gov/2015/10/biology-addiction.

39. Ibid.

40. A. Matthews-King, 'Alcohol and Tobacco by Far the Worst Drugs for Human Health, Global Review Finds', *The Independent* (11 May 2018), https://www.independent.co.uk/news/health/alcohol-drinking-smoking-drugs-addictive-health-worst-bad-cannabis-cocaine-amphetamines-opioids-a8345741.html.

41. Ibid.

42. D. F. Maron, 'How Opioids Kill', *Scientific American* (8 January 2018), https://www.scientificamerican.com/article/how-opioids-kill/.

43. Ibid.

44. N. Vega, 'Sean Parker on Facebook: We Created a Monster', *New York Post* (9 November 2017), https://nypost.com/2017/11/09/sean-parker-on-facebook-we-created-a-monster/.

45. D. Brooks, 'How Evil Is Tech?', *The New York Times* (20 November 2017), https://www.nytimes.com/2017/11/20/opinion/how-evil-is-tech.html.

46. Ibid.

47. L. Kim, 'Multitasking Is Killing Your Brain', *Huffington Post* (6 December 2017), https://www.huffingtonpost.com/larry-kim/multitasking-is-killing-your_b_9821244.html.

48. Ibid.

49. Ibid.

50. Ibid.

51. M. Gummer, 'The Positive Impact of Social Media and Technology on Society', *Pi Media* (6 February 2018), https://uclpimedia.com/online/the-positive-impact-of-social-media-and-technology-on-society.

52. D. Spector, 'How Beer Created Civilization', *Business Insider* (26 December 2013), https://www.businessinsider.com/how-beer-led-to-the-domestication-of-grain-2013-12.

53. Ibid.

54. Ibid.

55. J. P. Kahn, 'How Beer Gave Us Civilization', *New York Times* (15 March 2013), https://www.nytimes.com/2013/03/17/opinion/sunday/how-beer-gave-us-civilization.html.

56. S. Sloat, 'The "Stoned Ape" Theory Might Explain Our Extraordinary Evolution', *Inverse* (14 July 2017), https://www.inverse.com/article/34186-stoned-ape-hypothesis.

57. A. Simon-Lewis, 'Brain Scans Reveal How Psychedelic Drugs Create a "Higher State of Consciousness"', *Wired* (20 April 2017), https://www.wired.co.uk/article/psychedelic-drugs-found-to-cause-a-higher-state-of-consciousness.

58. E. Ekman and G. Agin.-Liebes, 'Can a Psychedelic Experience Improve Your Life?', *Greater Good Magazine* (23 October 2019), https://greatergood.berkeley.edu/article/item/can_a_psychedelic_experience_improve_your_life.

59. I. Sample, 'Psychedelic Drugs Induce "Heightened State Of Consciousness", Brain Scans Show', *The Guardian* (19 April 2017), https://www.theguardian.com/science/2017/apr/19/brain-scans-reveal-mind-opening-response-to-psychedelic-drug-trip-lsd-ketamine-psilocybin.

60. Ibid.

Chapter 11: Conscious and Unconscious

1. C. Baraniuk, 'The Enormous Power of the Unconscious Brain', *BBC Future* (16 March 2016), http://www.bbc.com/future/story/20160315-the-enormous-power-of-the-unconscious-brain.

2. B. K. Keim, 'Brain Scanners Can See Your Decisions Before You Make Them', *Wired* (13 April 2008), https://www.wired.com/2008/04/mind-decision/.

3. B. Hughes, 'Genius of the Modern World', *BBC* (11 November 2019), https://www.bbc.co.uk/programmes/b07ht3cd.

4. J. LeFanu, 'Wrong Image of Freud Has Entered the Subconscious: Dr James LeFanu', *The Telegraph* (14 May 2006), https://www.telegraph.co.uk/news/uknews/1518310/Wrong-image-of-Freud-has-entered-the-subconscious.html.

5. S. Pinker, 'The Blank Slate Controversy', *New York Times* (13 October 2002), https://www.nytimes.com/2002/10/13/books/chapters/the-blank-slate.html.

6. R. J. Richards, 'The Impact of German Romanticism on Biology in the Nineteenth Century', *University of Chicago* (2011), http://home.uchicago.edu/rjr6/articles/Idealism & biology.pdf.

7. M. Iseli, 'Thomas De Quincey and the Cognitive Prospects of the Unconscious', *European Romantic Review*, 24:325–33 (2013).

8. T. C. Gannon, Immortal Sea, Eternal Mind: Romanticism and the Unconscious Psyche, M.A. thesis, University of South Dakota (1979).

9. Ibid.

10. J. S. Hendrix, *Unconscious Thought in Philosophy and Psychoanalysis* (Palgrave Macmillan, 2015), p. 19.

11. Ibid.

12. S. Austin, 'Freud, Sigmund (1856–1939)', *Oxford Dictionary of National Biography* (2011), http://www.oxforddnb.com/view/10.1093/ref:odnb/9780198614128.001.0001/odnb-9780198614128-e-55514.

13. Ibid.

14. S. McLeod, 'Id, Ego and Superego', *Simply Psychology* (2019), https://www.simplypsychology.org/psyche.html.

15. Ibid.

16. K. Cherry, 'The Interpretation of Dreams by Sigmund Freud: History and Significance', *Very Well Mind* (7 April 2020), https://www.verywellmind.com/the-interpretation-of-dreams-by-sigmund-freud-2795855.

17. S. McLeod, 'Psychoanalysis', *Simply Psychology* (2019), https://www.simplypsychology.org/psychoanalysis.html.

18. K. Cherry, 'The Oedipus Complex in Children', *Very Well Mind* (14 May 2020), https://www.verywellmind.com/what-is-an-oedipal-complex-2795403.

19. Ibid.

20. B. Borrell, 'Oedipus Wrecked: Study Supporting the Mother of All Psychological Complexes Withdrawn', *Scientific American* (24 February 2009), https://www.scientificamerican.com/article/oedipus-complex-study-withdrawn/.

21. S. McLeod, 'Psychosexual Stages', *Simply Psychology* (2019), https://www.simplypsychology.org/psychosexual.html.

22. C. Sieczkowski, 'Unearthed Letter From Freud Reveals His Thoughts On Gay People', *Huffington Post* (7 December 2017), https://www.huffingtonpost.com/2015/02/18/sigmund-freud-gay-cure-letter_n_6706006.html.

23. M. Billig, 'The Dialogic Unconscious: Psycho-analysis, Discursive Psychology and the Nature of Repression', *Massey University* (2018), http://www.massey.ac.nz/~alock/virtual/p-a4.htm.

24. R. Young, 'Back to Bakhtin', *Cultural Critique*, 2:71–92 (1985–6).

25. M. Meares, *The Metaphor of Play: Origin and Breakdown of Personal Being* (Routledge, 2002), p. 38.

26. J. M. Quinodoz, *Reading Freud: A Chronological Exploration of Freud's Writings* (Routledge, 2004), p. 145.

27. S. Knapton, 'Traumatic Memories Really Can Be Repressed, But Also Restored, Say Scientists', *Daily Telegraph* (17 August 2015), https://www.telegraph.co.uk/news/science/science-news/11807695/Traumatic-memories-really-can-be-repressed-but-also-restored-say-scientists.html.

28. Ibid.

29. Ibid.

30. Ibid.

31. Ibid.

32. D. Wolman, 'The Split Brain: A Tale of Two Halves', *Nature* (14 March 2012), https://www.nature.com/news/the-split-brain-a-tale-of-two-halves-1.10213.

33. C. Profaci, 'Two Brains in One Head? The Story of the Split-Brain Phenomenon', *NeuWrite* (27 August 2015), https://neuwritesd.org/2015/08/27/two-brains-in-one-head-the-story-of-the-split-brain-phenomenon/.

34. B. Brogaard, 'Split Brains', *Psychology Today* (6 November 2012), https://www.psychologytoday.com/gb/blog/the-superhuman-mind/201211/split-brains.

35. D. A. Lienhard, 'Roger Sperry's Split Brain Experiments (1959–1968)', *The Embryo Project Encyclopedia* (27 December 2017), https://embryo.asu.edu/pages/roger-sperrys-split-brain-experiments-1959-1968.

36. B. Brogaard, 'Split Brains', *Psychology Today* (6 November 2012), https://www.psychologytoday.com/us/blog/the-superhuman-mind/201211/split-brains.

37. A. Novotney, 'Despite What You've Been Told, You Aren't "Left-Brained" or "Right-Brained"', *The Guardian* (16 November 2013), https://www.theguardian.com/commentisfree/2013/nov/16/left-right-brain-distinction-myth.

38. S. M. Kosslyn and G. Wayne Miller, 'Left Brain, Right Brain? Wrong', *Psychology Today* (27 January 2014), https://www.psychologytoday.com/us/blog/the-theory-cognitive-modes/201401/left-brain-right-brain-wrong.

39. S. Paulson, H. A. Berlin, E. Ginot, and G. Makari, 'Delving within: The New Science of the Unconscious', *Annals of the New York Academy of Sciences*, 1406:12–27 (2017).

40. Ibid.

41. Ibid.

42. Ibid.

43. Ibid.

44. S. Paulson, D. Barrett, K. Bulkeley, and R. Naiman, 'Dreaming: A Gateway to the Unconscious?', *Annals of the New York Academy of Sciences*, 1406:28–45 (2017).

Chapter 12: Sanity and Madness

1. H. J. Parkinson, 'How Do I Know If I Have a Mental Illness?', *The Guardian* (29 January 2016), https://www.theguardian.com/uk-news/2016/jan/29/how-do-i-know-if-i-have-a-mental-illness.

2. A. Head and J. Bond, 'We Need to Address the Socioeconomic Causes of Mental Health Issues If We Really Want to Tackle the Problem', *The Independent* (20 May 2018), https://www.

independent.co.uk/voices/mental-health-help-how-change-awareness-causes-treat-a8357741.html.

3. R. Bentall, 'Mental Illness Is A Result of Misery, Yet Still We Stigmatise It', *The Guardian* (26 February 2016), https://www.theguardian.com/commentisfree/2016/feb/26/mental-illness-misery-childhood-traumas.

4. D. Spence, 'The Psychiatric Oligarchs Who Medicalise Normality', *British Medical Journal*, 344:e3135 (2012).

5. B. J. Deacon, 'The Biomedical Model of Mental Disorder: A Critical Analysis of Its Validity, Utility, and Effects on Psychotherapy Research', *Clinical Psychology Review*, 33:846–61 (2013).

6. R. Prasad, 'Why US Suicide Rate Is On The Rise', *BBC News* (11 June 2018), https://www.bbc.co.uk/news/world-us-canada-44416727.

7. J. Christensen, 'Why the US Has the Most Mass Shootings', *CNN* (5 October 2017), https://edition.cnn.com/2015/08/27/health/u-s-most-mass-shootings/index.html.

8. P. M. Visscher, N. R. Wray, Q. Zhang, P. Sklar, M. I. McCarthy, M. A Brown, and J. Yang, '10 Years of GWAS Discovery: Biology, Function, and Translation', *American Journal of Human Genetics*, 101:5–22 (2017).

9. M. Balter, 'Schizophrenia's Unyielding Mysteries', *Scientific American*, 316:54–61 (2017).

10. O. D. Howes, R. McCutcheon, M. J. Owen, and R. M. Murray, 'The Role of Genes, Stress, and Dopamine in the Development of Schizophrenia', *Biological Psychiatry*, 81:9–20 (2017).

11. Ibid.

12. P. J. Harrison, 'Recent Genetic Findings in Schizophrenia and Their Therapeutic Relevance', *Journal of Psychopharmacology*, 29:85–96 (2015).

13. Ibid.

14. A. G. Diehl and A. P. Boyle, 'Deciphering ENCODE', *Trends in Genetics*, 32:238–49 (2016).

15. A. Park, 'Junk DNA—Not So Useless After All', *Time* (6 September 2012), http://healthland.time.com/2012/09/06/junk-dna-not-so-useless-after-all/.

16. Harrison, 'Recent Genetic Findings in Schizophrenia'.

17. E. Callaway, 'New concerns raised over value of genome-wide disease studies', *Nature News* (15 June 2017), https://www.nature.com/news/new-concerns-raised-over-value-of-genome-wide-disease-studies-1.22152.

18. Ibid.

19. Ibid.

20. P. M. Visscher, M. A. Brown, M. I. McCarthy, and J. Yang, 'Five Years of GWAS Discovery', *American Journal of Human Genetics*, 90:7–24 (2012).

21. M. Wolfe, 'What Is Crossing Over in Genetics?', *Sciencing* (18 September 2018), https://sciencing.com/crossing-over-genetics-6628252.html.

22. Ibid.

23. I. Lobo and K. Shaw, 'Discovery and Types of Genetic Linkage', *Nature Education*, 1:139 (2008).

24. J. M. Heather and B. Chain, 'The Sequence of Sequencers: The History of Sequencing DNA', *Genomics*, 107:1–8 (2016).

25. C. Zhuo, W. Hou, C. Lin, L. Hu, and J. Li, 'Potential Value of Genomic Copy Number Variations in Schizophrenia', *Frontiers in Molecular Neuroscience*, 10:204 (2017).

26. T. P. Rutkowski, J. P. Schroeder, G. M. Gafford, S. T. Warren, D. Weinshenker, T. Caspary, and J. G. Mulle, 'Unraveling the Genetic Architecture of Copy Number Variants Associated with Schizophrenia and Other Neuropsychiatric Disorders', *Journal of Neuroscience Research*, 95:1144–60 (2017).

27. H. Devlin, 'Radical New Approach to Schizophrenia Treatment Begins Trial', *The Guardian* (3 November 2017), https://www.theguardian.com/society/2017/nov/03/radical-new-approach-to-schizophrenia-treatment-begins-trial.

28. Ibid.

29. Ibid.

30. D. Rudacille, 'Attention Deficit, Autism Share Genetic Risk Factors', *Spectrum News* (22 August 2011), https://www.spectrumnews.org/news/attention-deficit-autism-share-genetic-risk-factors/.

31. M. G. Thompson, 'R.D. Laing and Anti-Psychopathology: The Myth of Mental Illness Redux', *Mad in America* (26 October 2013), https://www.madinamerica.com/2013/10/r-d-laing-anti-psychopathology-myth-mental-illness-redux/.

32. P. Gibney, 'The Double Bind Theory: Still Crazy-Making after All These Years', *Psychotherapy in Australia*, 12:48–55 (2006).

33. Ibid.

34. J. Diski, 'Rhythm Method', *London Review of Books*, 16:20–1 (1994).

35. Ibid.

36. Ibid.

37. T. Friedman and N. N. Tin, 'Childhood Sexual Abuse and the Development of Schizophrenia', *Postgraduate Medical Journal*, 83:507–8 (2007).

38. S. Bressert, 'Schizophrenia Symptoms', *Psych Central* (9 May 2020), https://psychcentral.com/disorders/schizophrenia/schizophrenia-symptoms/.

39. L. S. Vygotsky, 'Thought in Schizophrenia', *Archives of Neurology and Psychiatry*, 31:1067 (1934).

40. A. Kozulin, *Vygotsky's Psychology: A Biography of Ideas* (Harvester Wheatsheaf, 1990), pp. 226–8.

41. Ibid.

42. Ibid.

43. Ibid.

44. H. Werner and B. Kaplan, *Symbol Formation* (Wiley, 1963), p. 257.

45. Vygotsky, 'Thought in Schizophrenia'.

46. A. Kozulin, *Vygotsky's Psychology: A Biography of Ideas* (Harvester Wheatsheaf, 1990), p. 231.

47. Ibid., 229.

48. B. Rund, 'Attention, Communication, and Schizophrenia', *Yale Journal of Biology and Medicine*, 58:265–73 (1985).

49. Ibid.

50. Ibid.

51. L. S. Vygotsky, 'The Development of Higher Forms of Attention in Childhood', in J.V. Wertsch, (ed.), *The Concept of Activity in Soviet Psychology* (M. E. Sharpe, 1981), pp. 193–4.

52. Rund, 'Attention, Communication, and Schizophrenia'.

53. Ibid.

54. Ibid.

55. D. G. Smith, 'The Placenta Is Now a Suspect In Heightening Schizophrenia Risk', *Scientific American* (28 May 2018), https://www.scientificamerican.com/article/the-placenta-is-now-a-suspect-in-heightening-schizophrenia-risk/.

56. Ibid.

57. S. Peters, 'Researcher Acknowledges His Mistakes in Understanding Schizophrenia', *Mad in America* (26 January 2017), https://www.madinamerica.com/2017/01/researcher-acknowledges-mistakes-understanding-schizophrenia/.

58. S. McCarthy-Jones, 'The Concept of Schizophrenia Is Coming to an End—Here's Why', *The Independent* (4 September 2017), https://www.independent.co.uk/life-style/health-and-

families/healthy-living/concept-schizophrenia-coming-to-end-psychology-genetics-psychiatry-schizophrenia-a7925576.html.

59. Ibid.

60. C. Dodd, 'Open Dialogue: The Radical New Treatment Having Life-Changing Effects on People's Mental Health', *The Independent* (6 December 2015), https://www.independent.co.uk/life-style/health-and-families/health-news/open-dialogue-the-radical-new-treatment-having-life-changing-effects-on-peoples-mental-health-a6762391.html.

61. Ibid.

62. Ibid.

63. Ibid.

64. S. Malhotra and S. Sahoo, 'Rebuilding the Brain with Psychotherapy', *Indian Journal of Psychiatry*, 59:411–19 (2017).

Chapter 13: Depression and Anxiety

1. P. Allen, G. Couzens, A. Hall, J. Hall, and I. Bains, 'Horror as 16 German School Pupils from the Same Class All Die on Board Doomed Germanwings Flight—But They Nearly Missed It When One Student Forgot Their Passport', *Daily Mail* (24 March 2015), http://www.dailymail.co.uk/news/article-3009705/Horror-16-German-school-pupils-class-exchange-trip-lifetime-dead-board-doomed-Germanwings-flight.html.

2. 'Germanwings Plane Crash: Who Were the Victims?', *BBC News* (27 March 2015), https://www.bbc.co.uk/news/world-europe-32047560.

3. K. Willsher, L. Osborne, and A. Chrisafis, 'Police Search Homes of Andreas Lubitz, Co-pilot Blamed for Germanwings Crash', *The Guardian* (27 March 2015), https://www.theguardian.com/world/2015/mar/26/airlines-change-cockpit-rules-after-co-pilot-blamed-for-germanwings-crash.

4. Ibid.

5. H. Samuel, 'Germanwings Crash: Andreas Lubitz Searched Online for Suicide and Cockpit Doors', *The Telegraph* (2 April 2015), https://www.telegraph.co.uk/news/worldnews/germanwings-plane-crash/11512137/Germanwings-crash-Andreas-Lubitz-searched-online-for-suicide-and-cockpit-doors.html.

6. S. Shuster, 'German Privacy Laws Let Pilot "Hide" His Illness from Employers', *Time* (27 March 2015), http://time.com/3761895/germanwings-privacy-law/.

7. M. Rahim, 'Don't Blame Depression for the Germanwings Tragedy', *The Guardian* (27 March 2015), https://www.theguardian.com/commentisfree/2015/mar/27/depression-germanwings-tragedy-pilot-andreas-lubitz-mental-health.

8. Ibid.

9. P. J. Skerrett, 'Suicide Often Not Preceded by Warnings', *Harvard Health Blog* (5 August 2019), https://www.health.harvard.edu/blog/suicide-often-not-preceded-by-warnings-201209245331.

10. J. Rayner, 'Cottage Industry', *The Guardian* (6 February 2005), https://www.theguardian.com/lifeandstyle/2005/feb/06/foodanddrink.restaurants.

11. N. McGee, 'Niall McGee Didn't Believe in Depression—until Cancer Medication Put Him in a Suicidal Spiral', *The Globe and Mail* (28 May 2015), https://www.theglobeandmail.com/life/niall-mcgee-didnt-believe-in-depression-until-cancer-meds-put-him-in-a-suicidal-spiral/article24660218/.

12. Ibid.

13. Ibid.

14. K. Griffiths, 'Depressed Bradford Woman Took Own Life in France', *The Telegraph and Argus* (10 August 2016), http://www.thetelegraphandargus.co.uk/news/14673774.Depressed_Bradford_woman_took_own_life_in_France/.

15. M. Smith, 'Balancing Your Humors', *Psychology Today* (2 November 2013), https://www.psychologytoday.com/us/blog/short-history-mental-health/201311/balancing-your-humors.

16. L. Appignanesi, *Mad, Bad And Sad: A History of Women and the Mind Doctors from 1800 to the Present* (Virago Press, 2009), pp. 136–7.

17. L. Fitzpatrick, 'A Brief History of Antidepressants', *Time* (7 January 2010), http://content.time.com/time/health/article/0,8599,1952143,00.html.

18. Ibid.

19. Ibid.

20. A. Edemariam, '"I Don't Know Who I Am without It": The Truth about Long-Term Antidepressant Use', *The Guardian* (6 May 2017), https://www.theguardian.com/society/2017/may/06/dont-know-who-am-antidepressant-long-term-use.

21. Fitzpatrick, 'A Brief History of Antidepressants'.

22. S. Mutalik, 'A Short History of the SSRI', *Psychiatric Times* (8 December 2014), http://www.psychiatrictimes.com/psychopharmacology/short-history-ssri.

23. A. Khan and W. A. Brown, 'Antidepressants versus Placebo in Major Depression: An Overview', *World Psychiatry*, 14:294–300 (2015).

24. S. Boseley, 'Antidepressant Withdrawal Symptoms Severe, Says New Report', *The Guardian* (2 October 2018), https://www.theguardian.com/society/2018/oct/02/antidepressant-withdrawal-symptoms-severe-says-new-report.

25. S. Boseley, 'Antidepressants: Is There a Better Way to Quit Them?', *The Guardian* (22 April 2019), https://www.theguardian.com/lifeandstyle/2019/apr/22/antidepressants-is-there-a-better-way-to-quit-them.

26. H. Arkowitz and S. O. Lilienfeld, 'Is Depression Just Bad Chemistry?', *Scientific American* (1 March 2014), https://www.scientificamerican.com/article/is-depression-just-bad-chemistry/.

27. J. Dryden, 'Study Reverses Thinking on Genetic Links to Stress, Depression', *Washington University School of Medicine in St Louis* (4 April 2017), https://medicine.wustl.edu/news/study-reverses-thinking-genetic-links-stress-depression/.

28. Ibid.

29. K. Oved, A. Morag, M. Pasmanik-Chor, M. Rehavi, N. Shomron, and D. Gurwitz, 'Genome-Wide Expression Profiling of Human Lymphoblastoid Cell Lines Implicates Integrin Beta-3 in the Mode of Action of Antidepressants', *Translational Psychiatry*, 3:e313 (2013).

30. J. Chen, 'How New Ketamine Drug Helps with Depression', *Yale Medicine* (21 March 2019), https://www.yalemedicine.org/stories/ketamine-depression/.

31. Ibid.

32. S. Knapton, 'Depression Is a Physical Illness Which Could Be Treated with Anti-inflammatory Drugs, Scientists Suggest', *The Telegraph* (8 September 2017), https://www.telegraph.co.uk/science/2017/09/08/depression-physical-illness-could-treated-anti-inflammatory/.

33. Ibid.

34. Ibid.

35. J. R. Thorpe, 'What Causes Depression?', *Bustle* (31 May 2017), https://www.bustle.com/p/what-causes-depression-new-research-says-it-might-be-malfunctioning-brain-circuitry-54289.

36. Ibid.

37. Ibid.

38. M. MacGill, 'What Is Depression and What Can I do about It?', *Medical News Today* (22 November 2019), https://www.medicalnewstoday.com/kc/depression-causes-symptoms-treatments-8933.

39. H. Ledford, 'First Robust Genetic Links to Depression Emerge', *Nature* (15 July 2015), https://www.nature.com/news/first-robust-genetic-links-to-depression-emerge-1.17979.

40. Ibid.

41. G. Lu, J. Li, H. Zhang, X. Zhao, L. J. Yan, and X. Yang, 'Role and Possible Mechanisms of Sirt1 in Depression', *Oxidative Medicine and Cellular Longevity*, 2018:8596903 (2018).

42. E. J. Nestler and S. E. Hyman, 'Animal Models of Neuropsychiatric Disorders', *Nature Neuroscience*, 13:1161–9 (2010).

43. R. M. Henig, 'How Depressed Is That Mouse?', *Scientific American* (7 March 2012), https://www.scientificamerican.com/article/depression-how-depressed-is-mouse/.

44. Ibid.

45. Ibid.

46. G. W. Brown and T. O. Harris, *Social Origins of Depression: A Study of Psychiatric Disorder in Women* (Free Press, 1978) .

47. K. Oatley, 'Depression: Crisis without Alternatives', *New Scientist*, 29–31 (1984).

48. M. Smith, 'Social Psychiatry Could Stem the Rising Tide of Mental Illness', *The Conversation* (3 June 2020), https://theconversation.com/social-psychiatry-could-stem-the-rising-tide-of-mental-illness-138152.

49. R. Zamzow, 'Male Mice Pass Stress Signatures down to Their Pups', *Spectrum* (19 October 2015), https://www.spectrumnews.org/news/male-mice-pass-stress-signatures-down-to-their-pups/.

50. Ibid.

51. N. Rahhal, 'Stressed Men Have "Softer" Children: Study Finds Stress Affects Sperm Quality and Makes Offspring Less Resilient to Pressure', *Daily Mail* (13 November 2017), http://www.dailymail.co.uk/health/article-5078233/Father-s-stress-affect-sperm-children-study-shows.html.

52. Ibid.

53. S. Paulsson, 'A View of the Holocaust', *BBC History* (17 February 2011), http://www.bbc.co.uk/history/worldwars/genocide/holocaust_overview_01.shtml.

54. D. Millward, 'Holocaust Survivors Pass on Trauma to Their Children's Genes', *The Telegraph* (22 August 2015), https://www.telegraph.co.uk/news/worldnews/northamerica/usa/11817892/Holocaust-survivors-pass-on-trauma-to-their-childrens-genes.html.

55. Ibid.

56. S. Yasmin, 'No, Trauma Is Not Inherited', *Dallas News* (30 May 2017), https://www.dallasnews.com/news/debunked/2017/05/30/trauma-inherited.

57. J. Glausiusz, 'Doubts Arising about Claimed Epigenetics of Holocaust Trauma', *Haaretz* (30 April 2017), https://www.haaretz.com/science-and-health/.premium-doubts-arising-about-claimed-epigenetics-of-holocaust-trauma-1.5466710.

58. Yasmin, 'No, Trauma Is Not Inherited'.

Chapter 14: Normality and Diversity

1. S. McLeod, 'The Medical Model', *Simply Psychology* (2018), https://www.simplypsychology.org/medical-model.html.

2. P. Beresford, 'Towards a Social Model of Madness and Distress? Exploring What Service Users Say', *Joseph Rowntree Foundation* (22 November 2010), https://www.jrf.org.uk/report/towards-social-model-madness-and-distress-exploring-what-service-users-say.

3. B. Taylor, H. Jick, and D. MacLaughlin, 'Prevalence and Incidence Rates of Autism in the UK: Time Trend from 2004–2010 in Children Aged 8 Years', *BMJ Open*, 3:e003219 (2013).

4. U. Frith, 'Autism—Are We Any Closer to Explaining the Enigma?', *The Psychologist*, 27:744–5 (2014).

5. Ibid.

6. S. Baron-Cohen, 'Leo Kanner, Hans Asperger, and the Discovery of Autism', *The Lancet*, 386:1329–30 (2015).

7. S. Baron, '*Neurotribes* Review—the Evolution of Our Understanding of Autism', *The Observer* (23 August 2015), https://www.theguardian.com/books/2015/aug/23/neurotribes-legacy-autism-steve-silberman-book-review-saskia-baron.

8. Ibid.

9. K. Connolly, 'Hans Asperger Aided and Supported Nazi Programme, Study Says', *The Guardian* (19 April 2018), https://www.theguardian.com/world/2018/apr/19/hans-asperger-aided-and-supported-nazi-programme-study-says.

10. Baron, '*Neurotribes* Review'.

11. Ibid.

12. Ibid.

13. H. Schofield, 'France's Autism Treatment "Shame"', *BBC News* (2 April 2012), https://www.bbc.co.uk/news/magazine-17583123.

14. Ibid.

15. E. Anthes, 'Lab-Grown Neurons Showcase Effects of Autism Mutations', *Spectrum* (15 January 2018), https://www.spectrumnews.org/news/lab-grown-neurons-showcase-effects-autism-mutations/.

16. Ibid.

17. Ibid.

18. Ibid.

19. A. Sandoiu, 'Reversing Autism with a Cancer Drug', *Medical News Today* (27 June 2018), https://www.medicalnewstoday.com/articles/322274.php.

20. Ibid.

21. Ibid.

22. S. Boseley, 'Andrew Wakefield Case Highlights the Importance of Ethics in Science', *The Guardian* (24 May 2010), https://www.theguardian.com/society/2010/may/24/andrew-wakefield-analysis-ethics-science.

23. S. Boseley, 'WHO Warns over Measles Immunisation Rates as Cases Rise 300% across Europe', *The Guardian* (19 February 2018), https://www.theguardian.com/society/2018/feb/19/who-warns-over-measles-immunisation-rates-as-cases-rise-400-across-europe.

24. A. Buncombe, 'Andrew Wakefield: How a Disgraced UK Doctor Has Remade Himself in Anti-vaxxer Trump's America', *The Independent* (4 May 2018), https://www.independent.co.uk/news/world/americas/andrew-wakefield-anti-vaxxer-trump-us-mmr-autism-link-lancet-fake-a8331826.html.

25. Z. Drewett, 'Desperate Parents Trying to "Cure" Autism by Giving Children Bleach', *Metro* (28 January 2018), https://metro.co.uk/2018/01/28/desperate-parents-trying-cure-autism-giving-children-bleach-7266567/.

26. Ibid.

27. Ibid.
28. A. E. Cha, 'Study: Autism, Creativity and Divergent Thinking May Go Hand in Hand', *Washington Post* (25 August 2015), https://www.washingtonpost.com/news/to-your-health/wp/2015/08/25/study-autism-creativity-and-divergent-thinking-may-go-hand-in-hand/?utm_term=.e0136faa9746.
29. Ibid.
30. Ibid.
31. Ibid.
32. T. McVeigh, 'People with Autism and Learning Disabilities Excel in Creative Thinking, Study Shows', *The Guardian* (22 August 2015), https://www.theguardian.com/society/2015/aug/22/autism-creative-thinking-study.
33. N. Hinde, 'Just 16% of People with Autism Are in Full-Time Paid Work and It Needs to Change, Says Charity', *Huffington Post* (27 October 2016), https://www.huffingtonpost.co.uk/entry/16-percent-of-people-with-autism-are-in-full-time-paid-work_uk_5811e71be4b0ccfc9561da6f.
34. N. L. Pesce, 'Most College Grads with Autism Can't Find Jobs. This Group Is Fixing That', *Market Watch* (2 April 2019), https://www.marketwatch.com/story/most-college-grads-with-autism-cant-find-jobs-this-group-is-fixing-that-2017-04-10-5881421.
35. Hinde, 'Just 16% of People with Autism'.
36. J. Parrington, 'Early Learning', *Socialist Review* (October 1994), https://www.marxists.org/history/etol/newspape/socrev/1994/sr179/parrington.html.
37. J. E. Knox, 'The Changing Face of Soviet Defectology: A Study in Rehabilitating the Handicapped', *Studies in Soviet Thought*, 37:217–36 (1989).
38. Parrington, 'Early Learning'.
39. Ibid.
40. Ibid.
41. Ibid.
42. Ibid.
43. Ibid.
44. P. Wolfberg, K. Bottema-Beutel, and M. DeWitt, 'Including Children with Autism in Social and Imaginary Play with Typical Peers', *American Journal of Play*, 5:55–80 (2012).
45. Ibid.
46. Ibid.
47. Ibid.
48. Parrington, 'Early Learning'.
49. Wolfberg, Bottema-Beutel, and DeWitt, 'Including Children with Autism'.
50. Ibid.
51. Ibid.
52. Ibid.
53. K. Gander, 'World Bipolar Day: What Is Bipolar Disorder? What Are Its Symptoms?', *The Independent* (29 March 2016), https://www.independent.co.uk/life-style/health-and-families/features/world-bipolar-day-what-is-bipolar-disorder-and-what-are-its-symptoms-a6958426.html.
54. H. Thomson, 'Intelligence, Creativity and Bipolar Disorder May Share Underlying Genetics', *The Guardian* (19 August 2015), https://www.theguardian.com/science/2015/aug/19/intelligence-creativity-and-bipolar-disorder-may-share-underlying-genetics.
55. Ibid.

56. Ibid.

57. N. Wolchover, 'Why Are Genius and Madness Connected?', *Live Science* (2 June 2012), https://www.livescience.com/20713-genius-madness-connected.html.

58. Ibid.

59. Ibid.

60. Ibid.

61. Ibid.

62. H. J. Parkinson, 'Having a Mental Illness Doesn't Make You a Genius', *The Guardian* (10 June 2015), https://www.theguardian.com/commentisfree/2015/jun/10/mental-illness-study-bipolar-disorder-creativity.

Chapter 15: Crime and Punishment

1. S. Jenkins, 'Could Bradford be the Shoreditch of Yorkshire—or is it the Next Detroit?', *The Guardian* (3 May 2018), https://www.theguardian.com/cities/2018/may/03/could-bradford-be-the-shoreditch-of-yorkshire-or-is-it-the-next-detroit-.

2. C. Wilde, 'Bradford Wins Capital of Curry Crown for Sixth Year in Row', *Telegraph and Argus* (10 October 2016), http://www.thetelegraphandargus.co.uk/news/14791659.Bradford_wins_Capital_of_Curry_crown_for_sixth_year_in_row/.

3. M. Wainwright, 'Bradford Wheels out Brontes, Hockney, Priestley and Curry after Poor Reviews', *The Guardian* (24 February 2011), https://www.theguardian.com/travel/2011/feb/24/bradford-brontes-cinemas-curry-reviews.

4. J. Smith, 'The Yorkshire Ripper Was Not a "Prostitute Killer"—Now His Forgotten Victims Need Justice', *The Telegraph* (30 May 2017), https://www.telegraph.co.uk/women/life/yorkshire-ripper-not-prostitute-killer-forgotten-victims-need/.

5. Ibid.

6. M. Lockley, 'Black Panther Donald Neilson's Trail of Terror and Murder of Lesley Whittle 40 Years Ago Remembered by Top Cop', *Birmingham Mail* (17 January 2015), https://www.birminghammail.co.uk/news/midlands-news/black-panther-donald-neilsons-trail-8464470.

7. 'Legacy of Black Panther Murders', *BBC News* (27 January 2010), http://news.bbc.co.uk/local/shropshire/hi/people_and_places/history/newsid_8365000/8365884.stm.

8. J. Patterson, 'Why The Black Panther Can Hold Its Head Up High', *The Guardian* (27 January 2012), https://www.theguardian.com/film/filmblog/2012/jun/06/the-black-panther-donald-neilson.

9. Ibid.

10. Ibid.

11. M. Simon, 'Fantastically Wrong: The Scientist Who Seriously Believed Criminals Were Part Ape', *Wired* (12 November 2014), https://www.wired.com/2014/11/fantastically-wrong-criminal-anthropology/.

12. Ibid.

13. Ibid.

14. A. Rutherford, 'Why We Can't Blame "Warrior Genes" for Violent Crime', *New Statesman* (16 September 2016), https://www.newstatesman.com/2016/09/why-we-can-t-blame-warrior-genes-violent-crime.

15. S. Knapton, 'Violence Genes May Be Responsible for One in 10 Serious Crimes', *The Telegraph* (28 October 2014), https://www.telegraph.co.uk/news/science/science-news/11192643/Violence-genes-may-be-responsible-for-one-in-10-serious-crimes.html.

16. Ibid.
17. T. Adams, 'How to Spot a Murderer's Brain', *The Guardian* (12 May 2013), https://www.theguardian.com/science/2013/may/12/how-to-spot-a-murderers-brain.
18. Ibid.
19. Ibid.
20. Ibid.
21. J. Wells, 'Low Resting Heart Rate Linked to "Future Violent or Anti-social Behaviour"', *The Telegraph* (17 September 2015), https://www.telegraph.co.uk/men/the-filter/11863460/Low-resting-heart-rate-linked-to-future-violent-or-anti-social-behaviour.html.
22. Adams, 'How to Spot a Murderer's Brain'.
23. T. Wallace, 'Brain Lesions Contribute to Criminal Behaviour, Study Finds', *Cosmos* (20 December 2017), https://cosmosmagazine.com/social-sciences/brain-lesions-contribute-to-criminal-behaviour-study-finds.
24. Ibid.
25. Ibid.
26. Ibid.
27. R. Keller, 'Macabre World of the Warped Killer Who Inspired *Silence of the Lambs*—Featuring Soup Bowls Made from Human Skulls and Even Grislier "Trophies"', *Daily Mirror* (30 August 2017), https://www.mirror.co.uk/news/real-life-stories/grisly-world-warped-killer-who-11081022.
28. G. Paoletti, 'The Story of Serial Killer Edmund Kemper, Whose Story Is Almost Too Gross to Be Real', *All That's Interesting* (7 November 2018), https://allthatsinteresting.com/edmund-kemper.
29. 'Gary Heidnik', *Criminal Minds* (2018), http://criminalminds.wikia.com/wiki/Gary_Heidnik.
30. C. Carter, 'From Lonely Weakling to the Yorkshire Ripper: How Serial Killer Peter Sutcliffe Developed His Murderous Hatred', *Daily Mirror* (13 November 2020), https://www.mirror.co.uk/news/uk-news/lonely-weakling-yorkshire-ripper-how-11213299.
31. H. Mitchell and M. G. Aamodt, 'The Incidence of Child Abuse in Serial Killers', *Journal of Police and Criminal Psychology*, 20:40–7 (2005).
32. G. Tenbergen, M. Wittfoth, H. Frieling, J. Ponseti, M. Walter, H. Walter, K. M. Beier, B. Schiffer, and T. H. Kruger, 'The Neurobiology and Psychology of Pedophilia: Recent Advances and Challenges', *Frontiers in Human Neuroscience*, 9:344 (2015).
33. Ibid.
34. R. Sanders, 'Are Paedophiles' Brains Wired Differently?', *BBC News* (24 November 2015), https://www.bbc.co.uk/news/magazine-34858350.
35. Ibid.
36. J. Grayson, 'He Is a Paedophile, But That Does Not Make Him a Child Molester', *Huffington Post* (17 September 2017), https://www.huffingtonpost.co.uk/juliet-grayson/he-is-a-paedophile-but-th_b_12046562.html.
37. Ibid.
38. K. Gander, 'The Man Whose Brain Tumour "Turned Him into a Paedophile"', *The Independent* (24 February 2016), https://www.independent.co.uk/life-style/health-and-families/features/a-40-year-old-developed-an-obsession-with-child-pornography-then-doctors-discovered-why-a6893756.html.
39. Ibid.
40. Ibid.

41. C. Weaver, 'Are You Raising a Paedophile?', *Now to Love* (7 June 2016), https://www.nowtolove.com.au/news/real-life/nature-vs-nurture-are-you-raising-a-paedophile-10454.

42. Ibid.

43. Ibid.

44. L. Buchen, 'Neuroscience: In Their Nurture', *Nature* 467:146–8 (2010).

45. Ibid.

46. B. Weidmann, 'The Making of a Bully', *The Scientist* (25 January 2013), https://www.the-scientist.com/daily-news/the-making-of-a-bully-39884.

47. Ibid.

48. Ibid.

49. S. C. Johnson, 'The New Theory That Could Explain Crime and Violence in America', *Medium* (18 February 2014), https://medium.com/matter/the-new-theory-that-could-explain-crime-and-violence-in-america-945462826399.

50. Ibid.

51. Ibid.

52. Ibid.

53. Ibid.

54. W. Hirstein, 'What Is a Psychopath?', *Psychology Today* (30 January 2013), https://www.psychologytoday.com/us/blog/mindmelding/201301/what-is-psychopath-0.

55. Ibid.

56. S. A. Bonn, 'Serial Killer Myth #1: They're Mentally Ill or Evil Geniuses', *Psychology Today* (16 June 2014), https://www.psychologytoday.com/us/blog/wicked-deeds/201406/serial-killer-myth-1-theyre-mentally-ill-or-evil-geniuses.

57. S. Rose, 'From Split to Psycho: Why Cinema Fails Dissociative Identity Disorder', *The Guardian* (12 January 2017), https://www.theguardian.com/film/2017/jan/12/cinema-dissociative-personality-disorder-split-james-mcavoy.

58. Ibid.

59. Ibid.

60. J. Jitchotvisut, 'How Ted Bundy Got Away with So Many Murders, According to a Forensic Psychologist', *Insider* (28 January 2019), https://www.insider.com/ted-bundy-case-explained-forensic-psychologist-2019-1.

61. J. C. Motzkin, J. P. Newman, K. A. Kiehl, and M. Koenigs, 'Reduced Prefrontal Connectivity in Psychopathy', *Journal of Neuroscience*, 31:17348–57 (2011).

62. S. Vaknin, 'Serial and Mass Killers', *Mental Health Matters* (14 April 2009), https://mental-health-matters.com/serial-and-mass-killers/.

63. S. A. Bonn, 'Understanding What Drives Serial Killers', *Psychology Today* (15 September 2019), https://www.psychologytoday.com/gb/blog/wicked-deeds/201909/understanding-what-drives-serial-killers.

64. R. Zachary, 'Night Stalker: The Life and Death of Richard Ramirez', *Serial Killer* (14 October 2020), https://serialkillershop.com/blogs/true-crime/richard-ramirez-night-stalker.

65. V. Allan, 'Why We Must Understand Savile Psyche', *The Herald* (29 June 2014), https://www.heraldscotland.com/opinion/13167665.why-we-must-understand-savile-psyche/.

66. A. Kalia, '"We're All Competing for the Same Jobs": Life in Britain's Youngest City', *The Guardian* (5 February 2018), https://www.theguardian.com/cities/2018/feb/05/life-britain-youngest-city-bradford-uk-unemployment.

67. Ibid.

68. Ibid.

69. O. Burkeman, 'Voice of America', *The Guardian* (1 March 2002), https://www.theguardian.com/books/2002/mar/01/studsterkel.

70. V. Dodd, 'Rising Crime is Symptom of Inequality, Says Senior Met Chief', *The Guardian* (14 June 2018), https://www.theguardian.com/uk-news/2018/jun/14/rising-is-symptom-of-inequality-says-senior-met-chief.

71. Ibid.

72. C. Ferguson, 'Banking Is a Criminal Industry Because Its Crimes Go Unpunished', *Huffington Post* (16 July 2012), https://www.huffingtonpost.com/charles-ferguson/bank-crimes_b_1675714.html.

73. Ibid.

74. Ibid.

75. Ibid.

76. R. Schouten, 'Psychopaths on Wall Street', *Harvard Business Review* (14 March 2012), https://hbr.org/2012/03/psychopaths-on-wall-street.

Chapter 16: Class and Division

1. M. Kaufman, 'The Internet Revolution is the New Industrial Revolution', *Forbes* (5 October 2012), https://www.forbes.com/sites/michakaufman/2012/10/05/the-internet-revolution-is-the-new-industrial-revolution/.

2. P. Torres, 'It's the End of the World and We Know It: Scientists in Many Disciplines See Apocalypse, Soon', *Salon* (30 April 2017), https://www.salon.com/2017/04/30/its-the-end-of-the-world-and-we-know-it-scientists-in-many-disciplines-see-apocalypse-soon/.

3. J. Gibbons, 'Climate Change Is Reaching the Point of no Return', *The Times* (24 August 2018), https://www.thetimes.co.uk/article/climate-change-is-reaching-the-point-of-no-return-john-gibbons-5jtmj7jt6.

4. J. Vidal, '"Tip of the Iceberg": Is Our Destruction of Nature Responsible for Covid-19?', *The Guardian* (18 March 2020), https://www.theguardian.com/environment/2020/mar/18/tip-of-the-iceberg-is-our-destruction-of-nature-responsible-for-covid-19-aoe.

5. C. Bickerton, 'The Collapse of Europe's Mainstream Centre Left', *New Statesman* (1 May 2018), https://www.newstatesman.com/world/europe/2018/05/collapse-europe-s-mainstream-centre-left.

6. F. Rich, 'Ten Years after the Crash, We Are Still Living in the World It Brutally Remade', *New York Magazine* (5 August 2018), http://nymag.com/daily/intelligencer/2018/08/america-10-years-after-the-financial-crisis.html.

7. M. Savage, 'Richest 1% on Target to Own Two-Thirds of All Wealth by 2030', *The Observer* (7 April 2018), https://www.theguardian.com/business/2018/apr/07/global-inequality-tipping-point-2030.

8. J. Pickrell, 'Timeline: Human Evolution', *New Scientist* (4 September 2006), https://www.newscientist.com/article/dn9989-timeline-human-evolution/.

9. R. J. Bankoff and G. H. Perry, 'Hunter-Gatherer Genomics: Evolutionary Insights and Ethical Considerations', *Current Opinion in Genetics and Development*, 41:1–7 (2016).

10. J. Suzman, 'Why "Bushman Banter" Was Crucial to Hunter-Gatherers' Evolutionary Success', *The Guardian* (29 October 2017), https://www.theguardian.com/inequality/2017/oct/29/why-bushman-banter-was-crucial-to-hunter-gatherers-evolutionary-success.

11. Ibid.

12. Ibid.

13. Ibid.
14. A. Chua, 'How America's Identity Politics Went from Inclusion to Division', *The Guardian* (1 March 2018), https://www.theguardian.com/society/2018/mar/01/how-americas-identity-politics-went-from-inclusion-to-division.
15. J. Suzman, 'How Neolithic Farming Sowed the Seeds of Modern Inequality 10,000 Years Ago', *The Guardian* (5 December 2017), https://www.theguardian.com/inequality/2017/dec/05/how-neolithic-farming-sowed-the-seeds-of-modern-inequality-10000-years-ago.
16. B. Cunliffe, '*Against the Grain* by James C Scott Review—the Beginning of Elites, Tax, Slavery', *The Guardian* (25 November 2017), https://www.theguardian.com/books/2017/nov/25/against-the-grain-by-james-c-scott-review.
17. Ibid.
18. J. Diamond, 'The Worst Mistake in the History of the Human Race', *Discover Magazine* (1 May 1999), http://discovermagazine.com/1987/may/02-the-worst-mistake-in-the-history-of-the-human-race.
19. Cunliffe, '*Against the Grain* by James C Scott Review'.
20. Ibid.
21. Ibid.
22. G. E. M. De Sainte Croix, *The Class Struggle in the Ancient World* (Duckworth, 2001), p. 40.
23. M. Beard, 'A Radical, Short-Lived and Violent Experiment: the Origins of Democracy', *The Guardian* (29 April 2006), https://www.theguardian.com/commentisfree/2006/apr/29/comment.politics1.
24. Aristotle, 'Politics', *MIT Internet Classics Archive* (2009), http://classics.mit.edu/Aristotle/politics.1.one.html.
25. M. E. Price, 'Why We Think Monogamy Is Normal', *Psychology Today* (9 September 2011), https://www.psychologytoday.com/intl/blog/darwin-eternity/201109/why-we-think-monogamy-is-normal.
26. D. Ross, 'Feudalism and Medieval life', *Britain Express* (2020), https://www.britainexpress.com/History/Feudalism_and_Medieval_life.htm.
27. L. G. Bowman, 'The Paradox of the Declaration of Independence', *Aspen Institute* (1 July 2016), https://www.aspeninstitute.org/blog-posts/every-american-know-paradox-declaration-independence/.
28. J. M. C. Lustiger, 'Liberty, Equality, Fraternity', *First Things* (October 1997), https://www.firstthings.com/article/1997/10/002-liberty-equality-fraternity.
29. R. Cox, 'Inequality Gap: Growing Gulf between Rich and Poor Leaves 42 People with Same Wealth as World's 3.7bn Worst off', *The Independent* (22 January 2018), https://www.independent.co.uk/news/business/news/oxfam-report-worlds-richest-poorest-inequality-gap-wealth-2017-davos-wef-a8167676.html.
30. H. O'Shaughnessy, 'Chilean Coup: 40 Years Ago I Watched Pinochet Crush a Democratic Dream', *The Guardian* (7 September 2013), https://www.theguardian.com/world/2013/sep/07/chile-coup-pinochet-allende.
31. E. O'Carroll, 'Karl Marx: 10 Great Quotes on His Birthday', *Christian Science Monitor* (4 May 2012), https://www.csmonitor.com/Books/2012/0504/Karl-Marx-10-great-quotes-on-his-birthday/On-the-mainstream-media.
32. S. A. Diamond, 'Let's Talk about Loneliness: Alienation in a Linked Up Age', *Psychology Today* (21 February 2014), https://www.psychologytoday.com/us/blog/evil-deeds/201402/lets-talk-about-loneliness-alienation-in-linked-age.
33. J. Elster, *Karl Marx: A Reader* (Cambridge University Press, 2008), p. 39.

34. K. Moos, 'Working Out the Meaning of "Meaningful" Work', *Chronicle Vitae* (30 October 2014), https://chroniclevitae.com/news/781-working-out-the-meaning-of-meaningful-work.

35. G. Kennedy, 'Of Pins and Things', *Adam Smith Institute* (28 May 2012), https://www.adamsmith.org/blog/economics/of-pins-and-things.

36. T. Hindle, 'Mass Production', *The Economist* (20 October 2009), https://www.economist.com/news/2009/10/20/mass-production.

37. S. Ghosh, 'Peeing in Trash Cans, Constant Surveillance, and Asthma Attacks on the Job: Amazon Workers Tell Us Their Warehouse Horror Stories', *Business Insider* (5 May 2018), https://www.businessinsider.com/amazon-warehouse-workers-share-their-horror-stories-2018-4?r=US&IR=T.

38. Ibid.

39. J. Grove, 'Fixed-Term Now the Norm for Early Career Academics, Says UCU', *Times Higher Education* (14 April 2016), https://www.timeshighereducation.com/news/fixed-term-now-the-norm-for-early-career-academics-says-university-and-college-union-ucu.

40. S. Weale, 'Part-Time Lecturers on Precarious Work: "I Don't Make Enough for Rent"', *The Guardian* (16 November 2016), https://www.theguardian.com/uk-news/2016/nov/16/part-time-lecturers-on-precarious-work-i-dont-make-enough-for-rent.

41. L. Dodgson, 'Here's Why CEOs Often Have the Traits of a Psychopath', *Business Insider* (7 July 2017), http://uk.businessinsider.com/ceos-often-have-psychopathic-traits-2017-7.

42. J. Silver, Is Wall Street Full of Psychopaths?, *The Atlantic* (2012), https://www.theatlantic.com/health/archive/2012/03/is-wall-street-full-of-psychopaths/254944/

43. A. Rzepniknowska, 'Racism and Xenophobia Experienced by Polish Migrants in the UK before and after Brexit Vote', *Journal of Ethnic and Migration Studies*, 45:61–77, (2018).

44. M. Berry, I. Garcia-Blanco, and K. Moore, 'UK Press Is the Most Aggressive in Reporting on Europe's "Migrant" Crisis', *The Conversation* (14 March 2016), https://theconversation.com/uk-press-is-the-most-aggressive-in-reporting-on-europes-migrant-crisis-56083.

45. A. Bovey, 'Women in Medieval Society', *British Library* (30 April 2015), https://www.bl.uk/the-middle-ages/articles/women-in-medieval-society.

46. H. J. Sharman, E. K. Hunt, R. F. Nesiba, and B. Wiens-Tuers, *Economics: An Introduction to Traditional and Progressive Values* (M.E. Sharpe, 2008), p. 37.

47. R. Stites, *The Women's Liberation Movement in Russia: Feminism, Nihilsm, and Bolshevism, 1860–1930* (Princeton University Press, 1978), p. 160.

48. E. Goddard and J. Cox, 'Women's Suffrage: After 100 Years since Millions of Women Got the Vote around the World, How Do Their Rights Compare Now?', *The Independent* (6 February 2018), https://www.independent.co.uk/news/long_reads/women-suffrage-100-years-get-vote-right-uk-britain-ireland-us-world-countries-compare-switzerland-a8191506.html.

49. K. Parker, 'Despite Progress, Women Still Bear Heavier Load Than Men in Balancing Work and Family', *Pew Research Center* (10 March 2015), http://www.pewresearch.org/fact-tank/2015/03/10/women-still-bear-heavier-load-than-men-in-balancing-work-family/.

50. R. Solnit, 'A Broken Idea of Sex is Flourishing. Blame Capitalism', *The Guardian* (12 May 2018), https://www.theguardian.com/commentisfree/2018/may/12/sex-capitalism-incel-movement-misogyny-feminism.

51. A. Langone, '#MeToo and Time's Up Founders Explain the Difference between the 2 Movements—And How They're Alike', *Time* (22 March 2018), http://time.com/5189945/whats-the-difference-between-the-metoo-and-times-up-movements/.

52. R. Lewis, 'What Actually is a Belief? And Why Is It So Hard to Change?', *Psychology Today* (7 October 2018), https://www.psychologytoday.com/gb/blog/finding-purpose/201810/what-actually-is-belief-and-why-is-it-so-hard-change.

53. P. Agarwal, 'What Neuroimaging Can Tell Us about Our Unconscious Biases', *Scientific American* (12 April 2020), https://blogs.scientificamerican.com/observations/what-neuroimaging-can-tell-us-about-our-unconscious-biases/.

54. Ibid.

55. M. Haiken, 'More Than 10,000 Suicides Tied to Economic Crisis, Study Says', *Forbes* (12 June 2014), https://www.forbes.com/sites/melaniehaiken/2014/06/12/more-than-10000-suicides-tied-to-economic-crisis-study-says/.

56. J. Henley, '"Recessions Can Hurt, But Austerity Kills"', *The Guardian* (15 May 2013), https://www.theguardian.com/society/2013/may/15/recessions-hurt-but-austerity-kills.

57. Ibid.

58. J. Kanter and K. Manbeck, 'COVID-19 Could Lead to an Epidemic of Clinical Depression, and the Health Care System Isn't Ready for That, Either', *The Conversation* (1 April 2020), http://theconversation.com/covid-19-could-lead-to-an-epidemic-of-clinical-depression-and-the-health-care-system-isnt-ready-for-that-either-134528.

59. R. Pinto, M. Ashworth, and R. Jones, 'Schizophrenia in black Caribbeans Living in the UK: An Exploration of Underlying Causes of the High Incidence Rate', *British Journal of General Practice*, 58:429–434 (2008).

60. Ibid.

61. Ibid.

62. 'UK Life Blamed for Ethnic Schizophrenia', *BBC News* (14 July 2000), http://news.bbc.co.uk/1/hi/health/807945.stm.

63. T. Evans, 'Understanding the Gender Pay Gap in the UK', *Office for National Statistics* (17 January 2018), https://www.ons.gov.uk/employmentandlabourmarket/peopleinwork/earningsandworkinghours/articles/understandingthegenderpaygapintheuk/2018-01-17.

64. R. Neate, 'Global Pay Gap Will Take 202 Years to Close, Says World Economic Forum', *The Guardian* (18 December 2018), https://www.theguardian.com/world/2018/dec/18/global-gender-pay-gap-will-take-202-years-to-close-says-world-economic-forum.

65. S. Zacharek, E. Dockterman, and H. S. Edwards, 'TIME Person of the Year: The Silence Breakers', *Time* (5 March 2017), https://time.com/time-person-of-the-year-2017-silence-breakers/.

66. J. Ducharme, 'Any Type of Sexual Harassment Can Cause Psychological Harm, Study Says', *Time* (9 November 2017), http://time.com/5017072/sexual-harassment-psychological-damage/.

67. Ibid.

68. P. R. Albert, 'Why Is Depression More Prevalent in Women?', *Journal of Psychiatry and Neuroscience*, 40:219–21 (2015).

69. J. Bindel, 'Tell Me If You Still Think Prostitution Is Empowering after Hearing What The Buying Punters Have to Say', *The Independent* (19 September 2017), https://www.independent.co.uk/voices/sex-work-punters-what-do-they-think-prostitution-exploitation-rape-danger-a7889511.html.

70. Ibid.

71. M. Farley, 'Risks of Prostitution: When the Person Is the Product', *Journal of the Association for Consumer Research*, 3:97–108 (2018).

72. A. Travis, 'Yorkshire Ripper Mental Health Review Revives Central Issue of Trial', *The Guardian* (1 December 2015), https://www.theguardian.com/uk-news/2015/dec/01/yorkshire-ripper-peter-sutcliffe-mental-health-1981-trial.

73. L. Thornton, 'Yorkshire Ripper Killed EIGHT More Women, Claims the Ex-cop Who Interviewed Him More Than 30 Times', *Daily Mirror* (6 April 2017), https://www.mirror.co.uk/news/uk-news/yorkshire-ripper-linked-eight-more-10174613.

74. L. Wattis, 'The Social Nature of Serial Murder: The Intersection of Gender and Modernity', *European Journal of Women's Studies*, 243:381–93 (2016).

75. J. Bindel, 'Peter Sutcliffe Should Never Be Freed', *The Guardian* (2 March 2010), https://www.theguardian.com/commentisfree/2010/mar/02/peter-sutcliffe-hate-crimes-women.

Chapter 17: Resistance and Rebellion

1. W. F. Edgerton, 'The Strikes in Ramses III's Twenty-Ninth Year', *Journal of Near Eastern Studies*, 10:137–45 (1951).

2. H. Richardson, 'University Staff Suspend Strikes over Pensions', *BBC News* (13 April 2018), https://www.bbc.co.uk/news/education-43758015.

3. J. J. Mark, 'The First Labor Strike in History', *Ancient History Encylopedia* (4 July 2017), https://www.ancient.eu/article/1089/the-first-labor-strike-in-history/.

4. Ibid.

5. V. Blake, 'I'm Striking with University Colleagues as Our Pensions Are Being Destroyed', *The Guardian* (22 February 2018), https://www.theguardian.com/commentisfree/2018/feb/22/university-strike-pensions-retirement-security.

6. M. Coldrick, 'Margaret Thatcher and the Pit Strike in Yorkshire', *BBC News* (8 April 2013), https://www.bbc.co.uk/news/uk-england-22068640.

7. S. Milne, 'Now We See What Was Really at Stake in the Miners' Strike', *The Guardian* (12 March 2014), https://www.theguardian.com/commentisfree/2014/mar/12/miners-strike-gutting-unions-bob-crow.

8. M. Townsend, 'Miners' Strike: "All I Want is for Someone to Say: I'm Sorry"', *The Observer* (1 December 2012), https://www.theguardian.com/politics/2012/dec/02/miners-strike-orgreave-special-report.

9. A. Gillan, 'I Was Always Told I Was Thick. The Strike Taught Me I Wasn't', *The Guardian* (10 May 2004), https://www.theguardian.com/politics/2004/may/10/past.women.

10. D. Hayes, 'Thirty Years on: the Socialist Workers Party and the Great Miners' Strike', *International Socialism* (2 April 2014), http://isj.org.uk/thirty-years-on-the-socialist-workers-party-and-the-great-miners-strike/.

11. K. Kellaway, 'When Miners and Gay Activists United: The Real Story of the Film *Pride*', *The Guardian* (31 August 2014), https://www.theguardian.com/film/2014/aug/31/pride-film-gay-activists-miners-strike-interview.

12. Ibid.

13. M. Lehtonen, *The Cultural Analysis of Texts* (Sage Publications, 2000), p. 157.

14. Ibid.

15. J. Coles, J. Hearn, and C. Royle, 'UCU: This is a Dispute We Can Win', *Socialist Review* (April 2018), http://socialistreview.org.uk/434/ucu-dispute-we-can-win.

16. Ibid.

17. Ibid.

18. J. Spence and C. Stephenson, '"Side by Side with Our Men?" Women's Activism, Community, and Gender in the 1984–1985 British Miners' Strike', *International Labor and Working-Class History*, 75:68–84 (2009).

19. J. Rees and L. German, *A People's History of London* (Verso Books, 2012), p. 41.

20. Ibid.

21. Ibid.

22. M. Lavalette and G. Mooney, *Class Struggle and Social Welfare* (Routledge, 2013), p. 61.

23. Ibid.

24. A. Gorz, 'Reform and Revolution', *Socialist Register* (March 1968), https://socialistregister.com/index.php/srv/article/view/5272/0.

25. A. J. Rubin, 'May 1968: A Month of Revolution Pushed France Into the Modern World', *New York Times* (5 May 2018), https://www.nytimes.com/2018/05/05/world/europe/france-may-1968-revolution.html.

26. C. Hill, *The World Turned Upside Down: Radical Ideas during the English Revolution* (Penguin, 1991).

27. M. O. Grenby, 'Writing Revolution: British Literature and the French Revolution Crisis, a Review of Recent Scholarship', *Literature Compass*, 3:1351–85 (2006).

28. M. Cole, K. Levitin, and A.R. Luria, *The Autobiography of Alexander Luria: A Dialogue with the Making of Mind* (Psychology Press, 2010), p. 17.

29. C. A. Hartley and L. H. Somerville, 'The Neuroscience of Adolescent Decision-Making', *Current Opinion in Behavioral Sciences*, 5:108–15 (2015).

30. K. Marx, *The Eighteenth Brumaire of Louis Bonaparte* (1852), https://www.marxists.org/archive/marx/works/1852/18th-brumaire/ch01.htm.

31. G. Tiruneh, 'Social Revolutions: Their Causes, Patterns, and Phases', *SAGE Open* (18 September 2014), http://journals.sagepub.com/doi/full/10.1177/2158244014548845.

32. D. Saldanha, 'Antonio Gramsci and the Analysis of Class Consciousness: Some Methodological Considerations', *Economic and Political Weekly*, 23:11–18 (1988).

33. Ibid.

34. C. Hill, *The Century of Revolution: 1603–1714* (Routledge, 1980).

35. D. Sandbrook, 'The Man Who Wouldn't Be King', *New Statesman* (29 December 2010), https://www.newstatesman.com/society/2010/12/cromwell-god-essay-history.

36. T. Benn, 'The Levellers and the Tradition of Dissent', *BBC History* (17 February 2011), http://www.bbc.co.uk/history/british/civil_war_revolution/benn_levellers_01.shtml.

37. J. Rees, 'In Defence of October', *International Socialism*, 52:3–79 (Autumn 1991), https://www.marxists.org/history/etol/writers/rees-j/1991/xx/october.html.

38. C. Harris, 'Russia's February Revolution Was Led by Women on the March', *Smithsonian Magazine* (17 February 2017), https://www.smithsonianmag.com/history/russias-february-revolution-was-led-women-march-180962218/.

39. T. Vallance, 'Fresh Perspectives on the Levellers', *History Today* (6 June 2017), https://www.historytoday.com/reviews/fresh-perspectives-levellers.

40. S. Weissman, 'The Golden Era', *Jacobin Magazine* (18 December 2017), https://www.jacobinmag.com/2017/12/victor-serge-russian-revolution-bolsheviks.

41. B. J. Luskin, 'MRIs Reveal Unconscious Bias in the Brain', *Psychology Today* (7 April 2016), https://www.psychologytoday.com/gb/blog/the-media-psychology-effect/201604/mris-reveal-unconscious-bias-in-the-brain.

42. M. Rouault, J. Drugowitsch, and E. Koechlin, 'Prefrontal Mechanisms Combining Rewards and Beliefs in Human Decision-Making', *Nature Communications*, 10:301 (2019).

43. G. Walton, 'Théroigne de Méricourt, Heroine of the French Revolution', *Amazing Women in History* (2016), http://www.amazingwomeninhistory.com/theroigne-de-mericourt/.
44. Ibid.
45. Ibid.
46. H. Mantel, 'Rescued by Marat', *London Review of Books*, 14:15–16 (1992).
47. Ibid.
48. Walton, 'Théroigne de Méricourt, Heroine of the French Revolution'.
49. I. Ferguson, *Politics of the Mind: Marxism and Mental Distress* (Bookmarks, 2017), p. 30.
50. E. Fee and T. M. Brown, 'Freeing the Insane', *American Journal of Public Health*, 96:1743 (2006).
51. A. Rogers and D. Pilgrim, *Mental Health Policy in Britain: A Critical Introduction* (Palgrave Macmillan, 1996), p. 50.
52. B. Manning, *The English People and the English Revolution* (Bookmarks, 1995).
53. S. Roberts, 'Col. Thomas Rainborowe: "The Poorest He That Is in England Hath a Right to Live as the Greatest He"', *The History of Parliament* (18 April 2013), https://thehistoryofparliament.wordpress.com/2013/04/18/col-thomas-rainborowe-the-poorest-he-that-is-in-england-hath-a-right-to-live-as-the-greatest-he/.
54. R. Coalson, 'Gulags Were a Horrific Cornerstone of Stalinist Russia', *The Atlantic* (5 March 2013), https://www.theatlantic.com/international/archive/2013/03/gulags-were-a-horrific-cornerstone-of-stalinist-russia/273703/.
55. D. Sherry, *Russia 1917: Workers' Revolution and the Festival of the Oppressed* (Bookmarks, 2017), p. 219.
56. A. Kozulin, *Vygotsky's Psychology: A Biography of Ideas* (Harvester Wheatsheaf, 1990), pp. 239–40.
57. G. Dvorsky, 'How the Soviets Used Their Own Twisted Version of Psychiatry to Suppress Political Dissent', *Gizmodo* (4 September 2012), https://i09.gizmodo.com/how-the-soviets-used-their-own-twisted-version-of-psych-5940212.
58. Ibid.
59. Ibid.
60. R. D. Strous, 'Psychiatry during the Nazi Era: Ethical Lessons for the Modern Professional', *Annals of General Psychiatry*, 6:8 (2007).
61. T. Burns, *Our Necessary Shadow: The Nature and Meaning of Psychiatry* (Pegasus Books, 2014), p. 50.
62. G. Le Bon, *The Psychology of Revolutionary Crowds* (London: T Fisher Unwin, 1913).
63. Ibid.
64. Ibid.
65. J. Drury, 'Impact: From Riots to Crowd Safety', *The Psychologist* (February 2016), https://thepsychologist.bps.org.uk/volume-29/february/riots-crowd-safety.
66. C. T. Miéville, 'The Day That Shook the World', *Jacobin Magazine* (7 November 2017), https://jacobinmag.com/2017/11/october-revolution-china-mieville-bolsheviks.
67. N. Carlin, 'The Levellers and the Conquest of Ireland in 1649', *The Historical Journal*, 30:269–88 (1987).
68. J. Israel, *Revolutionary Ideas: An Intellectual History of the French Revolution from the Rights of Man to Robespierre* (Princeton University Press, 2014), pp. 182–3.
69. S. Sagall, 'Antisemitism and the Russian Revolution', *Socialist Review* (April 2018), http://socialistreview.org.uk/434/antisemitism-and-russian-revolution.
70. Ibid.
71. J. Rees, *The Leveller Revolution* (Verso, 2016), pp. 290–2.

72. J. Orr, *Marxism and Women's Liberation* (Bookmarks, 2015), pp. 94–5.

73. D. Sherry, *Russia 1917: Workers' Revolution and the Festival of the Oppressed* (Bookmarks, 2017), p. 185.

74. M. Bond, 'Mob Mentality', *Slate* (17 March 2015), https://slate.com/technology/2015/03/crowd-psychology-people-are-friendly-altruistic-happy-in-large-gatherings.html.

75. B. Trew, 'How Distaste of LGBT People in Egypt Has Turned into State-Sponsored Persecution', *The Independent* (17 May 2015), https://www.independent.co.uk/news/world/middle-east/how-distaste-of-lgbt-people-in-egypt-has-turned-into-state-sponsored-persecution-10256869.html.

76. M. L. Thomas, 'Fascism in Europe Today', *International Socialism*, 162:27–64 (2019).

77. P. Brown, in P. Brown (ed.), *Radical Psychology* (Tavistock Publications, 1973), pp. 244–56.

78. Ibid.

79. O. Nachtwey, *Germany's Hidden Crisis: Social Decline in the Heart of Europe* (Verso, 2018), p. 133.

80. Ibid.

81. S. Knapton, 'Nine in 10 People Would Electrocute Others If Ordered, Rerun of infamous Milgram Experiment Shows', *The Telegraph* (14 March 2017), https://www.telegraph.co.uk/science/2017/03/14/nine-10-people-would-electrocute-others-ordered-re-run-milgram/.

82. Ibid.

83. Ibid.

84. M. Stuchbery, 'What We Can Learn from the White Rose Siblings', *The Local De* (22 February 2019), https://www.thelocal.de/20190222/76-years-later-what-the-white-rose-siblings-teach-us.

85. O. B. Waxman, '"He Was Sent by God to Take Care of Us": Inside the Real Story behind *Schindler's List*', *Time* (7 December 2018), https://time.com/5470613/schindlers-list-true-story/.

86. S. B. Carroll, 'Deep Secrets and the Thrill of Discovery', *Quanta* (25 February 2016) https://www.quantamagazine.org/deep-secrets-and-the-thrill-of-discovery-20160225/.

Chapter 18: Music and Rhythm

1. D. Randall, *Sound System* (Pluto Books, 2017), p. 1.

2. B. Lawergren, 'The Origin of Musical Instruments and Sounds', *Anthropos*, 88:31–45 (1988).

3. S. T. Parker and K. E. Jaffe, *Darwin's Legacy: Scenarios in Human Evolution* (Altamira Press, 2008), p. 7.

4. 'Why Music?', *The Economist* (18 December 2008), https://www.economist.com/christmas-specials/2008/12/18/why-music.

5. Ibid.

6. Ibid.

7. Ibid.

8. Ibid.

9. R. D. Fields, 'The Power of Music: Mind Control by Rhythmic Sound', *Scientific American* (19 October 2012), https://blogs.scientificamerican.com/guest-blog/the-power-of-music-mind-control-by-rhythmic-sound/.

10. Ibid.

11. Ibid.

12. Ibid.

13. S. Szeles, '7 Theories on Why We Evolved to Love Music', *Nova* (21 May 2014), http://www.pbs.org/wgbh/nova/blogs/secretlife/blogposts/the-evolution-of-music/.

14. E. Landau, 'This Is Your Brain on Music', *CNN* (23 January 2018), https://edition.cnn.com/2013/04/15/health/brain-music-research/index.html.

15. Ibid.

16. Ibid.

17. Ibid.

18. Ibid.

19. J. Schulkin and G. B. Raglan, 'The Evolution of Music and Human Social Capability', *Frontiers in Neuroscience*, 8:1–13 (2014).

20. Ibid.

21. L. Marquand-Brown, 'Music and Human Evolution', *OUP Blog* (25 July 2017), https://blog.oup.com/2017/07/music-human-evolution/.

22. Ibid.

23. I. Morley, *The Prehistory of Music* (Oxford University Press, 2013), Chapter 2.

24. J. Richter and R. Ostovar, '"It Don't Mean a Thing if It Ain't Got That Swing"—An Alternative Concept for Understanding the Evolution of Dance and Music in Human Beings', *Frontiers in Human Neuroscience*, 10:485 (2016).

25. P. Gray, 'Play as a Foundation for Hunter-Gatherer Social Existence', *American Journal of Play*, 1:476–522 (2009).

26. P. Wiessner, 'Embers of Society: Firelight Talk among the Ju/'hoansi Bushmen', *Proceedings of the National Academy of Sciences USA*, 111:14027–35 (2014).

27. Morley, *The Prehistory of Music*.

28. Ibid.

29. Gray, 'Play as a Foundation for Hunter-Gatherer Social Existence'.

30. M. Del Nevo, *Art Music: Love, Listening and Soulfulness* (Transaction Publishers, 2013), p. 100.

31. F. Rohrer, 'The Devil's Music', *BBC News* (28 April 2006), http://news.bbc.co.uk/1/hi/magazine/4952646.stm.

32. D. Randall, *Sound System* (Pluto Press, 2017), pp. 98–9.

33. Rohrer, 'The Devil's Music'.

34. Ibid.

35. Ibid.

36. S. Behrman, 'From Revolution to Irrelevance: How Classical Music Lost Its Audience', *International Socialism*, 121 (2 January 2009), http://isj.org.uk/from-revolution-to-irrelevance-how-classical-music-lost-its-audience/.

37. Ibid.

38. P. McGarr, *Mozart: Overture to Revolution* (Redwords, 2001), pp. 20–2.

39. Ibid., 20–2.

40. Ibid., 69.

41. Behrman, 'From Revolution to Irrelevance'.

42. L. S. Vygotsky, 'Art and Life', in *The Psychology of Art* (1925), https://www.marxists.org/archive/vygotsky/works/1925/art11.htm.

43. S. Jeffries, 'Don Giovanni—Hero or Villain?', *The Guardian* (11 April 2012), https://www.theguardian.com/music/2012/apr/11/don-giovanni-hero-or-villain.

44. L. Howard, 'The Story and Backstory of Writing *The Magic Flute*', *Utah Opera* (19 February 2019), https://utahopera.org/explore/2019/02/the-story-and-backstory-of-writing-the-magic-flute/.

45. M. Lopez-Gonzalez and C. J. Limb, 'Musical Creativity and the Brain', *Cerebrum*, 2012:2 (2012).

46. N. Njoroge, 'Dedicated to the Struggle: Black Music, Transculturation, and the Aural Making and Unmaking of the Third World', *Black Music Research Journal*, 28:85–104 (2008).

47. M. Smith, *John Coltrane: Jazz, Racism and Resistance* (Redwords, 2003), pp. 30–1.

48. Ibid., 34–5.

49. A. Midgette, 'Knowing Mozart Better in the Evolution of the Piano', *New York Times* (30 August 2006), https://www.nytimes.com/2006/08/30/arts/music/knowing-mozart-better-in-the-evolution-of-the-piano.html.

50. Ibid.

51. A. Bateman, 'Beethoven and the Piano', *Music Unwrapped* (2020), http://www.musicunwrapped.co.uk/filemanager/Beethoven_Spread.pdf.

52. Ibid.

53. A. Lefkovitz, *Jimi Hendrix and the Cultural Politics of Popular Music* (Amazon Media, 2018), p. 9.

54. M. Clague, '50 Years Ago, Jimi Hendrix's Woodstock Anthem Expressed the Hopes and Fears of a Nation', *The Conversation* (14 August 2019), http://theconversation.com/50-years-ago-jimi-hendrixs-woodstock-anthem-expressed-the-hopes-and-fears-of-a-nation-120717.

55. 'Fifty Years of Jimi Hendrix', *The Economist* (14 December 2016), https://www.economist.com/prospero/2016/12/14/fifty-years-of-jimi-hendrix.

56. J. Storey, *Cultural Theory and Popular Culture: A Reader* (Pearson Education, 2006), p. 80.

57. J. Thompson, 'Top 20 Political Songs: "Imagine", John Lennon, 1971', *New Statesman* (25 March 2010), https://www.newstatesman.com/music/2010/03/lennon-imagine-political.

58. Ibid.

59. K. Macdonald, 'The Harmonic Magic behind the Timeless Beauty of John Lennon's "Imagine"', *Classic FM* (3 July 2020), https://www.classicfm.com/discover-music/harmonic-analysis-imagine-john-lennon/.

60. D. Gray, 'Drift Away', *Genius* (2020), https://genius.com/Dobie-gray-drift-away-lyrics.

61. D. Randall, *Sound System* (Pluto Press, 2017), pp. 66–7.

62. Ibid., 69.

63. Ibid., 69.

64. T. Ali and R. Blackburn, 'The Lost John Lennon Interview', *Counterpunch* (8 December 2005), https://www.counterpunch.org/2005/12/08/the-lost-john-lennon-interview/.

65. G. Campbell, 'Country Boy', *Genius* (2020), https://genius.com/Glen-campbell-country-boy-you-got-your-feet-in-la-lyrics.

66. G. Adams, 'Eminem: The Fall and Rise of a Superstar', *The Independent* (4 February 2009), https://www.independent.co.uk/arts-entertainment/music/features/eminem-the-fall-and-rise-of-a-superstar-1544787.html.

67. Eminem, 'Walk on Water', *Genius* (2020), https://genius.com/Eminem-walk-on-water-lyrics.

68. G. Cartwright, 'Richey Edwards', *The Guardian* (26 November 2008), https://www.theguardian.com/music/2008/nov/26/richey-edwards-manic-street-preachers.

69. Manic Street Preachers, '4st 7lb', *Genius* (2020), https://genius.com/Manic-street-preachers-4st-7lb-lyrics.

70. Cartwright, 'Richey Edwards'.

71. Ibid.

Chapter 19: Art and Design

1. L. Lennon, 'How a Dog Called Robot Helped Reveal Lascaux's Prehistoric Art Gallery', *Daily Telegraph* (11 September 2015), https://www.dailytelegraph.com.au/news/how-a-dog-called-robot-helped-reveal-lascauxs-prehistoric-art-gallery/news-story/717e1cb7dc68c1302f2575e5cdc67fe9.

2. Ibid.

3. R. Erlich, 'In Portugal, a Dam Threatens Prehistoric Rock Engravings', *Christian Science Monitor* (4 April 1995), https://www.csmonitor.com/1995/0404/04101.html.

4. A. M. Baptista and A. P. Fernandes Batarda, 'Rock Art and the Coa Valley Archaeological Park: A Case Study in the Preservation of Portugal's Prehistoric Rupestral Heritage', in P. Pettit, P. Bahn, and S. Ripoll (eds), *Palaeolithic Cave Art at Creswell Crags in European Context.* (Oxford University Press, 2007), pp. 263–79.

5. J. Marchant, 'A Journey to the Oldest Cave Paintings in the World', *Smithsonian Magazine* (January 2016), https://www.smithsonianmag.com/history/journey-oldest-cave-paintings-world-180957685/.

6. A. Kiely, 'The Origin Of The World's Art: Prehistoric Cave Painting', *Head Stuff* (12 September 2016), https://www.headstuff.org/culture/history/origin-worlds-art-prehistoric-cave-painting/.

7. M. Brooks, 'Drawing, Visualisation and Young Children's Exploration of "Big Ideas"', *International Journal of Science Education*, 31:319–41 (2009).

8. S. Brown, X. Gao, L. Tisdelle, S. B. Eickhoff, and M. Liotti, Naturalizing Aesthetics: Brain Areas for Aesthetic Appraisal across Sensory Modalities', *Neuroimage*, 58:250–8 (2011).

9. S. Brown and X. Gao, 'The Neuroscience of Beauty', *Scientific American* (27 September 2011), https://www.scientificamerican.com/article/the-neuroscience-of-beauty/.

10. A. Tucker, 'How does the Brain Process Art?', *Smithsonian Magazine* (November 2012), https://www.smithsonianmag.com/science-nature/how-does-the-brain-process-art-80541420/.

11. R. Chamberlain, I. C. McManus, N. Brunswick, Q. Rankin, H. Riley, and R. Kanai, 'Drawing on the Right Side of the Brain: A Voxel-Based Morphometry Analysis of Observational Drawing', *Neuroimage*, 96:167–73 (2014).

12. Tucker, 'How does the Brain Process Art?'

13. Chamberlain et al., 'Drawing on the Right Side of the Brain'.

14. Tucker, 'How does the Brain Process Art?'

15. C. Finn, 'Race to Save Lascaux Cave Art, Prehistoric "Sistine Chapel"', *Wired* (3 January 2008), https://www.wired.com/2008/01/race-to-save-la/.

16. A. Feltus, 'I Believe There Is No Progress in Art', *The Finch* (9 November 2018), http://thefinch.net/2016/11/09/alan-feltus-i-believe-there-is-no-progress-in-art/.

17. G. Horvath, E. Farkas, I. Boncz, M. Blaho, and G. Kriska, 'Cavemen Were Better at Depicting Quadruped Walking Than Modern Artists: Erroneous Walking Illustrations in the Fine Arts from Prehistory to Today', *PLoS One*, 7:e49786 (2012).

18. J. Jones, 'Bosch's Garden of Earthly Delights Shows a World Waking up to the Future', *The Guardian* (10 January 2017), https://www.theguardian.com/artanddesign/jonathanjonesblog/2017/jan/10/bosch-garden-of-earthly-delights-shows-a-world-waking-up-to-the-future.

19. H. Baan, 'Hieronymus Bosch: Visionary of Change', *Socialist Review* (March 2016), http://socialistreview.org.uk/411/hieronymus-bosch-visionary-change.

20. Editorial, 'Hieronymus Bosch Died 500 Years Ago, But His Art Will Still Creep You Out', *NPR Illinois* (26 June 2016), http://www.nprillinois.org/post/hieronymus-bosch-died-500-years-ago-his-art-will-still-creep-you-out.

21. Baan, 'Hieronymus Bosch: Visionary of Change'.

22. A. Denney, 'Bosch's Monsters Explained', *Another Man Magazine* (27 June 2018), http://www.anothermanmag.com/life-culture/10393/hieronymus-bosch-s-monsters-explained.

23. Jones, 'Bosch's Garden of Earthly Delights'.

24. J. Burleigh, 'Was Michelangelo's Artistic Genius a Symptom of Autism?', *The Independent* (1 June 2004), https://www.independent.co.uk/news/uk/this-britain/was-michelangelos-artistic-genius-a-symptom-of-autism-756718.html.

25. Ibid.

26. S. Shaikh and J. Leonard-Amodeo, 'The Deviating Eyes of Michelangelo's David', *Journal of the Royal Society of Medicine*, 98:75–6 (2005).

27. G. Vasari, *The Lives of the Artists* (Oxford University Press, 1991), pp. 277–8.

28. J. Molyneux, 'Michelangelo and Human Emancipation', *International Socialism*, 128 (14 October 2010), http://isj.org.uk/michelangelo-and-human-emancipation/.

29. Shaikh and Leonard-Amodeo, 'The Deviating Eyes of Michelangelo's David'.

30. 'What is Modernism?', *Saylor Organisation* (2011), https://www.saylor.org/site/wp-content/uploads/2011/05/Modernism.pdf.

31. C. Nineham, 'The Two Faces of Modernism', *International Socialism*, 64 (Autumn 1994), http://pubs.socialistreviewindex.org.uk/isj64/nineham.htm.

32. J. Jones, 'Pablo's Punks', *The Guardian* (9 January 2007), https://www.theguardian.com/culture/2007/jan/09/2.

33. J. Molyneux, 'A Revolution in Paint: 100 Years of Picasso's Demoiselles', *International Socialism*, 115 (2 July 2007), http://isj.org.uk/a-revolution-in-paint-100-years-of-picassos-demoiselles/.

34. E. Gibson, 'A Magical Metamorphosis of the Ordinary', *Wall Street Journal* (16 April 2011), https://www.wsj.com/articles/SB10001424052748703551304576261042931202326.

35. C. Temple, 'Visual Memory and What Picasso Was Really Seeing', *The Guardian* (9 October 2016), https://www.theguardian.com/lifeandstyle/2016/oct/09/visual-memory-and-what-picasso-was-really-seeing.

36. Ibid.

37. Jones, 'Pablo's Punks'.

38. Molyneux, 'A Revolution in Paint'.

39. Ibid.

40. Ibid.

41. C. Duncan, *The Aesthetics of Power* (Cambridge University Press, 1993), pp. 96–7.

42. 'Five Ways to Look at Malevich's Black Square', *Tate* (2020), https://www.tate.org.uk/art/artists/kazimir-malevich-1561/five-ways-look-malevichs-black-square.

43. Ibid.

44. C. Harrison and P. Wood, *Art in Theory 1900–2000: An Anthology of Changing Ideas* (Blackwell, 2003).

45. S. Zeki, 'Art and the Brain', *Daedalus*, 127:71–103 (1999).

46. 'Five Ways to Look at Malevich's Black Square'.

47. 'What is Modernism?'

48. P. Anderson, *A Zone of Engagement* (Verso, 1992), p. 36.

49. W. L. Adamson, 'Avant-Garde Modernism and Italian Fascism: Cultural Politics in the Era of Mussolini', *Journal of Modern Italian Studies*, 6:230–48 (2001).

50. J. Uglow, 'When Art Meets Power', *New York Review of Books* (8 March 2017), https://www.nybooks.com/daily/2017/03/08/when-art-meets-power-russia-revolution/.

51. Ibid.

52. C. Brandist, *Carnival Culture and the Soviet Modernist Novel* (Palgrave Macmillan, 1996), p. 52.

53. A. S. Shatskikh, *Vitebsk: The Life of Art* (Yale University Press, 2007), p. 118.

54. B. Nicholson, 'Where to begin with Sergei Eisenstein', *British Film Institute* (23 January 2018), https://www.bfi.org.uk/news-opinion/news-bfi/features/where-begin-sergei-eisenstein.

55. Ibid.

56. Ibid.

57. J. Hellerman, 'How "Parasite" Made One of the Greatest Montages Ever', *No Film School* (31 December 2019), https://nofilmschool.com/parasite-montage.

58. J. Vassilieva, 'Eisenstein/Vygotsky/Luria's Project: Cinematic Thinking and the Integrative Science of Mind and Brain', *Screening the Past* (December 2013), http://www.screeningthepast.com/2013/12/eisenstein-vygotsky-luria%E2%80%99s-project-cinematic-thinking-and-the-integrative-science-of-mind-and-brain/.

59. Ibid.

60. Ibid.

61. Ibid.

62. M. Wilmington, '"Nevsky's" Bloody Ballet for the Ages', *Los Angeles Times* (1 November 1987), https://www.latimes.com/archives/la-xpm-1987-11-01-ca-18046-story.html.

63. Vassilieva, 'Eisenstein/Vygotsky/Luria's Project'.

Chapter 20: Fact and Fiction

1. L. Reynolds, 'Girls in Jeans, Dads Helping around the House and the Beginning of the Urban Sprawl: How Peter and Jane Books from 60s and 70s Marked Changes in British Society', *Daily Mail* (11 March 2015), http://www.dailymail.co.uk/femail/article-2989622/Girls-jeans-dads-helping-house-beginning-urban-sprawl-Peter-Jane-books-60s-70s-marked-changes-British-society.html.

2. 'Work for the Idle Hands', *Telegraph and Argus* (5 January 2000), http://www.thetelegraphandargus.co.uk/news/8059971.Work_for_the_Idle_hands/.

3. R. Connolly, 'Room at the Top Revisited', *Radio Times* (29 March 2011), http://www.rayconnolly.co.uk/room-at-the-top-revisited/.

4. C. Flanagan, 'How Lolita Seduces Us All', *The Atlantic* (September 2018), https://www.theatlantic.com/magazine/archive/2018/09/how-lolita-seduces-us-all/565751/.

5. J. Self, 'William Golding: Pincher Martin', *Asylum* (26 November 2013), https://theasylum.wordpress.com/2013/11/26/william-golding-pincher-martin/.

6. M. A. Hector, 'Personality and Virginia Woolf', *Psychology Today* (23 August 2009), https://www.psychologytoday.com/us/blog/novels-poetry-and-psychology/200908/personality-and-virginia-woolf?amp.

7. D. Lodge, 'Sense and Sensibility', *The Guardian* (2 November 2002), https://www.theguardian.com/books/2002/nov/02/fiction.highereducation.

8. Ibid.

9. W. Somerset Maugham, *Cakes and Ale* (Vintage Classics, 2000), Preface.

10. K. Clark and M. Holquist, *Mikhail Bakhtin* (Harvard University Press, 1986), p. 240.

11. I. R. Makaryk, *Encyclopedia of Contemporary Literary Theory: Approaches, Scholars, Terms* (University of Toronto Press, 1993), p. 609.

12. A. Ananthaswamy, 'Ecstatic Epilepsy: How Seizures Can Be Bliss', *New Scientist* (22 January 2014), https://www.newscientist.com/article/mg22129531-000-ecstatic-epilepsy-how-seizures-can-be-bliss/.

13. Ibid.

14. Clark and Holquist, *Mikhail Bakhtin*, p. 245.

15. P. J. Leithart, 'The Monologue beyond the Dialogue', *First Things* (24 March 2017), https://www.firstthings.com/web-exclusives/2017/03/the-monologue-beyond-the-dialogue.

16. M. Berman, *All That Is Solid Melts into Air* (Verso, 1982).

17. N. Collins, 'How Wuthering Heights Caused a Critical Stir When First Published in 1847', *The Telegraph* (22 March 2011), https://www.telegraph.co.uk/culture/tvandradio/8396278/How-Wuthering-Heights-caused-a-critical-stir-when-first-published-in-1847.html.

18. A. Green, 'Lockwood's Cruelty in Wuthering Heights', *Owlcation* (8 February 2018), https://owlcation.com/humanities/Lockwoods-Cruelty-in-Wuthering-Heights.

19. N. Rogers, 'Review: Chartism and Class Struggle', *Labour/Le Travail*, 19:143–51 (1987).

20. T. Chaffin, 'Frederick Douglass's Irish Liberty', *The New York Times* (25 February 2011), https://opinionator.blogs.nytimes.com/2011/02/25/frederick-douglasss-irish-liberty/.

21. C. Fowler, 'Was Emily Brontë's Heathcliff black?', *The Conversation* (25 October 2017), https://theconversation.com/was-emily-bront-s-heathcliff-black-85341.

22. D. Bryfonski, *Class Conflict in Emily Bronte's Wuthering Heights* (Greenhaven Press, 2011), p. 104.

23. K. Hughes, 'The Strange Cult of Emily Brontë and the "Hot Mess" of Wuthering Heights', *The Guardian* (21 July 2018), https://www.theguardian.com/books/2018/jul/21/emily-bronte-strange-cult-wuthering-heights-romantic-novel.

24. T. Eagleton, 'Emily Brontë and the Great Hunger', *The Irish Review* 12:108–19 (1992).

25. M. P. Asl, 'Recurring Patterns: Emily Brontë's Neurosis in Wuthering Heights', *International Journal of Education and Literacy Studies*, 2:46–50 (2014).

26. R. Onion, 'A School Progress Report for the Brontë Sisters', *Slate* (22 July 2014), http://www.slate.com/blogs/the_vault/2014/07/22/history_of_the_bront_family_school_report_assessing_their_progress_at_cowan.html.

27. B. Barnett, 'Branwell Bronte: The Mad, Bad and Dangerous Brother of Charlotte, Emily and Anne', *The Independent* (17 September 2017), https://www.independent.co.uk/news/long_reads/branwell-bronte-emily-charlotte-anne-family-haworth-yorkshire-a7940396.html.

28. Asl, 'Recurring Patterns'.

29. Ibid.

30. Ibid.

31. E. Bronte, *Wuthering Heights* (Penguin, 1965), p. 11.

32. Ibid.

33. Ibid., 12.

34. Ibid., 11.

35. C. Tóibín, 'Pure Evil—Colm Tóibín on The Turn of the Screw', *The Guardian* (3 June 2006), https://www.theguardian.com/books/2006/jun/03/fiction.colmtoibin.

36. M. Norris, 'Investigating Ambiguity: Sources of Insanity in "The Turn of the Screw" (P8)', *Medium* (12 May 2016), https://medium.com/@mrnorris/investigating-ambiguity-sources-of-insanity-in-the-turn-of-the-screw-p8-551975c0c08c.

37. B. Leithauser, 'Ever Scarier: On "The Turn of the Screw"', *The New Yorker* (29 October 2012), https://www.newyorker.com/books/page-turner/ever-scarier-on-the-turn-of-the-screw.

38. J. Sexton, 'A Non-Apparitionist Reading of The Turn of the Screw', *BC Open Textbooks* (2014), https://opentextbc.ca/englishliterature/wp-content/uploads/sites/27/2014/05/Non-apparitionist-reading-of-Turn-of-Screw.pdf.

39. Norris, 'Investigating Ambiguity'.

40. M. A. Nichols, 'The Victims of Jack the Ripper', *The History Press* (1 September 2011), https://www.thehistorypress.co.uk/articles/the-victims-of-jack-the-ripper/.

41. Norris, 'Investigating Ambiguity'.

42. Leithauser, 'Ever Scarier'.

43. W. K. Penny, 'Shattered Eden: Subjectivity and the Fall out of Language in Henry James's The Turn of the Screw', *Literary Imagination*, 18:255–73 (2016).

44. Ibid.

45. Leithauser, 'Ever Scarier'.

46. Ibid.

47. J. Carey, *William Golding: The Man Who Wrote Lord of the Flies* (Faber and Faber, 2009), pp. 190–205.

48. L. Surette, 'A Matter of Belief: Pincher Martin's Afterlife', *Twentieth Century Literature*, 40:205–25 (1994).

49. D. Peter, 'William Golding's *Pincher Martin*: A Study of Self and Its Terror of Negation', *McMaster University* (December 1982), https://macsphere.mcmaster.ca/bitstream/11375/10954/1/fulltext.pdf.

50. Surette, 'A Matter of Belief'.

51. R. McCrum, 'William Golding's Crisis', *The Observer* (11 March 2012), https://www.theguardian.com/books/2012/mar/11/william-golding-crisis.

52. M. N. Singh, 'Golding's Pincher Martin: Monomania Caused Moral Degradation in Modern Man', *International Journal on Studies in English Language and Literature*, 4:22–6 (2016).

53. H. Goyal, 'Predicament of the Protagonist in William Golding's Pincher Martin', *Research Journal of English Language and Literature*, 5:285–91 (2017).

54. Ibid.

55. Y. Nir and G. Tononi, 'Dreaming and the Brain: From Phenomenology to Neurophysiology', *Trends in Cognitive Science*, 14:88–100 (2010).

56. P. Iyer, 'The Butler Didn't Do It Again', *Times Literary Supplement* (28 April 1995), https://www.the-tls.co.uk/articles/public/the-unconsoled/.

57. M. Kakutani, 'From Kazuo Ishiguro, A New Annoying Hero', *New York Times* (17 October 1995), https://www.nytimes.com/1995/10/17/books/books-of-the-times-from-kazuo-ishiguro-a-new-annoying-hero.html.

58. Ibid.

59. S. Jordison, 'The Unconsoled Deals in Destruction and Disappointment', *The Guardian* (27 January 2015), https://www.theguardian.com/books/booksblog/2015/jan/27/kazuo-ishiguro-reading-group.

60. Ibid.

61. C. Quarrie, 'Impossible Inheritance: Filiation and Patrimony in Kazuo Ishiguro's The Unconsoled', *Studies in Contemporary Fiction*, 55:138–51 (2014).

62. Jordison, 'The Unconsoled Deals'.

63. A. H. Fairbanks, 'Ontology and Narrative Technique in Kazuo Ishiguro's "The Unconsoled"', *Studies in the Novel*, 45:603–19 (2013).

64. Ibid.

Chapter 21: Science and Technology

1. M. A. Weitekamp, '"We're Physicists": Gender, Genre and the Image of Scientists in The Big Bang Theory', *Journal of Popular Television* 3:75–92 (2015).

2. M. Brooks, 'Why the Scientist Stereotype Is Bad for Everyone, Especially Kids', *Wired* (15 June 2012), https://www.wired.com/2012/06/opinion-scientist-stereotype/.

3. J. Viegas, 'Wild Chimps Upgrade Termite-Fishing Tool', *NBC News* (4 March 2009), http://www.nbcnews.com/id/29509302/ns/technology_and_science-science/t/wild-chimps-upgrade-termite-fishing-tool/.

4. C. Q. Choi, 'Human Evolution: The Origin of Tool Use', *Live Science* (11 November 2009), https://www.livescience.com/7968-human-evolution-origin-tool.html.

5. K. Marx, *Capital*, volume I (Swan Sonnenschein, Lowrey, & Co, 1867), Chapter 7, https://www.marxists.org/archive/marx/works/1867-c1/ch07.htm.

6. S. Brenner, 'Life Sentences: Detective Rummage Investigates', *Genome Biology*, 3: comment1013.1011–comment1013.1012 (2002).

7. A. Bradford, 'What Is a Scientific Hypothesis? Definition of Hypothesis', *Live Science* (26 July 2017), https://www.livescience.com/21490-what-is-a-scientific-hypothesis-definition-of-hypothesis.html.

8. A. Ossola, 'Scientists Are More Creative Than You Might Imagine', *The Atlantic* (12 November 2014), https://www.theatlantic.com/education/archive/2014/11/the-creative-scientist/382633/.

9. J. Cutraro, 'How Creativity Powers Science', *Science News for Students* (24 May 2012), https://www.sciencenewsforstudents.org/article/how-creativity-powers-science.

10. Ibid.

11. Ibid.

12. Ibid.

13. Ibid.

14. B. Griffith, 'Radioactivity Discovered', *Royal Society of Chemistry* (1 November 2008), https://eic.rsc.org/feature/radioactivity-discovered/2020216.article.

15. Ibid.

16. J. Parrington, 'Lise Meitner—A Scientific Pioneer Who Overcame Prejudice', *Socialist Worker* (17 December 2013), https://socialistworker.co.uk/art/37090/Lise+Meitner+++a+scientific+pioneer+who+overcame+prejudice.

17. Ibid.

18. Ibid.

19. Ibid.

20. 'Ancient Mesopotamian Civilizations', *Khan Academy* (20 October 2018), https://www.khan-academy.org/humanities/world-history/world-history-beginnings/ancient-mesopotamia/a/mesopotamia-article.

21. Ibid.

22. M. Shuttleworth, 'War Machines of Archimedes', *Explorable* (2 September 2011), https://explorable.com/archimedes-war-machines.

23. A. Jogalekar, 'Falsification and Its Discontents', *Scientific American* (24 January 2014), https://blogs.scientificamerican.com/the-curious-wavefunction/falsification-and-its-discontents/.

24. J. Naughton, 'Thomas Kuhn: The Man Who Changed the Way the World Looked at Science', *The Guardian* (19 August 2012), https://www.theguardian.com/science/2012/aug/19/thomas-kuhn-structure-scientific-revolutions.

25. R. Ford Denison and K. Muller, 'The Evolution of Cooperation', *The Scientist* (31 December 2016), https://www.the-scientist.com/features/the-evolution-of-cooperation-34284.

26. J. Carey, 'No Need for Geniuses: Revolutionary Science in the Age of the Guillotine by Steve Jones', *Sunday Times* (27 March 2016), https://www.thetimes.co.uk/article/no-need-for-geniuses-revolutionary-science-in-the-age-of-the-guillotine-by-steve-jones-w29mmpd39.

27. J. Parrington, 'You Can Lead a Sperm to Ovum', *The Guardian* (25 July 2002), https://www.theguardian.com/science/2002/jul/25/medicalresearch.medicalscience.

28. L. W. Swanson and J. W. Lichtman, 'From Cajal to Connectome and Beyond', *Annual Review of Neuroscience*, 39:197–216 (2016).

29. M. I. Kaiser, 'The Limits of Reductionism in the Life Sciences', *History and Philosophy of the Life Sciences*, 33:453–76 (2011).

30. A. Blunden, 'The Germ Cell of Vygotsky's Science', in *Vygotsky and Marx: Towards a Marxist Psychology* (Routledge, 2017), pp. 132–45.

31. E. R. Scerri, 'The Evolution of the Periodic System', *Scientific American* (21 January 2011), https://www.scientificamerican.com/article/the-evolution-of-the-periodic-system/.

32. C. Orzel, 'What Has Quantum Mechanics Ever Done for Us?', *Forbes* (13 August 2015), https://www.forbes.com/sites/chadorzel/2015/08/13/what-has-quantum-mechanics-ever-done-for-us/.

33. D. Overbye, 'A Century Ago, Einstein's Theory of Relativity Changed Everything', *New York Times* (24 November 2015), https://www.nytimes.com/2015/11/24/science/a-century-ago-einsteins-theory-of-relativity-changed-everything.html.

34. D. Greenbaum and M. Gerstein, 'Illuminating the Genome's Dark Matter', *Cell*, 163:1047–8 (2015).

35. Ibid.

36. Ibid.

37. C. S. Powell, 'Relativity versus Quantum Mechanics: The Battle for the Universe', *The Guardian* (4 November 2015), https://www.theguardian.com/news/2015/nov/04/relativity-quantum-mechanics-universe-physicists.

38. Ibid.

39. Ibid.

40. J. V. Wertsch, *Vygotsky and the Social Formation of Mind* (Harvard University Press, 1985), p. 158.

41. L. S. Vygotsky, *Thought and Language* (MIT Press, 1962), p. 318.

Chapter 22: Mind and Meaning

1. E. Palermo, 'The Origins of Religion: How Supernatural Beliefs Evolved', *Live Science* (5 October 2015), https://www.livescience.com/52364-origins-supernatural-relgious-beliefs.html.

2. B. N. Thompson, 'Theory of Mind: Understanding Others in a Social World', *Psychology Today* (3 July 2017), https://www.psychologytoday.com/us/blog/socioemotional-success/201707/theory-mind-understanding-others-in-social-world.

3. Palermo, 'The Origins of Religion'.

4. H. C. Peoples, P. Duda, and F. W. Marlowe, 'Hunter-Gatherers and the Origins of Religion', *Human Nature*, 27:261–82 (2016).

5. R. Gray, 'Cave Fires and Rhino Skull Used in Neanderthal Burial Rituals', *New Scientist* (28 September 2016), https://www.newscientist.com/article/mg23230934-800-cave-fires-and-rhino-skull-used-in-neanderthal-burial-rituals/.

6. Ibid.

7. 'Buddhism', *History* (22 July 2020), https://www.history.com/topics/religion/buddhism.

8. S. Grogan, '"A Solace to a Tortured World…" – The Growing Interest in Spiritualism during and after WW1', *World War I Centenary* (2014), http://ww1centenary.oucs.ox.ac.uk/author/suzannegrogan.

9. Peoples et al., 'Hunter-Gatherers and the Origins of Religion'.

10. Ibid.

11. L. Wade, 'To Foster Complex Societies, Tell People a God Is Watching', *Science* (4 March 2015), http://www.sciencemag.org/news/2015/03/foster-complex-societies-tell-people-god-watching.

12. C. Seawright, 'Shesmu, Demon-God of the Wine Press, Oils and Slaughterer of the Damned', *Tour Egypt* (2013), http://www.touregypt.net/featurestories/shesmu.htm.

13. J. Sharman, 'Vatican Rushes to Deny Reports Pope Thinks Hell Doesn't Exist and Sinning Souls Just "Disappear"', *The Independent* (30 March 2018), https://www.independent.co.uk/news/world/europe/pope-francis-hell-does-not-exist-catholic-vatican-full-quote-a8281041.html.

14. Ibid.

15. K. Marx, 'A Contribution to the Critique of Hegel's Philosophy of Right: Introduction', *Deutsch-Französische Jahrbücher* (1844), https://www.marxists.org/archive/marx/works/1843/critique-hpr/intro.htm.

16. F. Engels, 'On the History of Early Christianity', *Die Neue Zeit* (1894), https://www.marxists.org/archive/marx/works/1894/early-christianity/index.htm.

17. S. McIntire and W. E. Burns, *Speeches in World History* (Facts on File, 2009), p. 104.

18. M. Empson, *Kill All The Gentlemen: Class Struggle and Change in the English Countryside* (Bookmarks, 2018), p. 31.

19. A. Saville, 'The 17th Century English Scientific Revolution', *The Queen's College, Oxford* (2020), https://www.queens.ox.ac.uk/scientific%20revolution.

20. K. Thomas, *Religion and the Decline of Magic: Studies in Popular Beliefs in Sixteenth and Seventeenth Century England* (Penguin, 1991).

21. J. Walker, 'An Introduction To John Bunyan's The Pilgrim's Progress', *Banner of Truth* (15 September 2005), https://banneroftruth.org/uk/resources/articles/2005/an-introduction-to-john-bunyans-the-pilgrims-progress/.

22. R. McCrum, 'The 100 Best Novels: No 1—The Pilgrim's Progress by John Bunyan (1678)', *The Guardian* (23 September 2013), https://www.theguardian.com/books/2013/sep/23/100-best-novels-pilgrims-progress.

23. Ibid.

24. Ibid.

25. R. Hampson, 'What You Didn't Know about King's "Dream" Speech', *USA TODAY* (12 August 2013), https://eu.usatoday.com/story/news/nation/2013/08/12/march-on-washington-king-speech/2641841/.

26. Ibid.

27. N. Morgan, 'The Story behind Martin Luther King's "I Have a Dream"', *Forbes* (16 January 2012), https://www.forbes.com/sites/nickmorgan/2012/01/16/the-story-behind-martin-luther-kings-i-have-a-dream/.

28. A. Rutherford, '"There Is Grandeur in This View of Life"', *The Guardian* (15 February 2008), https://www.theguardian.com/commentisfree/2008/feb/15/thereisgrandeurinthisviewoflife.

29. B. Reese, 'Interview with Christof Koch', *GigaOm* (25 May 2018), https://gigaom.com/2018/05/25/interview-with-christof-koch/.

30. S. McLeod, 'Behaviorist Approach', *Simply Psychology* (2020), https://www.simplypsychology.org/behaviorism.html.

31. G. Johnson, 'What Really Goes On in There', *New York Times* (10 November 1991), https://archive.nytimes.com/www.nytimes.com/books/98/12/06/specials/dennett-consciousness.html?mcubz=1.

32. S. Ritchie, 'Do Daniel C. Dennett's Memes Deserve to Survive?', *The Spectator* (4 March 2017), https://www.spectator.co.uk/2017/03/do-daniel-c-dennetts-memes-deserve-to-survive/.

33. R. Dawkins, *The Selfish Gene: 30th Anniversary Edition* (Oxford University Press, 2006), p. 192.

34. Ibid.

35. J. Hughes, 'Meme Theory: Do We Come up with Ideas or Do They, in Fact, Control Us?', *The Independent* (13 July 2012), https://www.independent.co.uk/life-style/gadgets-and-tech/features/meme-theory-do-we-come-up-with-ideas-or-do-they-in-fact-control-us-7939077.html.

36. A. Ripstein, 'Commodity Fetishism', *Canadian Journal of Philosophy*, 17:733–48 (1987).

37. C. Koch, 'Is Consciousness Universal?', *Scientific American* (1 January 2014), https://www.scientificamerican.com/article/is-consciousness-universal/.

38. J. White, 'Is the Universe Self-aware?', *Episcopal Café* (21 June 2017), https://www.episcopalcafe.com/44054-2/.

39. J. Piper, 'In the Beginning Was the Word', *Desiring God* (21 September 2008), https://www.desiringgod.org/messages/in-the-beginning-was-the-word.

40. T. Metzinger, *Being No One: The Self-Model Theory of Subjectivity* (MIT Press, 2004), p. 66.

41. H. James, 'The Art of Fiction', *Longman's Magazine* (1884), https://public.wsu.edu/~campbelld/amlit/artfiction.html.

42. M. J. Campion, 'How the World Loved the Swastika—until Hitler Stole It', *BBC News* (23 October 2014), https://www.bbc.co.uk/news/magazine-29644591.

43. Ibid.

44. L. Boissoneault, 'The True Story of the Reichstag Fire and the Nazi Rise to Power', *Smithsonian Magazine* (21 February 2017), https://www.smithsonianmag.com/history/true-story-reichstag-fire-and-nazis-rise-power-180962240/.

45. W. Bennett, 'F is for Fascism', *Socialist Review* (November 2007), http://socialistreview.org.uk/319/f-fascism.

46. P. Bradshaw, 'Amelie', *The Guardian* (5 October 2001), https://www.theguardian.com/film/2001/oct/05/1.

47. V. N. Voloshinov, *Marxism and the Philosophy of Language* (Harvard University Press, 1986), p. 13.

Chapter 23: Mind and Machine

1. M. A. Boesler, 'Young Neil Armstrong Recounts His Experience on the Moon', *Business Insider* (25 August 2012), https://www.businessinsider.com/neil-armstrong-1970-bbc-interview-2012-8?IR=T.

2. E. Siegel, 'The Most Impossible Technology from "Star Trek"', *Forbes* (22 July 2016), https://www.forbes.com/sites/startswithabang/2016/07/22/the-most-impossible-technology-from-star-trek/.

3. Y. Chen, Y. Niu, and W. Ji, 'Genome Editing in Nonhuman Primates: Approach to Generating Human Disease Models', *Journal of Internal Medicine*, 280:246–51 (2016).

4. S. Zhang, 'China Is Genetically Engineering Monkeys with Brain Disorders', *The Atlantic* (8 June 2018), https://www.theatlantic.com/science/archive/2018/06/china-is-genetically-engineering-monkeys-with-brain-disorders/561866/.

5. Ibid.

6. D. Cyranoski, 'China Is Positioning Itself as a World Leader in Primate Research', *Nature*, 532:300–2 (2016).

7. Z. Rahim, 'Scientist Injects Human Genes into Monkeys' Brains and Makes Them More Intelligent, Study Says', *The Independent* (12 April 2019), https://www.independent.co.uk/news/science/monkey-human-genes-brains-china-dna-intelligence-study-a8866616.html.

8. H. Brewis, 'Chinese Scientists Create Super Monkeys by Injecting Brains with Human DNA', *Evening Standard* (12 April 2019), https://www.standard.co.uk/news/world/chinese-scientists-create-super-monkeys-by-injecting-brains-with-human-dna-a4116916.html.

9. Ibid.

10. J. Hamill, 'Chinese Scientists Create Super-intelligent Monkeys by Injecting Human DNA into Their Brains', *Metro* (12 April 2019), https://metro.co.uk/2019/04/12/chinese-scientists-create-super-intelligent-monkeys-injecting-human-dna-brains-9183475/.

11. M. E. Coors, J. J. Glover, E. T. Juengst, and J. M. Sikela, 'The Ethics of Using Transgenic Non-human Primates to Study What Makes Us Human', *Nature Reviews in Genetics*, 11:658–62 (2010).

12. I. Kelava and M. A. Lancaster, 'Dishing out Mini-brains: Current Progress and Future Prospects in Brain Organoid Research', *Developmental Biology*, 420:199–209 (2016).

13. M. Molteni, 'Mini Brains Just Got Creepier—They're Growing Their Own Blood Vessels', *Wired* (3 April 2018), https://www.wired.com/story/mini-brains-just-got-creepiertheyre-growing-their-own-veins/.

14. L. P. Hickey, '"The Brain in a Vat" Argument', *Internet Encyclopedia of Philosophy* (2020), https://www.iep.utm.edu/brainvat/.

15. I. Nathan, 'Empire Essay: The Matrix Review', *Empire* (1 January 2000), https://www.empireonline.com/movies/empire-essay-matrix/review/.

16. Ibid.

17. T. Robey, 'Ghost In the Shell Review: Scarlett Johansson's Soulful Action Spectacle Proves the Purists Wrong', *The Telegraph* (31 March 2017), https://www.telegraph.co.uk/films/0/ghost-review-scarlett-johanssons-soulful-action-spectacle-proves/.

18. J. Falconer, 'What is the Technological Singularity?', *The Next Web* (19 June 2011), https://thenextweb.com/insider/2011/06/19/what-is-the-technological-singularity/.

19. N. Ismail, 'AI: The Greatest Threat in Human History?', *Information Age* (20 October 2016), https://www.information-age.com/ai-greatest-threat-human-history-123462789/.

20. A. Sulleyman, 'AI Is Highly Likely to Destroy Humans, Elon Musk Warns', *The Independent* (24 November 2017), https://www.independent.co.uk/life-style/gadgets-and-tech/news/elon-musk-artificial-intelligence-openai-neuralink-ai-warning-a8074821.html.

21. R. Cellan-Jones, 'Stephen Hawking Warns Artificial Intelligence Could End Mankind', *BBC News* (2 December 2014), https://www.bbc.co.uk/news/technology-30290540.

22. M. Weisberger, 'Why Does Artificial Intelligence Scare Us So Much?', *Live Science* (8 June 2018), https://www.livescience.com/62775-humans-why-scared-of-ai.html.

23. T. Friend, 'How Frightened Should We Be of A.I.?', *The New Yorker* (7 May 2018), https://www.newyorker.com/magazine/2018/05/14/how-frightened-should-we-be-of-ai.

24. Ibid.
25. M. Zastrow, '"I'm in Shock!" How an AI Beat the World's Best Human at Go', *New Scientist* (9 March 2016), https://www.newscientist.com/article/2079871-im-in-shock-how-an-ai-beat-the-worlds-best-human-at-go/.
26. C. Woodford, 'Neural Networks', *Explain That Stuff* (17 June 2020), https://www.explainthatstuff.com/introduction-to-neural-networks.html.
27. I. Sample, '"It's Able to Create Knowledge Itself": Google Unveils AI That Learns on Its Own', *The Guardian* (18 October 2017), https://www.theguardian.com/science/2017/oct/18/its-able-to-create-knowledge-itself-google-unveils-ai-learns-all-on-its-own.
28. Ibid.
29. Ibid.
30. J. Sokol, 'Why Artificial Intelligence Like AlphaZero Has Trouble with the Real World', *Quanta Magazine* (21 February 2018), https://www.quantamagazine.org/why-alphazeros-artificial-intelligence-has-trouble-with-the-real-world-20180221/.
31. Ibid.
32. J. C. Baillie, 'Why AlphaGo Is Not AI', *Spectrum* (17 March 2016), https://spectrum.ieee.org/automaton/robotics/artificial-intelligence/why-alphago-is-not-ai.
33. M. Bennett-Smith, 'Stephen Hawking: Brains Could Be Copied to Computers to Allow Life after Death', *Huffington Post* (25 September 2013), https://www.huffingtonpost.co.uk/entry/stephen-hawking-brains-copied-life-after-death_n_3977682.
34. A. Regalado, 'Nectome Will Preserve Your Brain, But You Have to Be Euthanized First', *MIT Technology Review* (13 March 2018), https://www.technologyreview.com/s/610456/a-startup-is-pitching-a-mind-uploading-service-that-is-100-percent-fatal/.
35. D. B. Kirtley, 'Futurist Says We'll Use Lasers to Beam Our Minds Into Space Someday Soon', *Wired* (1 March 2014), https://www.wired.com/2014/03/geeks-guide-michio-kaku/.

Epilogue: A Twenty-First-Century Mind

1. F. Thibaut, 'The Mind-Body Cartesian Dualism and Psychiatry', *Dialogues in Clinical Neuroscience*, 20:3 (2018).
2. N. Barber, 'The Blank Slate Controversy', *Psychology Today* (21 September 2016), https://www.psychologytoday.com/gb/blog/the-human-beast/201609/the-blank-slate-controversy.
3. H. Rolston, 'Skyhooks and Cranes', *Metanexus* (1 June 1999), http://www.metanexus.net/skyhooks-and-cranes/.
4. J. Kemsley, 'The Safety Zone', *The Safety Zone* (September 2011), http://cenblog.org/the-safety-zone/2011/09/chemical-oscillations-the-belousov-zhabotinsky-reaction/.
5. Ibid.
6. H. Wen and Z. Liu, 'Separating Fractal and Oscillatory Components in the Power Spectrum of Neurophysiological Signal', *Brain Topography*, 29:13–26 (2016).
7. H. Wang, 'Modeling Neurological Diseases with Human Brain Organoids', *Frontiers in Synaptic Neuroscience*, 10:15 (2018).
8. N. Williams and R. N. Henson, 'Recent Advances in Functional Neuroimaging Analysis for Cognitive Neuroscience', *Brain and Neuroscience Advances*, 2:1–4 (2018).
9. A. Kozulin, *Vygotsky's Psychology: A Biography of Ideas* (Harvester Wheatsheaf, 1990), pp. 225–33.
10. A. M. Moe and N. M. Docherty, 'Schizophrenia and the Sense of Self', *Schizophrenia Bulletin*, 40:161–8 (2014).

11. R. Ostrow, 'Crowd Control for the Mind: Coping with Inner Voices', *The Australian* (15 December 2016), https://www.theaustralian.com.au/life/health-wellbeing/crowd-control-for-the-mind-coping-with-inner-voices/news-story/103ea4e51b8bb3c5ee13a83ff2927e54.

12. P. de Sousa, W. Sellwood, A. Spray, C. Fernyhough, and R. P. Bentall, 'Inner Speech and Clarity of Self-Concept in Thought Disorder and Auditory-Verbal Hallucinations', *Journal of Nervous Mental Disorders*, 204:885–93 (2016).

13. C. Nordqvist, 'Understanding the Symptoms of Schizophrenia', *Medical News Today* (23 April 2020), https://www.medicalnewstoday.com/articles/36942.php.

14. L. A. Sass and J. Parnas, 'Schizophrenia, Consciousness, and the Self', *Schizophrenia Bulletin*, 29:427–44 (2003).

15. S. Matthews, 'Forget Anti-depressants, Doctors Should Be Able to Prescribe Music, Arts and Writing Courses to Help Patients Suffering with the Blues, Claims GP', *Daily Mail* (11 July 2018), http://www.dailymail.co.uk/health/article-5918013/Doctors-able-prescribe-music-depressed-patients-claims-GP.html.

16. Ibid.

17. J. Doward, 'Psychiatrists under Fire in Mental Health Battle', *The Observer* (12 May 2013), https://www.theguardian.com/society/2013/may/12/psychiatrists-under-fire-mental-health.

18. H. Fraad, 'Profiting from Mental Ill-Health', *The Guardian* (15 March 2011), https://www.theguardian.com/commentisfree/cifamerica/2011/mar/15/psychology-healthcare.

19. Ibid.

20. D. Joravsky, 'The Mechanical Spirit: The Stalinist Marriage of Pavlov to Marx', *Theory and Society*, 4:457–77 (1977).

21. H. Katschnig, 'Are Psychiatrists an Endangered Species? Observations on Internal and External Challenges to the Profession', *World Psychiatry*, 9:21–8 (2010).

22. 'What is the Difference between Clinical Psychology and Psychiatry?', *Psychology School Guide* (2020), https://www.psychologyschoolguide.net/guides/difference-between-clinical-psychology-and-psychiatry/.

23. W. Self, 'Psychiatrists: The Drug Pushers', *The Guardian* (3 August 2013), https://www.theguardian.com/society/2013/aug/03/will-self-psychiatrist-drug-medication.

24. J. Nicolas, 'Why Pretend Social Work Is about Social Justice? It's Not', *The Guardian* (20 October 2015), https://www.theguardian.com/social-care-network/2015/oct/20/why-pretend-social-work-is-about-social-justice-its-not.

25. J. Larson, 'Blaming Parents: What I've Learned and Unlearned as a Child Psychiatrist', *Scientific American* (15 April 2011), https://blogs.scientificamerican.com/guest-blog/blaming-parents-what-ive-learned-and-unlearned-as-a-child-psychiatrist/.

26. K. Pickles, 'Diabetes is the Fastest Modern Health Crisis after the Number of Cases Doubles in the Last 20 Years with 3.7 Million People Being Diagnosed', *Daily Mail* (27 February 2018), http://www.dailymail.co.uk/health/article-5438433/Diabetes-fastest-modern-health-crisis-cases-double.html.

27. S. Boseley, 'Expert Revered for Painstaking Work That Proved Link between Smoking and Cancer', *The Guardian* (8 December 2006), https://www.theguardian.com/science/2006/dec/08/cancer.uk.

28. L. Surugue, 'Schizophrenia Is a "Modern" Disease, Developing after Humans Diverged from Neanderthals', *International Business Times* (16 August 2016), https://www.ibtimes.co.uk/schizophrenia-modern-disease-developing-after-humans-diverged-neanderthals-1576327.

29. Ibid.
30. D. E. Bloom, '7 Billion and Counting', *Science*, 333:562–9 (2011).
31. 'Life Expectancy of the World Population', *Worldometer* (2020), worldometers.info/demographics/life-expectancy/.
32. V. Cable, 'Vince Cable Speech: Capitalism in Crisis', *Liberal Democrats* (7 June 2018), https://www.libdems.org.uk/capitalism_in_crisis.
33. M. El-Erian, 'Is Stagnation the "New Normal" for the World Economy?', *The Guardian* (3 February 2016), https://www.theguardian.com/business/2016/feb/03/is-stagnation-the-new-normal-for-the-world-economy.
34. D. Campbell, 'Stress and Social Media Fuel Mental Health Crisis among Girls', *The Guardian* (23 September 2017), https://www.theguardian.com/society/2017/sep/23/stress-anxiety-fuel-mental-health-crisis-girls-young-women.
35. Ibid.
36. Ibid.
37. J. Stiglitz, 'Joseph Stiglitz: Why We Have to Change Capitalism', *The Telegraph* (23 January 2010), https://www.telegraph.co.uk/finance/newsbysector/banksandfinance/7061058/Joseph-Stiglitz-Why-we-have-to-change-capitalism.html.
38. J. Bennet, '"We Need an Energy Miracle"', *The Atlantic* (November 2015), https://www.theatlantic.com/magazine/archive/2015/11/we-need-an-energy-miracle/407881/.
39. E. Green, 'The Existential Dread of Climate Change: How Despair about Our Changing Climate May Get in the Way of Fixing It', *Psychology Today* (13 October 2017), https://www.psychologytoday.com/us/blog/there-is-always-another-part/201710/the-existential-dread-climate-change.
40. A. Marr, 'Andrew Marr: British Politics Is Broken—the Centre Cannot Hold', *New Statesman* (23 March 2015), https://www.newstatesman.com/politics/2015/03/andrew-marr-british-politics-broken-centre-cannot-hold.
41. I. Sample, 'Scientists Find Genetic Mutation That Makes Woman Feel No Pain', *The Guardian* (28 March 2019), https://www.theguardian.com/science/2019/mar/28/scientists-find-genetic-mutation-that-makes-woman-feel-no-pain.
42. Ibid.
43. Ibid.
44. Ibid.
45. Ibid.
46. Ibid.
47. N. Remnick, 'At 100, Still Running for Her Life', *New York Times* (22 April 2016), https://well.blogs.nytimes.com/2016/04/22/at-100-still-running-for-her-life/.
48. Ibid.
49. Ibid.
50. Ibid.
51. N. Jayapalan, *History of English Literature* (Atlantic, 2001), p. 9.
52. O. J. Benedictow, 'The Black Death: The Greatest Catastrophe Ever', *History Today* (3 March 2005), https://www.historytoday.com/archive/black-death-greatest-catastrophe-ever.
53. M. Ibeji, 'Black Death: Political and Social Changes', *BBC History* (17 February 2011), http://www.bbc.co.uk/history/british/middle_ages/blacksocial_01.shtml.
54. J. Neumann and J. Dettwiller, 'Great Historical Events That Were Significantly Affected by the Weather: Part 9, the Year Leading to the Revolution of 1789 in France (II)', *Bulletin American Meteorological Society*, 71:33–41 (1990).

55. P. Ames, 'Portugal Celebrates the World's Coolest Coup', *Global Post* (24 April 2014), https://www.pri.org/stories/2014-04-24/portugal-celebrates-worlds-coolest-coup.

56. I. Birrell, 'Greta Thunberg Teaches Us about Autism as Much as Climate Change', *The Guardian* (23 April 2019), https://www.theguardian.com/commentisfree/2019/apr/23/greta-thunberg-autism.

INDEX OF NAMES

For the benefit of digital users, indexed terms that span two pages (e.g., 52–53) may, on occasion, appear on only one of those pages.

Morsi, Mohamed 295
Mozart, Wolfgang Amadeus 2, 312–3, 315, 386
Mubarak, Hosni 295
Mukhopadhyay, Tito 229
Muotri, Alysson 90
Murphy, Tim 13
Murray, Robin 200
Musk, Elon 403
Mussolini, Benito 293, 296, 335
Myshkin, Prince Lev Nikolayevich (The Idiot) 344

Nabokov, Vladimir 340
Nachtwey, Oliver 296–7
Neilson, Donald 240–1
Nelson, Charles 136
Neo (The Matrix) 401
Neubauer, Simon 76
Newton, Isaac 365–6, 383

O'Keefe, John 117
Obama, Barack 13
Ottesen Kennair, Leif Edward 275

Pääbo, Svante 126
Pagel, Mark 27
Park, In-hyun 101
Parker, Sean 165
Parkinson, Hannah Jane 237
Paşca, Sergiu 102
Patterson, Penny 43–4
Pavlov, Ivan 9, 14–15
Pelaprat, Etienne 140
Penny, William 352
Piaget, Jean 49–51
Picasso, Pablo 2, 140, 325, 330–3
Pilgrim, David 290
Pinel, Philippe 290
Pinker, Steven 30
Pinochet, Augusto 267
Pitt, Brad 117, 119

Plato 109, 153, 311
Popova, Lyubov 336
Popper, Karl 365
Pritchard, Jonathan 189
Provokiev, Sergei 338
Pryer, Anthony 311
Puni, Ivan 336
Putnam, Hilary 400

Queen of the Night (The Magic Flute) 313
Quint, Peter (The Turn of the Screw) 350–2
Quiroga, Rodrigo Quian 117–8

Radulovic, Jelena 179–80
Rahim, Masuma 205
Raine, Adrian 243–4
Rainsborough, Thomas 291
Ramesses III 278
Ramirez, Richard 254
Ramón y Cajal, Santiago 63–4
Randall, Dave 303, 317–8
Raskolinov, Rodion Romanovich (Crime and Punishment) 345
Ravidat, Marcel 320
Razzaque, Russell 202
Reeves, Aaron 273
Rego, Paula 2
Reich, Wilhelm 296
Reicher, Stephen 294
Reis Mesquita, Giovana 160
Reiss, Allan 149
Ren, Bing 82
Renoir, Pierre-Auguste 333
Rizzolatti, Giacomo 67
Robert-Fleury, Tony 290
Rodchenko, Alexander 336
Rogers, Anne 290
Rose, Steven 30
Rubens, Peter Paul 333
Rund, Bjorn 198–9
Rusakov, Dmitri 112
Rutherford, Ernest 362

INDEX OF SUBJECTS

For the benefit of digital users, indexed terms that span two pages (e.g., 52–53) may, on occasion, appear on only one of those pages